Science

Teacher's Guide

6

W0051264

Deborah Roberts

Terry Hudson

Alan Haigh

Geraldine Shaw

OXFORD
UNIVERSITY PRESS

Great Clarendon Street, Oxford, OX2 6DP, United Kingdom

Oxford University Press is a department of the University of Oxford. It furthers the University's objective of excellence in research, scholarship, and education by publishing worldwide. Oxford is a registered trade mark of Oxford University Press in the UK and in certain other countries.

British Library Cataloguing in Publication Data

Data available

ISBN 978-1-38-201737-4

12

Paper used in the production of this book is a natural, recyclable product made from wood grown in sustainable forests. The manufacturing process conforms to the environmental regulations of the country of origin.

Printed in Great Britain by CPI Group (UK) Ltd., Croydon CR0 4YY

Acknowledgements

The publisher and authors would like to thank the following for permission to use photographs and other copyright material:

Cover: Artwork by Blindsalida. **Photos: Page xii:** Rawpixel.com/ Shutterstock; **page xvii(tl):** Evgeny Karandaev/Shutterstock.

Table on page vi: From *Theory of Learning* by Julie Cotton; © Kogan Page 1995. Reproduced with permission of the Licensor through PLSclear.

Every effort has been made to contact copyright holders of material reproduced in this book. Any omissions will be rectified in subsequent printings if notice is given to the publisher.

The manufacturer's authorised representative in the EU for product safety is Oxford University Press España S.A. of El Parque Empresarial San Fernando de Henares, Avenida de Castilla, 2 – 28830 Madrid (www.oup.es/en or product.safety@oup.com). OUP España S.A. also acts as importer into Spain of products made by the manufacturer.

Contents

Introduction

We are living in an ever-changing world, where the way we work, live, learn, communicate and relate to one another is constantly shifting. In this climate, we need to instill in our learners the skills to equip them for every eventuality so they are able to overcome challenges, adapt to change and have the best chance of success. To do this, we need to evolve beyond traditional teaching approaches and foster an environment where students can start to build lifelong learning skills for success. Students need to learn how to learn, how to problem solve, be agile and work flexibly. Going hand-in-hand with this is the development of self-awareness and mindfulness through the promotion of wellbeing to ensure students learn the socio-emotional skills to succeed.

Teaching and learning with *Oxford International Primary Science*

This series is based on the English National Curriculum Programme of Study for Primary Science. The books for each year (or stage) follow the scheme and meet all the learning objectives – including working scientifically. Each lesson includes the learning objectives from the curriculum and summary of the key teaching points. A full mapping grid identifying the unit and lesson where each objective can be found is available online at https://www.oxfordowl.co.uk/. Objectives are written in student-friendly language in the Student Book.

The teaching units in the series are flexible: they can be adapted as you see fit to meet the needs of your students. Each unit stands alone and can be taught in any order.

The books are designed for students aged 5 to 11. Each year has a **Student Book**, a **Workbook** and a **Teacher's Guide**. There are also numerous online resources and sources of support: these include further examples and support for formative and summative assessment, and can be found at: https://www.oxfordowl.co.uk/. Underpinning the rationale for the series is the strong belief that science provides a way of thinking and working. It helps us to make sense of the world we live in and provides intellectual skills that help us in all curriculum areas and in life.

This introduction shows how to use the resources to develop your students' scientific knowledge, skills and understanding.

This series has seven main aims:

1 To deliver scientific knowledge and facts

2 To deliver scientific understanding

3 To deliver scientific methods of enquiry

4 To deliver scientific thinking and reasoning

5 To help students understand the development of science and its uses in context in the world around them

6 To support the wellbeing of students

7 To give students a global outlook

1 Scientific knowledge and facts

The Student Book and Workbook introduce concepts in a logical sequence and ensure that new ideas are introduced sensitively. Key scientific concepts and ideas are explained. Students are then asked to discuss and apply their new knowledge.

2 Scientific understanding

Knowledge without understanding is only useful for recall. Understanding moves to a deeper intellectual level and enables students to think and apply that knowledge. Effective learning requires students to develop appropriate attitudes, skills and enthusiasm, and this can be encouraged by good teaching and exciting resources. This means students can gain an understanding of the principles and practice of science.

The knowledge, or content, in this series is based on the English National Curriculum. Each book has five units.

Though this is not a theoretical book, it is important to consider underpinning ideas that have informed good practice in the classroom:

- Teaching approach
- Cognitive style
- Active learning

Teaching approach

The kind of teaching strategies used are key to achieving understanding. Telling and giving students information is important but usually only improves students' short-term memory of scientific facts. This is often called 'passive learning' as students are not intellectually engaged in the process themselves.

Teaching and learning can either be teacher-centred (didactic) or student-centred (heuristic). Good teaching is a combination of these.

Advantages		*Advantages*
• Clear learning objectives • Teacher can demonstrate a professional approach, e.g. presentations • Teacher is seen as 'expert' • Fewer problems with classroom management and behaviour		• Can be motivating and powerful • Develops a range of skills • Learning is relevant • Encourages creativity and problem solving • Student has a say in the learning
Didactic	⟵⟶	**Heuristic**
Disadvantages		*Disadvantages*
• May build on inaccurate knowledge • May not be motivating • Does not develop skills • Does not give students responsibility • Limited by the teacher		• May not deal with underlying principles – too pragmatic • If only existing skills are learned, this approach may not encourage questioning of existing approaches • Lack of structure may confuse • Classroom management may be problematic

Advantages and disadvantages of teacher-centred and student-centred approaches (Cotton, J., 1995, *The theory of learning: an introduction,* Kogan Page, London) © Kogan Page 1995. Reproduced with permission of the Licensor through PLSclear.

Cognitive style

Cognitive style is a student's personal and preferred way of organising and representing information. The cognitive style, or way of thinking, impacts on how our students see and make sense of the world.

Cognitive styles can be split into four types:

1 Wholists like to see the whole picture when learning – the big picture.

2 Analysts prefer to get down to the detail and look at only one or two details.

3 Verbalisers welcome chances to talk through problems.

4 Imagers see mental pictures when dealing with information.

Most people are a combination of all of these but have a preference for one or two. These cognitive styles may have a major impact on the processing of information.

We need to be aware of the possible cognitive styles of our learners and ensure that our approach balances all four styles.

Lev Vygotsky stressed the importance of social relationships between the teacher and learners, and amongst learners. He stressed that language and discussion are key to development.

Benjamin Bloom proposed an 'educational taxonomy' identifying different learning 'domains':

• Cognitive (knowledge)

• Affective (attitudes)

• Psychomotor (skills)

The domain can be seen as a ladder that starts with remembering and proceeds to more complex tasks such as analysing. The 'rungs' of the ladder are:

1 Recall data

2 Understand

3 Apply (use)

4 Analyse

5 Synthesise

6 Evaluate

This ladder helps teachers to devise tasks, sequences of tasks and questions appropriate to the level of thinking of their students. Considering this will help develop your students' ability to think and reason.

Active learning

Active learning approaches encourage students to engage with tasks and to develop skills that they may not develop during teacher-led, didactic lessons. Active learning combines a number of models of teaching growing out of 'discovery learning' ideas.

Whole-class instruction can be as effective as individual instruction, especially in terms of the time students spend on tasks. The key to effective teaching is the appropriate selection of approaches at any particular time and with any particular group. The table opposite summarises common teaching approaches. It does not indicate poor teaching on the left (teacher-centred) and good teaching on the right (student-centred).

Possible active learning strategies

- Group discussion (talking and listening)
- Active reading
- Active writing
- Presentation
- Role-play and drama
- Information technology
- Visits, visitors and field trips
- Data handling
- Problem solving
- Video and audio tape recording
- Games and simulations

Questioning and group work are very important. These will be dealt with in more detail on pages ix–xi.

The Student Books present ideas in a range of ways – written, diagrams, charts, tables and photographs. The lessons contain a rich variety of learning and teaching approaches, such as individual reading and writing, paired and small group discussion work, whole-class discussion and activity, problem solving, investigations, research activities, presentations, surveys and review and reflection. In addition, suggestions for other activities, such as field trips and educational visits, are included in the Teacher's Guide.

3 Scientific methods of enquiry and working scientifically

This series promotes scientific enquiry and closely follows the working scientifically objectives in the English National Curriculum. Students are encouraged to use and reflect on the different ways that scientists work and think, which have produced the knowledge, theories and laws of science over the last 1000 years. It is based on 'empiricism'; arriving at knowledge and understanding through observation and experiment.

Scientists progress by observation and questioning what they see and already know. From this they develop hypotheses which they test by experiments and develop new knowledge. This will be further explored in the section 'How to be a Scientist' on pages 2–3 of this Guide and in the student resources.

Science teaches students to think in a structured way that is good for analysing and solving problems. However, students should understand that science is also a creative human endeavour. Imagination is vital to scientific progress.

The books in this series allow students to develop their skills to work scientifically by addressing each of the appropriate scientific enquiry processes at each stage. Students are encouraged to plan and carry out full-scale investigations in the later stages and, as such, apply the skills learned earlier.

4 Scientific thinking and reasoning

It is essential to encourage students to think and reason for themselves. Thinking and reasoning are important life skills. The abilities to think, reason and research make students independent learners who can interpret and understand new ideas more quickly. Unfortunately, this aspect of education is often neglected.

In this series, the ability to think and reason will be encouraged, nurtured, practised and assessed at each level. Scientists use deductive logical thinking to make sound inferences which take them from the known to discover the unknown. They use reason and argument based on fact and evidence to prove their case. Allowing students to experience these processes promotes their curiosity and enthusiasm. 'Discovery learning' approaches allow students to experience the thrill of finding out.

Resist the temptation to provide answers, solutions and too much support for your students. We hope that the learning activities within the books, and the support provided in the Teacher's Guide, will help you to create a learning environment where at times students can plan, find out and learn new ideas themselves – with you as a guide and facilitator. Allow them time to think and discuss ideas before gently guiding those who need support.

Teacher-centred learning	Student-centred learning
Teacher exposition	Group work
Accent on competition	Accent on cooperation
Whole-class teaching	Resource-based learning
Teacher responsible for learning	Student more responsible for learning
Teacher providing knowledge	Teacher as guide/facilitator
Students seen as empty vessels which need filling	Students have ownership of ideas and work
Subject knowledge valued	Process skills are valued
Teacher-imposed discipline	Self-discipline
Teacher and student roles emphasised	Students seen as source of knowledge and ideas
Teacher decides the curriculum	Students involved in curriculum planning
Passive student roles	Students actively involved in learning
Limited range of learning styles and activities	Wide range of learning styles employed

Select a variety of approaches to promote active learning

5 Science in context

It is vital to link what students learn in the classroom to the real world. This makes their learning relevant and helps them to relate new ideas to their own experiences:

- Stress that science involves an ongoing process of change and improvement in ideas. Explain that our ideas about science are built on earlier ideas. Point out that people in the past could only use what they knew at the time to make sense of the world. Sometimes this meant they put forward ideas that scientists now know are not correct. For example, many people thought that the world was flat, and the Sun orbited the Earth.

- Emphasise that some early thinkers created ideas that are still remarkably similar to our modern ideas. For example, over 2200 years ago Aristarchus suggested the Earth orbits the Sun. Democritus stated that matter is made of smaller particles more than 2300 years ago. Even our understanding of forces, based on Isaac Newton's laws of motion, were proposed by Philoponus over 1500 years ago.

- Explain that science theories develop when a person or a team puts forward new ideas. If other scientists test these ideas and agree then the idea becomes a part of science theory. It might be changed later with new evidence. This is how ideas develop.

- Explain that developing new technologies and materials also helps form new science ideas. For example, until the invention of the microscope 500 years ago, scientists could not see microorganisms and did not know they existed. Improvements in telescopes have resulted in changing ideas about the stars and even our nearest planets. Modern materials have allowed spacecraft and computers to be made.

The activities in each lesson provide you with many opportunities to relate the science content and processes to the real world. Whenever possible, take students out to see examples of science being used in the real world – such as on farms, in factories or even at an airport. You can invite people in to talk about their jobs and how they use science – for example, doctors, vets, farmers, gardeners and builders.

6 Wellbeing

The Student Book and Workbook provide opportunities for teachers to consider the vital importance of wellbeing and to weave this into their teaching. The enquiry-based approach encourages curiosity and helps students to think about the world around them.

Wellbeing does not mean feeling happy all of the time. Making mistakes, feeling challenged and even being confused at times can help to develop resilience.

The resources support wellbeing directly by:

- **Providing questions and science facts** to challenge and engage students. They can reflect on prior learning and apply new skills.

For example, students are asked to think back to their work on components of an electrical circuit in earlier lessons and list those they have used to build a circuit.

- **Encouraging active science.** This means an active brain and also an active body. Students learn better and make better progress when they are physically active in lessons.

For example, students play the role of particles by arranging themselves into a close packed pattern to represent solids, move slightly to represent liquids and then move around the room to model the particles in gases.

- **Promoting group work.** Collaborative work is used throughout the resources so that students have opportunities to develop their collaborative skills. This growth through practice develops confidence and happiness.

There are opportunities for group work and collaboration in every lesson. For example, students work together to decide on key questions to ask a visiting health profession and work in groups regularly to plan and carry out investigations and surveys.

- **Presenting 'stretch zone' challenges** to encourage students to develop thinking skills and welcome challenge. They will become more familiar with moving away from a 'comfort' zone into the 'stretch zone' without worrying.

For example, students are asked to apply their understanding of habitats and survey techniques to hypothesise why some animals are found in certain locations but not others.

- **Offering mindful moments.** These provide opportunities for students to pause and re-focus their attention.

For example, the end-of-unit questions in the Student Book and self-review statements in the Workbook offer ideal opportunities for students to think about their learning.

Teachers are encouraged to develop the following approaches:

- Providing praise with a growth mindset. Teachers should work to praise the process rather than the intelligence or marks. Giving positive feedback on how something is being done is highly effective. This includes praising effort, perseverance, resilience, teamwork, strategies, etc.

- Discussing and evaluating mistakes. Learning always involves making mistakes. Students should not fear or worry about mistakes. They should see them as opportunities to learn.

7 A global outlook

The *Oxford International Primary Science* resources are designed to address the idea that academic lifelong success is the result of both academic performance and emotional wellbeing. As educators we want to prepare our students for a workplace that is unknown to us. Ideas and activities identify areas where students can develop skills while feeling safe and confident enough to apply themselves to the content of the lessons. Skills are separated into distinct categories designed to provide the opportunities to develop key lifelong skills. Students are inspired by images and information that result in curiosity and wonder. Students become confident problem solvers by taking risks that also develop creative skills. Real-world skills are encouraged through carefully designed projects and activities. Students are introduced to project management and aspects of literacy, for example financial and functional literacy. There is an emphasis on carrying out research and careful analysis of the information that they find in addition to their own findings and data. Students have opportunities to develop interpersonal skills through communication and relationship building. They are encouraged to voice their ideas through discussion activities. The projects particularly allow students to take part in leadership roles and the responsibility that comes with this. The resources address the students' self-development skills through critical thinking, ethics and motivation. There are a number of sections throughout the scheme where students are introduced to ethical and sensitive issues.

Wellbeing is an area that is emphasised in the resources with a desire to address mental health issues, supporting learners in and out of school. Students are encouraged to care for their own minds by promoting mindfulness and to manage stress more effectively. Students should become more optimistic about their lives and the world around them. They are encouraged to care for their bodies with an emphasis on being active and eating healthily. Students are encouraged through a number of activities to build and maintain relationships and friendships with family and others. They learn how to communicate confidently with a range of people and connect through kindness and thoughtful behaviours. Students are more conscious about the world beyond their immediate environment but know they have a valued role and place, resulting in becoming better and more responsible lifelong citizens from an early age.

Teaching techniques for this series

Science learning is made up of:

- Remembering science facts
- Gaining scientific knowledge
- Developing science skills
- Developing science understanding

Facts are important but being told facts does not ensure knowledge and understanding. Working out science problems and engaging with scientific processes is much more likely to help students develop understanding. This is why applying scientific skills – doing science rather than remembering it – is vital.

Think about the question below:

Question: Who was the scientist who discovered the force of gravity?

Answer: Isaac Newton.

Knowing the answer to this question takes the learner no further. It demands no higher order thinking skills and does not help with solving any other problems. However, if a student understands Newton's theory of gravity and motion, they can start to explain and predict how things move, float and fly. Understanding enables a student to apply knowledge and solve problems and furthers their learning.

This series aims to provide science facts and knowledge but also science understanding. Certain strategies are better at teaching understanding than others.

Effective questioning is the key

Students can learn to understand by listening and reading. This is only possible if they have acquired advanced learning techniques and have sufficient background knowledge and understanding in which to fit any new ideas. That is why you can enhance your understanding, for example about your teaching, through these approaches, but this is not true for young learners. For inexperienced and less skilled students, the teacher enables them to progress from memory recall to deeper understanding. This series focuses on teaching and learning approaches that promote understanding. Science facts and knowledge are covered, of course, otherwise there would be a lack of context and content but the activities are also designed to develop thinking and learning skills.

Research tells us that teachers ask up to 400 questions per day. That can be 30 per cent of teaching time. It is clear then that time spent improving our questioning techniques will have an important impact on learning.

To give you some idea about the complexity of questioning you may wish to think about your own practice:

- Why you are asking a question
- What type of questions you are going to ask
- When you are going to ask questions
- How you are going to ask questions
- Who you are going to ask questions to
- How you expect the questions to be answered
- How you will respond if the person does not understand the question
- How you will react to an inappropriate or wrong answer
- How you will react to an appropriate answer
- How long you will wait for an answer

As teachers we ask questions for a number of reasons:

- To get attention
- To check students are paying attention
- To check understanding
- To reinforce or revise a topic
- To increase understanding
- To encourage thinking
- To develop a discussion

Bloom describes six levels of thought process:

- Knowledge
- Comprehension
- Application
- Analysis
- Synthesis
- Evaluation

We need to ask questions that encourage deeper thinking. If we only ask questions at the knowledge end of the spectrum, we will not encourage students to analyse or synthesise new ideas.

We also need to think about the nature and style of our questions. Two major categories are closed and open.

Closed questions

These tend to have only one or a limited range of correct answers. They require factual recall. They are useful for whole-group question and answer sessions, to quickly check learning or refresh memory or as a link to new work. Examples include:

Question: What is the boiling point of water at sea level?

Answer: 100°C.

Question: What are the three stages of the water cycle?

Answer: Water; clouds; rain.

These are very good for knowledge recall but are generally non-productive regarding anything else.

Open questions

These may have several possible answers and it may be difficult to decide which are correct. They are used to develop understanding and encourage students to think about issues and ideas. They encourage students to think and manipulate information and are much more complex. We are not looking for a single right answer; we are looking for what the student thinks may be the right answer. Once the teacher gets the student thinking, then the teacher can use this information to move the learning on towards the right answer, while promoting understanding at the same time.

Examples include:

Question: Where do you think the water in rain clouds comes from?

Answer: Any answer will have a little 'rightness' in it that the teacher can use. The student may answer 'From the sea.'

The teacher then can follow several further lines of enquiry to extend the learning. For example, the teacher could ask, 'Do you know of any other places the water might have come from?' Or 'How do you think that the water got into the clouds?'

These follow up 'how' and 'why' questions encourage students to think more deeply about the science and their understanding of the key ideas and principles.

Open questions require students to make links between ideas and apply knowledge – they often require students to be logical and imaginative. They require a longer time to think and answer than closed questions, and may lead onto wider discussion and debate.

Question series

Closed and open questions can be linked together to form a series. A series must be well planned but can lead to much improved understanding. Start with a few relatively easy factual closed questions and move towards more open questions. This is known as 'agenda building'. At the same time, you can move from individual to paired and then to small group discussion as the questions become more open and demand higher-level thinking.

In this series we promote an 'enquiring classroom' where closed questions are used, but also open questions which promote enquiring minds.

Question and answer techniques – some tips

1 The 'don't lead students down dark tunnels' technique

Students need to know where they are going before they start their learning journey, so tell them. For example:

'Today we are going to learn how magnets react to each other.'

This gives students the big idea on which they can hang the information that follows and make sense of it.

2 The 'ask students what they think' technique

Students usually lack the confidence to answer questions like:

'How did the water from the sea get into the clouds?'

Unless they are confident they know the correct answer, they will probably be reluctant to answer because they are afraid of failure. However, rephrase the question and say:

'How do you think that the water might have got from the sea up into the clouds?'

Then you are giving them permission to try even if they are not sure they are correct. In this way you do not always get the same students volunteering answers and you can give other students confidence.

3 The 'praise all answers' technique

To encourage students to share their thinking and suggestions, we have to value and thank them for their efforts. We may say, 'Good try but not quite there yet. Let me see if I can help you – do you think it could have something to do with the heat of the Sun?'

There is usually an element of correctness in most students' attempts which we can praise.

4 The 'teach from students' answers' technique

In one Student Book there is a question which asks:

'Why does a watermelon contain 600 seeds?'

As a student, I do not know the answer and I am afraid of failure.

If the teacher asks students what they think and values their answers, then this productive line of enquiry will help take a student and the rest of the class to the correct answer as well as an understanding.

Select the element of correctness from the student and then expand and explain to help all the rest of the students understand. For example:

Student: I think it is because they want to grow lots of new watermelons!

Teacher: Good answer. It is all about germination and new life. However, do you think that when one watermelon sheds its seeds, there will be 600 new watermelons springing up beside it?

Student: No, not all 600!

Teacher: Good, that's important. You have told me that not all of them survive. How many do you think survive?

The teacher leads and expands and informs the student's answers to arrive at the understanding of 'producing many, so that a few can survive to carry on the species'. Because students have been actively involved in this journey they will not only remember; they will understand.

5 The 'do not let students struggle' technique

If you find that you are asking questions and the answers are nowhere near what you are looking for, then give students the answer or suggest a choice of answers. Without this, the progress of the lesson is halted and students and teacher get frustrated; move the lesson on. Tell them and expand and explain your answer.

6 The 'right answer' technique

If you get the right answer, then all is good, or is it? Only the person who has given the answer understands why it is correct so you need to expand and explain, so that the rest of the class can share in that student's understanding.

Teacher: Good answer, what made you think of that?

Teacher: I see what you mean; you made the connection between the boiling kettle and steam and thought the Sun's heat did the same with the sea only it is invisible. Well done!

Whole-class or group work

Whole-class question and answer methods work very well and highly structured whole-class activities can help to keep students on task. However, maximum contributions and participation are usually encouraged in small groups.

Group work can help students to learn more effectively. They can learn science better and cooperative learning can help social cohesion, motivation and improvements in self-esteem. The student who is shy is more likely to contribute to a discussion with another person or one or two other people than volunteer ideas in front of the whole class. In addition, by sharing work, students can cover more ground more quickly. The small group is also a good forum for generating creativity.

Another advantage is that small group work frees the teacher from having to be at the front and leading the whole class. The teacher can move around the room and direct attention and support when and where it is most needed. Individual needs can be better met in this way.

Some specific examples of group work are described below.

1 Short, informal discussions

These are sometimes called 'Buzz groups'. They are very useful as they do not need any structure and can be used at any time. Simply ask pairs or small groups to look at a picture or think about an issue and then give them a few minutes to share their ideas. There are numerous examples of these in the resources, linked to questions for the buzz groups to consider.

2 Think-pair-share

Students think about something individually for a few minutes and then work with a partner to compare their ideas. Finally, the pair present their ideas to the class.

3 Circle of voices

This works with larger groups of students (four, five or six). Students take it in turns to speak about their ideas on a topic or question. No one else is allowed to speak so this helps to develop social and listening skills.

4 1-2-4

Students think about an issue or carry out a task individually – for example, make a list of animals and plants they have seen – and then work with a partner to compare lists and discuss a slightly more complicated question – for example, which animals eat other animals? Finally, pairs join to form fours. They share ideas and then work together on a final task – for example, make some food chains.

5 Jigsaw

Students work alone to become experts on an aspect – for example, different habitats – then join back together to share their expertise and answer a larger question – for example, how are animals adapted to habitats?

Differentiation

Differentiation is closely linked to inclusion: ensuring all students have access to the curriculum. This means that learning and teaching approaches must consider individual needs. Not all learners will learn at the same pace or in the same ways.

Approaches supported by the resources are:

1 **Modifying content.** At times we can adjust the content for some learners to provide sufficient support or adequate challenge. Examples are support or challenge Workbook activities and stretch zone tasks in the Student Book. This is often called differentiation by task.

2 **Differentiating expected outcomes.** This allows all students to tackle the same tasks but outcomes are differentiated – usually in terms of 'All students should … ', 'Most students will … ' and 'Some students may … '. These differentiated outcomes are given in each lesson section of this Teacher's Guide. This is often called differentiation by outcome.

3 **Differentiating the process.** This means providing more or less support as students are carrying out a task. Advice on this is in each Teacher's Guide lesson section and also there are additional support activities that can be given to some students. For example, investigation support pages.

4 **Questioning.** This is a very effective way of differentiating work. Use questions to check progress and decide when extra support or challenge is needed. Questions in the resources are designed to progress from low on Bloom's taxonomy (remember and understand) towards higher levels (analyse and evaluate).

5 **Varied approaches to assessment.** The resources include a wide range of assessment methods. These include verbal, written and drawn responses and individual and collaborative assessments. There are also differentiated questions ranging from easier introductory questions to more challenging ones.

Assessment

Assessment is an essential part of learning. Without being able to check progress, teachers and students will not be able to identify areas of strength and areas in need of development.

Assessment can be classified as either formative or summative.

Formative assessment takes place during learning and is used to address issues as they arise. This means learning and teaching can be modified during lessons to better meet the needs of learners. Feedback to students is ongoing.

Each activity within the Student Book and Workbook provides opportunities for formative assessment and feedback. This can be through teachers listening to discussions or presentations, observing the outputs of investigations and through assessing outcomes such as posters, reports and leaflets. Individual questions in discussion tasks can be used to monitor understanding and identify misconceptions. These can be addressed as they are noted. Some of these are noted in the Review and reflect sections in the lesson guidance in this Teacher's Guide.

Summative assessment is used to measure or evaluate student progress at the end of a process – for example, when a unit is completed or at the end of a year. Summative assessment compares students' attainment against a standard or benchmark.

The 'What have I learned?' features at the end of each unit can be used for summative assessment. Teachers can record which questions each student is answering correctly and use this to measure individual attainment. It can also indicate how well the class is progressing though the work. In this way, the assessment can inform individual interventions (extra support for a student) or whole-class interventions (reviewing work that is not well understood).

There is also a Quiz Yourself section in the Workbook, which contains questions from across the year. These quiz questions and activities are intended to encourage students to reflect on their learning and to reinforce their developing knowledge about scientific concepts in a fun way. They could also be used for formative or summative assessment.

Each activity – group and individual – can be assessed through observation and questioning and progress notes. Written or drawn responses for each activity, especially in the Workbook, can be assessed/marked using the school's marking policy and unit, end-of-term and end-of-year judgements made about individual and class progress.

Feedback is a crucial aspect of assessment. This should be as positive and encouraging as possible (see the wellbeing section on page viii) and identify clear targets. Involve students in assessment and target setting. This is encouraged through the 'formal' self-assessment statements in the Workbook at the end of each unit, where students are asked to rate their level of confidence. Assessment is done with learners not done to learners.

Teaching Primary Maths and Science through English: identifying the challenges and providing the support

The challenges

Ministries of Education at both local and national level are increasingly adopting the policy of English Medium Instruction (EMI), for either one or two subjects or across the whole curriculum. The rationale for doing so varies according to the local context, but improving the levels of achievement in English is usually an important factor.

In international schools it is likely that students do not share a mother tongue with each other or perhaps the teacher. English is, therefore, chosen as the medium for instruction to level the playing field and to provide the opportunity to develop proficiency in an international language.

This does not mean that the maths or science teacher is now being asked to replace the English teacher, or to have the same skills or knowledge of English (though in many primary schools one teacher may indeed teach both). It does mean that the science or maths teacher needs to become more language aware.

This raises significant challenges, including:

- the teacher's knowledge of English
- students' level of English (which may vary considerably in international schools)
- resources which provide appropriate language support
- assessment tools which ensure that it is the content and not the language which is being tested
- differentiation which acknowledges different levels of proficiency in both language and content.

Meeting the challenges positively

Perhaps lack of confidence in their own English proficiency is one of the most common concerns among teachers. However, while it is a factor, success in EMI is not necessarily linked to the teachers' proficiency in the second language. Teachers who have English as their mother tongue may well lack the sensitivity to, or awareness of, the language that a non-native speaker has acquired through learning and studying the second language. Developing this awareness and demonstrating it in both materials and method is the key to effective EMI.

Classroom language/Teacher Talk

Often non-native-speaker teachers are more concerned about their ability to run and manage the whole class in English than they are about the actual teaching of the maths or science concepts. The resources or textbook should help them with the latter. However, this use of English in the class is very important as it provides exposure to the second language, which plays a valuable role in language acquisition. The Teacher Talk for purposes such as checking attendance and collecting homework does not have to be totally accurate or accessible to students.

When teaching the science concepts, however, it is essential that the Teacher Talk is comprehensible. Some basic strategies to ensure this include:

- simplify your language
- use short simple sentences and project your voice
- paraphrase as necessary
- use visuals, the board, gestures and body language to clarify meaning
- repeat as necessary
- plan before the lesson
- prepare clear simple instructions and check understanding.

Creating a language-rich environment

Primary teachers often excel at providing a colourful and engaging physical environment for students. In the EMI classroom this becomes even more important. Posters, Word walls, lists of key structures, students' work and English signs and notices all provide a backdrop which provides the opportunity for exposure and language acquisition.

Planning

When planning the teacher needs to identify what the Language Demands are. This means thinking about what language students will need to understand or produce, and deciding how best to scaffold the learning to ensure that language does not become an obstacle to understanding the concept. This involves providing Language Support and goes beyond the familiar strategy of identifying key vocabulary.

Support for listening and reading

Listening and reading are receptive skills, requiring understanding rather than production of language.

Here are some suggestions for approaching such tasks.

If you are asking your students to listen to or read texts in English, ask yourself the following questions when you are planning the unit:

1 Do I need to teach any vocabulary before they listen/read?

2 How can I prepare them for the content of the text so that they are not listening 'cold'?

3 Can I provide visual support to help them understand the key content?

4 How many times should I ask them to read/listen?

5 What simple question can I set before they listen/read for the first time to focus their attention?

6 How can I check more detailed understanding of the text? Can I use a graphic organiser (e.g. tables, charts and diagrams) or gap-fill task to reduce the Language Demands?

7 Do I need to differentiate the task for those students who find reading/listening difficult?

8 Could I make the tasks interactive (e.g. jigsaw reading i.e. when students access different information before coming together, and information share)?

9 How am I going to check their answers and give feedback?

Support for speaking and writing

Speaking and writing are productive skills and may need more language input from the teacher, who has to decide what language students will need to complete the task and how best to provide this. When you plan to use a task which requires students to produce English (speak or write), you need to think about how to help them do this.

This means that you have to think in detail about what language the task requires (Language Demands) and what strategies you will use to help them use English to perform the task (Language Support).

You need to ask yourself the following questions:

1 What vocabulary does the task require? (LD)

2 Do I need to teach this before they start? How? (LS)

3 What phrases/sentences will they need?
Think about the language for learning maths/science: e.g. predicting and comparing. What structures do they need for these language functions? (LD)

4 While I am monitoring this task, is there any way I can provide further support for their use of English (especially for the weaker students)? (LS)

5 What language will students need to use at the feedback stage (e.g. when they present their task)? Do I need to scaffold this? (LD, LS)

Teaching vocabulary and structures

Vocabulary

Learning the key maths and science vocabulary is central to EMI and 'learning' means more than simply understanding the meaning. Knowing a word also involves being able to pronounce it accurately and use it appropriately. Below is a list of strategies which could be useful:

- Avoid writing the list of vocabulary on the board at the start of the unit and 'explaining' it. The vocabulary should be introduced as and when it arises in the unit. This helps students associate the word or phrase with the concept and context.
- Record the vocabulary clearly on the board and check that you are confident with the pronunciation and spelling.
- Give students a chance to say the word once they have understood it. The most efficient way to do this is through repetition drilling.
- Use visuals whenever possible to reinforce students' understanding of the word.
- Ensure students are recording the vocabulary systematically in their glossaries and, if possible, use a Word wall which lists the vocabulary under unit/topic headings.
- Remember to recycle and revise the vocabulary.

Structures

In order for students to talk or write about their maths/science they will need to go beyond vocabulary: they will also need to use those phrases and sentence frames which a particular task requires. For example, they may need the following expressions in maths and science:

X is the same as Y.

The sides are the same length.

The next number in the sequence.

I predict that X will happen.

If X happens, then Y happens.

The next step is …

The teacher needs to build up these banks of common maths/science phrases and encourage students to record them. This is an important part of identifying the Language Demands and providing the necessary support. The teacher does not have to focus on grammar here as the language can be taught as 'chunks' rather than specific grammatical structures.

How to use the language support in the classroom

The study of science involves becoming familiar with an extensive and specific vocabulary. This is sometimes referred to as the language of science. Add to this the fact that for many students the language of instruction – English – is not their first language, and it is clear that we need to be sensitive to the use of language and language support.

The Student Book supports language development by clearly identifying key words in the Word clouds on the WOW pages and making bold the key content and enquiry words for each lesson. The interactive glossary is also of vital importance in helping students understand the language. The Student Book topic pages also combine words with pictures as this is most effective in helping students understand the meanings of words. The linking of image to word is an essential factor in language development.

  How to Support Non-English Speakers

Repetition is also very important and the Student Book introduces and reinforces words by showing the words and asking students to use them in their discussions and answers.

Each section of the teaching notes linked to a particular activity or lesson also provides specific language support. Detailed and specific advice is provided for each key word and other words vital for scientific literacy. A range of strategies are suggested, including card sorts and card games, Word walls, team games to define or explain words, use of similar words to explain meaning and exploration of the origins of words.

Key principles underpinning language support in this series are:

- Words should be introduced and explained carefully.
- The word should be explained in context.
- Repetition is vital.
- Words should be linked to pictures or actions.
- Students should develop their own glossaries.
- The learning of vocabulary should be fun.
- Language should not be a barrier to learning.

Not all students will understand ideas and concepts at the same rate and there is likely to be variation in language skills. The Student Book pages are set out to be easy to follow and use, but there are also suggestions for further work and activities within each unit of this Teacher's Guide. These will help you to differentiate the learning and provide alternative learning opportunities. You should find the advice about pair and group work particularly valuable in helping you to meet individual needs.

In addition, the Workbook contains a wide range of support activities and suggestions for extension work and home learning.

Component Overview

The Student Books

The Student Books are textbooks for students to read and use. The Student Books include everything you need to deliver the course to your students, guide their activities and assess their progress.

Student Book	Typical student age range
Student Book 1	Age 5–6
Student Book 2	Age 6–7
Student Book 3	Age 7–8
Student Book 4	Age 8–9
Student Book 5	Age 9–10
Student Book 6	Age 10–11

 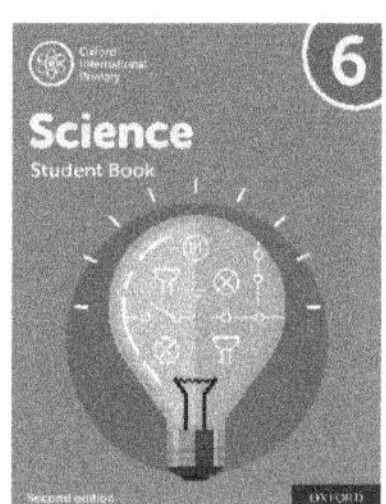

The Workbooks

Every activity in the Student Book is supported by an associated activity in the Workbook. The activities can be used as homework, but you may choose to use these additional practice exercises to supplement the class work.

Workbook	Typical student age range
Workbook 1	Age 5–6
Workbook 2	Age 6–7
Workbook 3	Age 7–8
Workbook 4	Age 8–9
Workbook 5	Age 9–10
Workbook 6	Age 10–11

 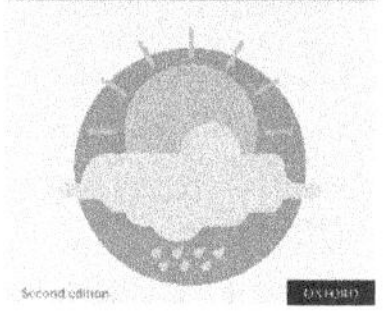  
   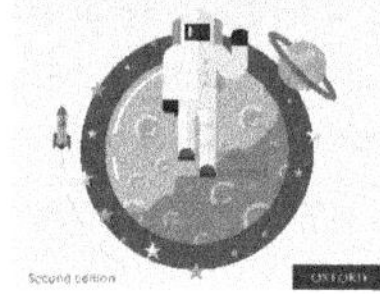 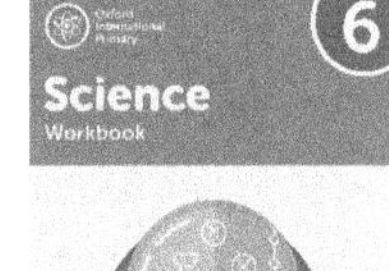

The Teacher's Guides

Each Teacher's Guide includes:

- An introduction with advice about delivering science and using the Student Books and Workbooks.
- A brief lesson plan for every lesson in each Student Book and Workbook.
- Model answers to the activities and investigations; and answers to the assessment activities.

There are six Teacher's Guides:

  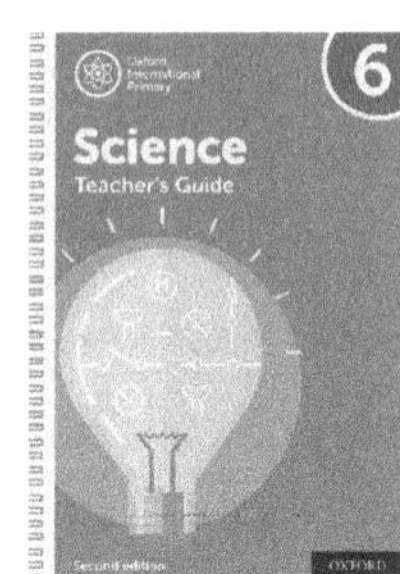

Digital resources

Interactive eBooks

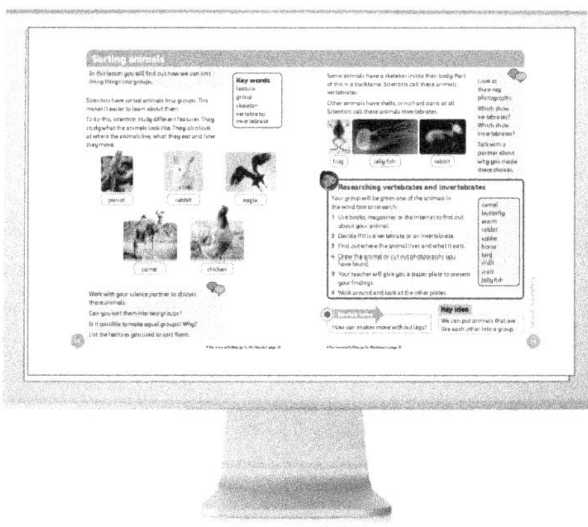

For the teacher:

Teachers can access the Student Books, Workbooks and Teacher's Guides online in eBook format, on the Oxford Owl website (www.oxfordowl.co.uk).

The enhanced eBooks show the course content on screen, making it easier for teachers to deliver engaging lessons.

For the students:

Teachers can allocate an eBooks version of the Student Book to the students for use at home. The Student eBooks include interactive activities, animations, and audio of all the key vocabulary.

Downloadable assessment materials

The downloadable assessments offer additional opportunities to assess, monitor and support students' progress. The assessments included are:

- end-of-unit tests – cover the content of each unit
- end-of-year practice papers – cover a sample of objectives from the year
- three transition papers – one to be taken at the end of Year 2 and two in Year 6. To support movement to the next phase of learning, the questions cover objectives from the preceding years.

Every test/paper comes with everything you need to assess and record progress including:

- objective coverage, transcripts (for lower years) and teaching advice
- answers, mark schemes and guidance on assessment.

Oxford Primary Illustrated Science Dictionary

The *Oxford Primary Illustrated Science Dictionary* gives comprehensive coverage of the key science terminology students use in the course. Each entry is in alphabetical order and, along with a clear and straightforward definition, has a fun and informative colour illustration or diagram to help explain the meaning. The dictionary is suitable for students with English as an additional language.

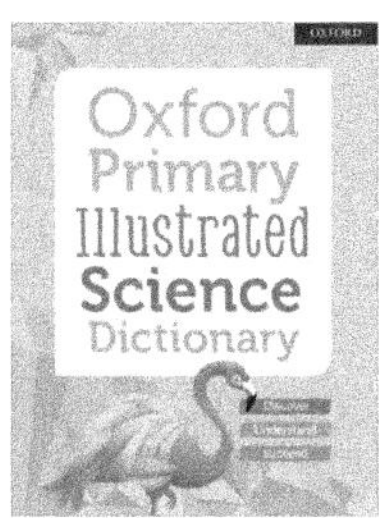

The curriculum

The *Oxford International Curriculum* offers a new approach to teaching and learning focused on wellbeing, which places joy at the heart of the curriculum and develops the global skills students need for their future academic, personal and career success.

Through six subjects; English, maths, science, computing, wellbeing and global skills projects, the *Oxford International Curriculum* offers a coherent and holistic approach to ensure continuity and progression across every student's educational journey, equipping them with the skills to shape their own future. Through this approach, we can help your students discover the joy of learning and develop the global skills they need to thrive in a changing world.

Unit starter

Student Book

Learning goals are stated clearly in every unit.

The Word cloud presents key words introduced in the unit. These are included in the write-in glossary for students to complete.

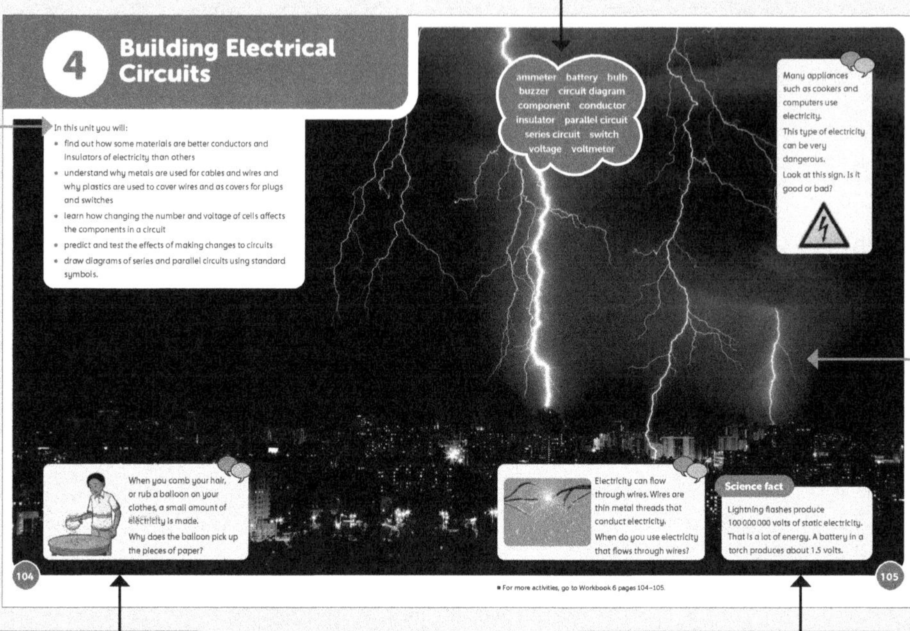

The introductory spread is bright and colourful to spark interest in young students.

Discussion activities allow students to develop communication skills.

Science fact boxes engage students to think about how science has developed or is used in everyday life.

Workbook

Every unit provides Language support throughout.

Each spread matches the Student Book lesson and consolidates learning.

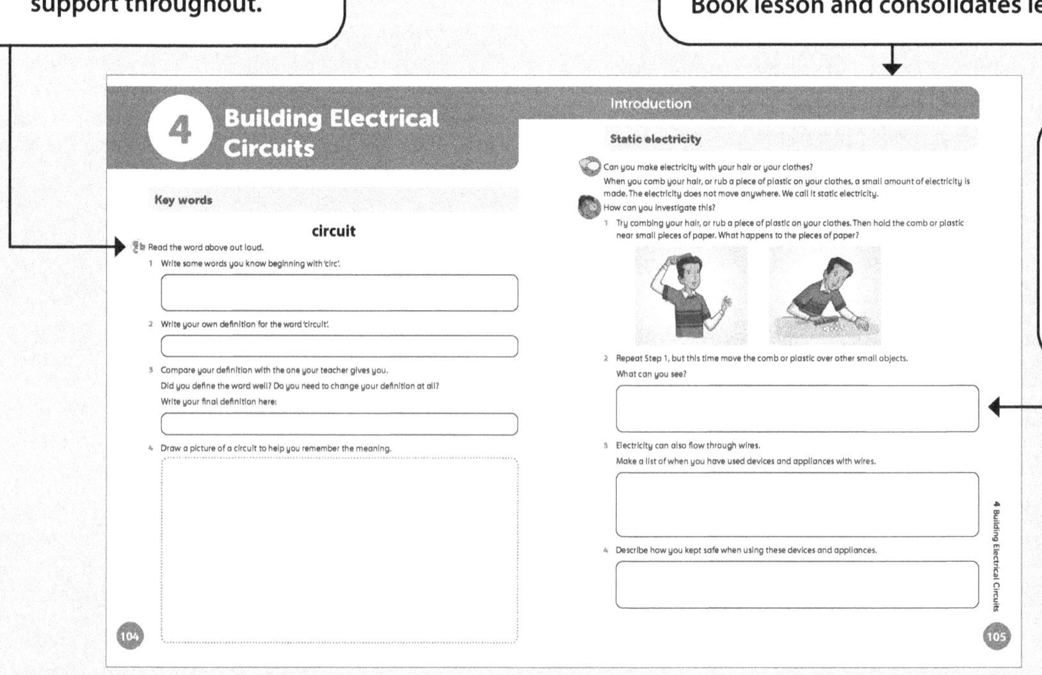

Write-in format allows students to record their work at school and at home.

Lesson pages

Student Book

Key word boxes show the main science vocabulary for the lesson.

Learning objectives for the lesson are clearly set out at the start and summarised in the Key idea box at the end.

Investigations engage students to work scientifically.

Warning boxes prompt students to identify risks and to learn how to keep themselves and others safe during practical work.

Think back boxes remind students of prior learning.

The student-friendly text is accessible for English language learners. Step-by-step instructions guide students through the activities they will undertake.

Be a scientist boxes help students to develop their skills to work scientifically.

Workbook

Carefully scaffolded activities promote deep learning.

Illustrations engage and support students to learn English.

Stretch zone activities challenge the most confident students.

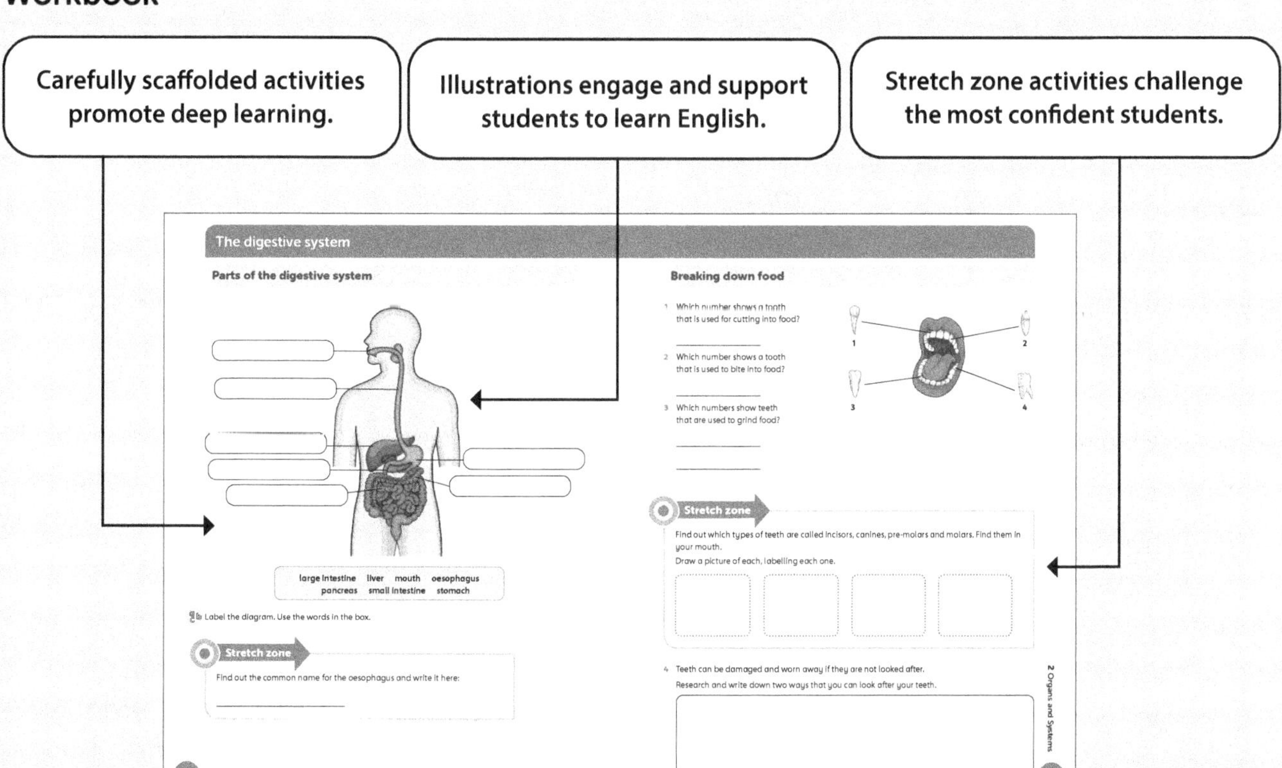

What have I learned? pages

Student Book

Students' progress is assessed through the questions and tasks at the end of each unit.

Be a scientist questions encourage application of science knowledge and skills.

Workbook

The review pages include a reminder of all the topics learned in the unit.

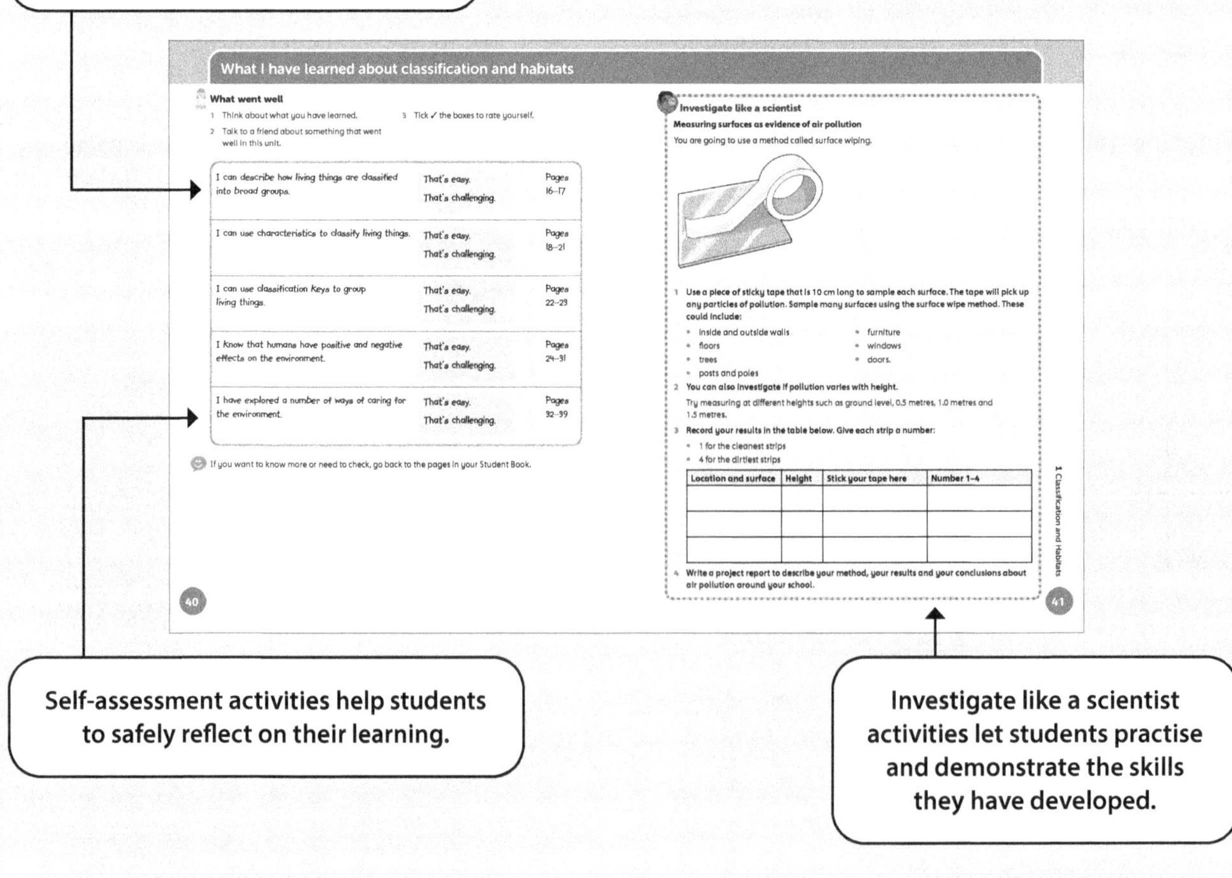

Self-assessment activities help students to safely reflect on their learning.

Investigate like a scientist activities let students practise and demonstrate the skills they have developed.

Teaching Notes

These pages are designed to support you in teaching your students how to tackle scientific enquiry, i.e. investigative approaches within science lessons. The structure of the pages closely follows the layout of the pages in the Student Book so you can guide students stage by stage through the process.

The diagram shows the important ideas about scientific enquiry presented to students.

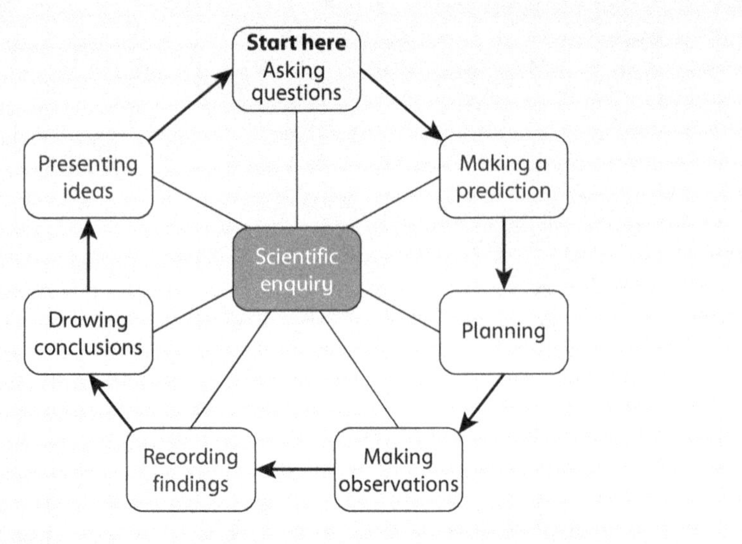

Teaching techniques

Students now have five years' experience of carrying out investigative tasks. They have developed their skills to become more confident of the steps to carry out effective scientific investigations. In Year 5 they learned more about the types of variables and how to plan and carry out fair tests. They carried out more accurate measurements using a variety of devices and learned to analyse more complex data.

This year, students will develop their investigation skills further by considering different types of investigations that develop out of a more sophisticated way of asking questions. They will plan in more detail, be more independent learners and take more responsibility for their safety and the safety of others.

The approach to scientific enquiry adopted throughout this scheme is to encourage students to be more responsible for identifying what should be investigated and how. Students are encouraged to reflect on the outcomes and explain what their results tell them. This can be summarised as:

WHAT am I going to investigate?

HOW am I going to investigate this?

WHAT do my results tell me?

It is important to realise that students do not have to carry out full investigations all of the time. You can concentrate on one or two phases of the scientific enquiry process. For example, present students with results from secondary sources and ask them to make sense of them. Or plan an investigation and discuss this but do not actually carry out the investigation. This is a good way to develop scientific enquiry skills. However, it is also important to allow students to put these together and carry out full investigations. This is when they are being scientists.

Investigations are best done within the context of the science ideas being studied at the time. They should never be a bolt-on, additional activity. Students need sufficient background knowledge to make sense of the investigation. The scientific enquiry ideas within the scheme start with a suitable stimulus. It is important to set the context within which students can ask questions. They need enough information to understand the basic scientific ideas and formulate questions to be asked, but not too much so that all curiosity and discovery is removed. A misconception is that an enquiry-based approach leaves students to find out everything. This is not the case. Good enquiry-based learning allows students to take steps of discovery on their own from a basis of confident understanding of concepts. This scheme is designed to support you in achieving this.

Work through the 'Being a Good Scientist' pages on pages 6–13 of the Student Book with students. This is an introductory lesson or two at the start of the year. Having set the scene, you may wish to then work through the planning grid on the next pages of this Teacher's Guide. Students can then use this as they work through the rest of the scientific enquiry phases.

Allow students to read through the Student Book text. Ask them to talk about the investigative process in the diagram. Explain that it is the same as in Year 5 but they will be using it to plan and carry out more detailed investigations. Next, work through the sections of the process explained on the next few pages of the Student Book and ask students to talk about the discussion questions.

Asking questions

The key to effective scientific enquiry is to encourage students to ask questions. All good research is based on a question. It is useful to suggest that students start questions with words such as 'which', 'what', 'do' and 'does' but at this age they will realise there are other questions that start with other words, such as 'is'. They are given some questions that would fit in with the example investigation. These are split into:

- Verification questions – such as 'Do all plants have flowers?'

Explain that they do not need to know much about plants to find out the answer to this question. The answer to verification questions will often be 'yes' or 'no'.

- Theory questions – such as 'Why does breathing rate increase during exercise?'

Point out that students will need to know some background science to tackle this question. They will need to know something about air, breathing, the lungs and circulation. The answer cannot be 'yes' or 'no' and will require students to link their answer to some science theory. The key to this type of question is to suggest WHY something happens.

- Experimental questions – such as 'Can a mirror make its own light?'

These are questions that are answered through carrying out a fair test. Some prior knowledge is needed and students will need to design tests that can be repeated by other people – they are testable. The emphasis is not on why but on WHAT.

At any time when you are not focusing on investigative work you can still develop questioning skills by modelling good practice. When undertaking any sort of scientific work keep asking students 'What would happen if …' or 'Why does this …' types of question.

Making a prediction

It is here that you can encourage students to discuss their ideas about what they think will happen in an investigation. Stress that a prediction is more than a guess. Students should use what they already know to help them. To stop a prediction being a guess, students should try to give a reason. Help students to understand this by setting up a number of situations when they can provide you with a prediction. Include practice in predicting at the start of any practical work in science.

Planning

Students should be able to design their own plans for investigations. Talk them through the prompt ideas and questions in the planning box on page 6 of the Student Book and emphasise the need to keep everyone safe. You can then read through the text about planning on page 8. Point out that when planning, two key questions are:

- What will you keep the same?
- What will you change?

Write down the three types of investigations so students see them in isolation (descriptive, comparative and experimental). Ask them to read the description of each.

Stress what a causal effect is. This is a new term for students, though they will have experienced it in investigations and in life. Also, remind students that the factors that are kept the same or changed are known as 'variables'. Ask them to explain what independent, dependent and control variables are. They can check the text on page 9 to help them.

Also talk to students about the equipment they will need and remind them that it is a good idea to discuss and share plans before starting. Ensure that they understand that secondary sources of information include books, magazines, information leaflets and the internet.

Making observations

This phase relies on observation skills and often the accurate use of measurement. For example, in the investigations used in the Student Book and the Workbook students carry out many different observations such as plant and animal characteristics, soil erosion, pH colour changes, litter composition, pulse and breathing rates, starch testing, reflection, refraction and investigating circuits. Talk to students about the need for accuracy and repeated measurements.

Recording findings

Remind students that there are many different ways to record their results. They are now experienced in using tables and charts but point out they will be designing their own tables and producing more complicated charts and graphs. Stress that a table keeps all of their results neat and tidy. It can help them to see patterns. Students can also use their results to make a chart or graph and point out that the independent variable is written along the x (horizontal) axis and the dependent variable is shown up the y (vertical) axis.

Drawing conclusions

Encourage students to look at their results carefully. Emphasise the need to look for patterns. Ask students to consider if any of their results were unusual. Discuss the value of repeating investigations to check how accurate the results are. Also encourage students to share their results to see if they are the same as other groups in the class. Ask them if their prediction was correct.

At the end of every investigation ask students if they can think of any improvements. This is an important part of scientific enquiry. Also ask if the investigation made them think of any other questions. Stress that good scientific enquiry always leads onto other questions. These can lead onto more investigations.

Presenting ideas

Encourage students to present their ideas in a variety of ways. Not only does this help them to develop a range of skills but it also models how science is communicated in real-word contexts. Point out the 'Tips for presenting ideas' section on page 13 of the Student Book and allow students to keep referring to it during the year.

Scientific enquiry planning grid

Asking questions

What am I trying to investigate? What is my question?

Making a prediction

What do I already know that will help me to decide what will happen?

I think what will happen is …

My reason is that …

Planning

Which type of test will I need to carry out?

The equipment I will need is …

What is the independent variable?

What is the dependent variable?

What I am going to measure is …

What are the control variables?

What I am going to do is …

I will be careful of …

My drawing of what I will set up:

Making observations

What observations and measurements should I make?

How can I make my observations accurate?

Which measuring devices can I use?

Recording findings

What is the best way to record data?

Will I use a table, chart or graph?

Will I draw diagrams or take photographs?

Drawing conclusions

Can I see any patterns?

Are any results unusual?

Do the results support my prediction?

How could I make my investigation more accurate?

Presenting ideas

How should I present my ideas to others?

Have I used scientific language and illustrations?

Does my work lead to other questions to study?

1 Classification and Habitats

In this unit students will:

- describe how living things are classified into broad groups according to common observable characteristics, and based on similarities and differences, including microorganisms, plants and animals
- give reasons for classifying plants and animals based on specific characteristics
- use classification keys based on observable characteristics
- find out how humans have positive (good) and negative (bad) effects on the environment
- learn about a number of ways of caring for the environment.

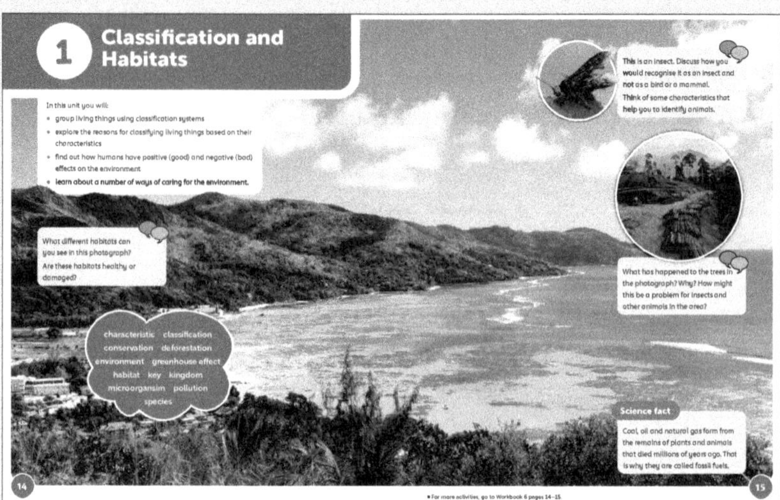

Supporting activities are in Workbook 6 pages 14–15.

Getting started

This unit explores how the characteristics of plants and animals are used to classify them. Students will review their knowledge of living and non-living things and use and design classification keys. They will learn about the classification of microorganisms and research some examples of each type. Students then move onto consider ways that human activities can have a harmful effect on the environment. They learn about air pollution, the greenhouse effect, and damage caused by quarrying and deforestation. They are encouraged to investigate water pollution and waste disposal and to consider recycling and reuse as ways of reducing waste. They will survey litter and consider ways of reducing litter. Students learn about renewable and non-renewable sources of energy. Finally, students carry out enquiry-based studies to learn about ways to encourage others to care for the environment.

Science in context

Use the lessons in this unit to encourage students to learn more about the importance of grouping and identifying living things in real life. Allow students to survey and investigate plants and animals in their local area. Take them out to see living things in the wild, in parks and on farms. Encourage students to find out about people who use keys as part of their work and invite them into school to talk about how they use classification and identification. Students can also survey the local area to consider how human activity is damaging habitats and to look at ways that pollution and litter are being managed. You could arrange a visit to see alternative energy being used and invite a person from the local power company into school to talk about how energy is generated and used in the area.

Scientific enquiry skills

Scientific enquiry skills for this unit focus on observation skills and the setting up of fair tests. Remind students of the different variables they must consider during investigations: independent (what they change); dependent (what they measure); and control (what they keep the same). Students further develop their skills of collecting, recording and presenting data. They will consider the accuracy and validity of the observations and measurements they take and be encouraged to evaluate their work. Students are asked to present their results in a variety of ways, including drawings, bar charts, graphs and tables. Students should be encouraged to use computer technologies to help in collecting and presenting data.

You can use the Investigation master sheet on pages 4–5 of this Teacher's Guide to support investigative work. This provides prompts and structure to support students in planning and carrying out fair tests and recording and drawing conclusions about their findings.

Resources

Student Book: writing materials; materials to make leaflets; variety of seeds; petri dishes; rulers; sets of different types of plastic forceps; timer; access to an outdoor area suitable for a plant and animal survey; identification keys for common species found in a local habitat; graph paper; clear plastic bottles with a section cut out; soil; twigs and sticks; grass seed; measuring jugs; water; watering cans; small cups or bottles; five water samples with different pH values; pH indicator strips; materials to make posters; access to an outdoor area suitable for a litter survey; gloves; large sheets of paper; pre-prepared packages of information about litter, saving energy, acid rain; additional stimulus materials to capture the imagination of students; access to the internet or books on littering, saving energy, acid rain; access to specialist speakers where possible; magazines with suitable images to cut and stick on the leaflets or posters.

Workbook: access to an outdoor area, including pond if possible; hand lenses or microscopes; cameras (optional); writing materials; materials to make leaflets; variety of seeds; petri dishes; rulers; sets of different types of plastic forceps; access to the internet or books about birds; large sheets of paper; access to the internet or local newspapers; materials to make leaflets or booklets; pieces of dowelling or skewers; clear plastic bags or plastic food wrap; thermometers; five water samples with varying degrees of turbidity; materials to make posters; variety of clean waste materials (such as plastic bottles, cardboard, yoghurt pots); string; sticky tape; glue; access to an outdoor area suitable for a litter survey; gloves; large plastic bags; white kitchen towel; tissue paper or pieces of clean fabric; access to various surfaces, e.g. floors, inside and outside walls, trees, furniture.

Key words for unit

Bold words are in the Word cloud and are included in the glossary.

acid   acid rain   alkali   atmosphere   **characteristic**   class   **classification**   classification key   climate change   **conservation**   **deforestation**   endangered species   energy consumption   **environment**   erosion   extinct   fossil fuel   **greenhouse effect**   group   **habitat**   **key**   **kingdom**   landfill   litter   **microorganism**   non-renewable energy   pH scale   **pollution**   quarrying   recycle   reduce   renewable energy   reuse   sample   solar panel   **species**   variable   waste

Scientific enquiry key words

Plan and/or carry out enquiries to answer questions

Make predictions

Recognise and control variables

Make observations

Take measurements, using equipment accurately

Record data and results

Analyse data, notice patterns and group or classify things

Report and present findings

Draw conclusions and give explanations

Identify causal relationships

Language support

At this age, students are likely to be able to read independently and look up any words they find unfamiliar, so encourage them to use scientific dictionaries. Start the unit by reading out the words in the Word cloud. Ask students to discuss each word and define those they are familiar with. This monitoring of prior knowledge is vital and can help you enormously in setting the level of work in the first few lessons. For every unit it is worth creating a Word wall so students see the words often and can become familiar with them. Students are old enough to make their own word cards for display so this could be an early task. Have daily quizzes about the words – point to one and ask, 'What does this word mean? Use it in a sentence.'

Remind students that their Student Book has a glossary at the back and they should add definitions of key words as they progress through the unit. This could be a regular end-of-lesson task or starter to encourage recall.

The activities on pages 14 and 15 of the Workbook are designed to support language development as they encourage students to think about some of the key words.

Repeat any new words regularly and use them in context. Students should listen, say, read and then write the words. They can also make a list of new non-science specific words they use in lessons. Some of these are listed in the individual lesson notes that follow.

Remember that the eBook has examples of key words being pronounced. Use this at the start of lessons and return to it at the end.

You could create a science library in your room. Collect resources such as science books and magazines, science encyclopaedias and dictionaries. You can also collect or download specific information about topics for each lesson or activity and make a booklet of these like a class magazine. Students will enjoy helping with the production of these small information booklets and they can include some of their own work. You will find these specific booklets very valuable support for lessons and especially Stretch zone activities.

Unit at a glance

The key teaching points for students in this unit are:

- to introduce the unit objectives
- to introduce the learning outcomes
- to engage students with the content of the unit
- to review and build on prior learning and understanding of the topics.

In the next lesson, students will learn more about classification systems and the five-kingdom classification of living things.

The purpose of this introductory lesson is for students to start thinking about and reviewing prior knowledge of habitats and the characteristics of living things. These introductory pages show a range of different habitats, an insect, and a forest with trees being cut down to prompt recall of earlier work and any general knowledge of plants. The photographs are used as a starting point and as a prompt for discussions.

Read through the key words and then allow students time to enjoy looking over the page before you start the sequence of discussion tasks. A suggested sequence follows below.

Arrange students into pairs or small groups of three or four for discussion work. A useful strategy is to start with students in pairs and then move pairs together to make small groups so students can share their ideas and discussions with others.

What different habitats can you see in this photograph? Are these habitats healthy or damaged?

Ask students to look carefully at the main photograph. They can discuss it, make a list of any habitats they see, and then decide if the habitats look healthy.

> **Possible response:** *Students should identify a range of habitats including forest/woodland; farmland; beach; town/urban and marine/sea. They all look healthy, though students may point out that habitats have been altered to build houses and create farmland.*

This is an insect. Discuss how you would recognise it as an insect and not as a bird or a mammal. Think of some characteristics that help you to identify animals.

Ask students to look carefully at the insect on page 15. Remind them to use their observation skills and to look carefully at every part of the insect.

> **Possible response:** *Students may observe the antennae, eyes, type of wing and segmented body as different from birds and mammals. Detailed insect anatomy is not needed.*

What has happened to the trees in the photograph? Why? How might this be a problem for insects and other animals in the area?

Allow students to work with a partner or with their small group to talk about the photograph showing deforestation. Point out that they will have to use prior knowledge of habitats as well as their observation skills. Ask volunteers to share their ideas with the class.

> **Possible response:** *The trees have been cut down to use for many possible things. Students may list some of the following: buildings; fuel; furniture; fences or making paper. This is a problem as the habitats for living things have been removed so they will lose where they live and their food supply. Some may point out that the soil can also wash or blow away.*

Science fact Coal, oil and natural gas form from the remains of plants and animals that died millions of years ago. That is why they are called fossil fuels.

Read out the Science fact or ask a volunteer to read it out. Ask students if they have seen or used any fossil fuels. Remind them that oil is used to make many things – such as vehicle fuels (diesel and gasoline/petrol), lubricating oils and plastics – so it is likely students have used fossil fuels even if they do not use coal or gas for heating and cooking.

Workbook activities

Key words (page 14)

Explain that students have to find some of the key words for the unit hidden in the wordsearch. As an extra challenge there is one word in the wordsearch that is not listed in the box. Ask them to circle the word in the grid, write the word down and add a definition.

> **Answer:** *Students should find and circle the following words: acid rain, conservation, deforestation, environment, habitat, litter, pollution, recycle. The missing word is key (row 13, columns 1–3). A key is used to identify living things through asking and answering questions.*

Test your key word memory! (page 15)

Ask students to look back at the key words on page 14 of the Student Book for 30 seconds. Tell them to then close the book and try the activity. If they need to, allow them to look back at the Student Book for another 30 seconds. They then ask a partner to check their answers. Encourage students to say the words out loud to their partners.

> **Answer:** *environment; species; conservation; pollution; key; kingdom; greenhouse effect; deforestation*

Extra activities

1 **Computing link:** Students can research three habitats that are found in their region. They can download pictures or draw large versions of the habitats and list some of the animals and plants found there. They can display these in the room.

2 **Computing link:** Students can research uses of wood or stone and locate any examples of forestry or quarrying in their area. They can then design and produce a poster in class or as homework.

Classification systems

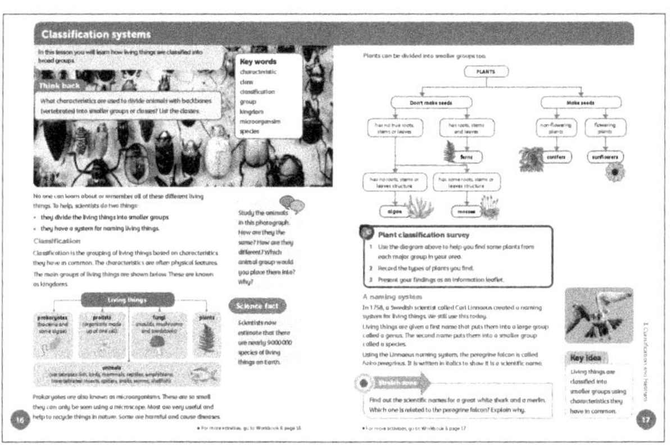

Supporting activities are in Workbook 6 pages 16–17.

Getting started

In this lesson students will learn that living things can be classified into five kingdoms. They will learn about using some major characteristics of plants to help to classify them into smaller groups. They will apply their new knowledge of plant groups to carry out a survey to find and observe examples.

Language support

To help develop language skills and review key words you can write the words that students will have heard of before (characteristic, class, classification, group, microorganism and species) on the board and ask students to tell their partner what each word means. They will have heard the word 'kingdom' before but in a different context. Explain that 'kingdom' is used to describe a huge group of living things. Point out that plants and animals are separate kingdoms.

Resources

Student Book: writing materials; materials to make leaflets.

Workbook: access to an outdoor area, including pond if possible; hand lenses or microscopes; cameras (optional); writing materials; materials to make leaflets.

Key words

characteristic   class   classification   group   kingdom   microorganism   species

Other words in the lesson

conifer   falcon   fern   fungi (moulds, mushrooms, toadstools)   genus   moss   prokaryotes (algae and bacteria)   protists

Scientific enquiry key words

Plan and/or carry out enquiries to answer questions

Make observations

Record data and results

Analyse data, notice patterns and group or classify things

Report and present findings

Lesson at a glance

The key teaching points for students in this lesson are:

- living things can be classified into five kingdoms
- the living things within a kingdom can be further classified into smaller groups
- physical features, called characteristics, are used to group similar animals.

In the next lesson, students will learn more about how animals are classified according to their characteristics.

Think back: What characteristics are used to divide animals with backbones (vertebrates) into smaller groups or classes? List the classes.

Ask students to think back to their work in earlier years about vertebrates. They can discuss some of the characteristics and then make a list to share with the class.

Answer: The classes are fish, amphibian, reptile, bird and mammal. Characteristics include skin covering, whether they lay eggs, how they feed their young, how they breathe, and if they are cold blooded or warm blooded.

Study the animals in this photograph. How are they the same? How are they different? Which animal group would you place them into? Why?

Allow students to work with a partner to talk about the photograph showing the different insects (beetles). Encourage them to observe similarities and differences and decide which animal group they belong to.

Answer: The animals are insects (beetles). Similarities include: many have legs, antennae, hard coverings (wing cases), eyes, and are made up of head and body. Differences include: size variation, colour variation, and length of legs and antennae.

Ask a volunteer to read the text under the photograph to the class. Point out that classification is sorting living things into similar groups. Let them study the diagram of the five-kingdom classification system. Ask them to talk about any of the kingdoms they have heard about before.

Science fact Scientists now estimate that there are nearly 9 000 000 species of living things on Earth.

Read out the Science fact or ask a volunteer to read it out. Ask students why scientists cannot know about every type of living thing, especially when there are nine million different species (types).

Students will need support in understanding prokaryotic organisms (microorganisms). Ask them to read the text at the bottom of page 16 and explain that prokaryotic organisms have a very simple, single-celled structure. To illustrate how small they are you could point out that prokaryotic cells, such as bacteria, can be 100 times smaller than a human cell. You could also ask them to look at 1 millimetre on a ruler and imagine that they could lay 1000 cells between the graduations. Tell students that they will be studying some prokaryotes in the next unit when they learn more about diseases, but also let them know that microorganisms play a vital role in recycling materials in the environment and we use them in making many foods – such as yoghurt, cheese and vinegar.

Point out the plant key at the top of page 17. Talk through the classification questions with students. You can explain that this is a dichotomous key – each question has two possible answers or branches. Di- means two. Tell them they should always answer the question then follow the line or branch to the next question until they arrive at the answer.

Investigation: Plant classification survey

The worksheet on page 17 of the Workbook supports this investigation. Allow students to work in small groups of three or four. Explain that they are going to use the key they have just studied to help them find some plants from each major group in your area. They will record the types of plants they find and present their findings as an information leaflet.

Ask students to read the text beneath the investigation box. They will learn about Carl Linnaeus, who created a naming system for living things. Explain that we still use this today. Point out that living things are given a first name that puts them into a large group called a genus. The second name puts them into a smaller group called a species. Use the peregrine falcon, *Falco peregrinus*, as an example and explain why it is written in italics.

Stretch zone: Find out the scientific names for a great white shark and a merlin. Which one is related to the peregrine falcon? Explain why.

Ask students to work with a partner and allow them access to the internet, books or wildlife magazines. They can write down the names and compare their findings with another pair.

Key idea

Living things are classified into smaller groups using characteristics they have in common.

Read through the key idea or ask a volunteer to read it out to the class. Ask students to tell a partner the name of a group of living things that is smaller than a kingdom.

Workbook activities

Identifying the correct kingdom (page 16)

Point out the drawings of the examples of living things from the five main groups of living things. Explain that students have to decide which kingdom each of these living things belongs to and record their answers in the table. They can look at the Student Book for support. They can then answer the questions about vertebrates and flowering plants.

Answer: Prokaryotes = staphylococcus, spirillum; protists = amoeba, Euglena; fungi = mushroom, mould; plants = fern, hibiscus, seaweed; animals = fish, crab, worm, camel. The vertebrates are fish and camel. The flowering plant is hibiscus.

Plant classification survey (page 17)

This activity supports the investigation on page 17 of the Student Book. Ask students to use the plant classification key in the Student Book.

Warning! Stay with your group. Do not touch any plants without permission from an adult. Discuss why this is important.

Read out the warning and ask students to discuss why the advice is so important. Ask them how they will keep themselves and other students safe.

Arrange for students to visit a local park, garden or botanical centre to look for different plants. If there is a pond nearby, let them study pond water through a hand lens or microscope. Ask students to draw or take photographs of any plants they observe and help them to find some plants from each major group. Remind them to use the key to help them. Students should record the types of plants they find in the table provided. They can present their findings as an information leaflet.

Possible response: This will vary depending on the area but students should find many examples of flowering and non-flowering plants (conifers, ferns, mosses) and, if you take them to the shore or a pond, they may find algae.

Ask students to test a partner by saying the name of a major group of living things – starting with the five kingdoms – and their partner has to describe it. They can take it in turns until all of the kingdoms have been covered. They can then do the same for the main groups of plants.

After this you can allow some quiet time for students to reflect on what they answered well in the lesson and to think of one way they could remember better the different types of plants – algae, mosses, ferns, conifers and flowering plants.

Conclude the lesson by asking students to complete question 1 in the 'What have I learned about classification and habitats?' activity on page 40 of the Student Book and the first statement on page 40 of the Workbook. (See also the teaching notes in this Teacher's Guide, pages 43–45.)

Extra activities

1 Students can make a poster about plant groups by collecting samples of different plants (ferns, conifers, flowering plants, algae and mosses), pressing them, and then sticking them onto poster paper with labels.

2 **Computing link:** Ask students to work in a small group to research one of the five kingdoms in more detail. They can produce a short scientific article about the living things found in the group. They can download photographs of the living things and describe some examples of where they are found.

Differentiation

Supporting: Large versions of the five-kingdom classification key can be displayed around the room to help students become familiar with it.

Consolidating: Display photographs of animals, flowering plants, non-flowering plants, fungi, protists and prokaryotes and regularly ask students to stand up and point to them as you say the words.

Extending: Students could research other classifications, such as the six- and seven-kingdom classifications.

Differentiated outcomes	
All students	should be able to state that living things can be classified into kingdoms using their characteristics
Most students	will be able to name and describe the five kingdoms and list some characteristics used to classify them
Some students	may be able to explain that there are other classification systems including six- and seven-kingdom systems

Using characteristics to classify animals

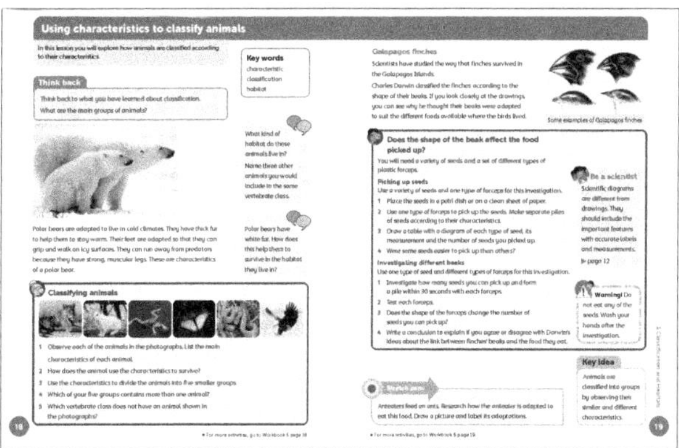

Supporting activities are in Workbook 6 pages 18–19.

Getting started

In this lesson students will study some of the specific physical characteristics used to classify the members of the animal kingdom into smaller groups. They will relate these characteristics to how animals are adapted to their habitats and investigate an example involving the shape and size of bird beaks.

Language support

Students will be familiar with the words 'characteristic', 'classification' and 'habitat' but it is worth reviewing this. Write the words on the board and then write down the phrases 'what it looks like', 'where an animal lives' and 'the grouping of living things'. Ask students to link each word to a meaning. Listen to their suggestions and then draw thick lines to show the answers.

Resources

Student Book: variety of seeds; petri dishes; rulers; sets of different types of plastic forceps; timer.

Workbook: variety of seeds; petri dishes; rulers; sets of different types of plastic forceps; access to the internet or books about birds.

> **Key words**
>
> characteristic   classification   habitat
>
> **Other words in the lesson**
>
> adapted   climate   finch   forceps   predators   vertebrate

Lesson at a glance

The key teaching points for students in this lesson are:

- animals have physical characteristics that help them to adapt to their habitats
- these characteristics can be used to classify animals into groups.

In the next lesson, students will learn about characteristics used to classify plants and microorganisms.

Think back: Think back to what you have learned about classification. What are the main groups of animals?

Students can work with a partner. Ask them to list the main groups of animals. Remind them some will have a backbone and an internal skeleton and some will not. If you wish, they could divide these groups further into smaller groups or give examples.

> **Answer:** *The two main groups are invertebrates and vertebrates. Students can further classify vertebrates into fish, amphibians, reptiles, birds and mammals. They may mention insects, spiders, snails, worms, jellyfish and shellfish as examples of invertebrates.*

Ask students to study the photograph at the top of page 18. They can also read the text that accompanies this.

 **What kind of habitat do these animals live in? Name three other animals you would include in the same vertebrate class.**

Students can work with a partner or in a small group of three or four. If you choose the latter, they could stay in their group for the discussion task and investigation that follows. Point out the background as well as the animals.

> **Answer:** *The polar bears live in cold habitats. Other examples include any mammals such as camels, cows, horses, seals and elephants.*

 **Polar bears have white fur. How does this help them to survive in the habitat they live in?**

Ask students to think back to earlier work in Year 4 on adaptations and predator–prey relationships. Encourage them to think about how animals might approach prey without being seen, or how they might hide from predators. Ask for volunteers to share their ideas with the class.

> **Answer:** *The white fur helps polar bears to be camouflaged so they can approach prey without being seen. The fur is also warm and waterproof to help them to stay warm.*

 ## Investigation: Classifying animals

The worksheet on page 18 of the Workbook supports this investigation. Allow students to work with their partner or in a small group. They will list the main characteristics of each animal and think about how the animal uses the characteristics to survive. They then divide the animals into five smaller groups based on the characteristics. Ask them to determine which of their five groups contains more than one animal and also work out which vertebrate class does not have an animal shown in the photographs.

> **Possible response:** *The characteristics should include: nature of body covering (fur, scales or feathers for example), shape, legs or not, wings or not, types of wings, and antennae or not. These various physical characteristics are used to help the animal stay alive as the coverings can protect an animal and help it to hide, the legs and wings help it to move, and the ears, antennae and eyes help it to sense the surroundings. The mammal group has two members. Fish are not represented.*

After completing the investigation ask students to read through the information about Galapagos finches at the top of page 19 and encourage them to look at the pictures of the finches. Ask, 'Which finch do you think is adapted to eat small seeds? Which finch might be adapted to eat larger seeds or nuts?'

 ## Investigation: Does the shape of the beak affect the food picked up?

Warning! Do not eat any of the seeds. Wash your hands after the investigation.

Read out the safety warning and ask students to discuss why the rules are important.

Next, explain that they are going to use a variety of seeds and a set of different types of plastic forceps to model bird food and beaks. The worksheet on page 19 of the Workbook supports this investigation.

To investigate picking up seeds, ask students to follow the instructions and find out how many of the different sized seeds they can pick up using only one type of forceps. Make sure they make separate piles of seeds according to their characteristics. They should draw a table with a diagram of each type of seed, its measurement (length), and the number of seeds they picked up. Ask students to decide if some of the seeds were easier to pick up than others.

To investigate different beaks, ask students to use one type of seed and different types of forceps for this investigation. They should count how many seeds they can pick up in 30 seconds with each type of forceps. They need to form a pile of each seed picked up. After they have tested each type of forceps, they can analyse their results and find out if the shape of the forceps changes the number of seeds they can pick up.

After both investigations ask students to write a conclusion to explain if they agree or disagree with Darwin's ideas about the link between finches' beaks and the food they eat.

> **Possible response:** *Students should find that small forceps are better at picking up small seeds and large forceps are better at picking up larger seeds. This matches the adaptations of beak sizes.*

Be a scientist: Scientific diagrams are different from drawings. They should include the important features with accurate labels and measurements. (Student Book, page 12)

Read out the Be a scientist information or ask a volunteer to read it out. Point out that when they draw the seeds they are recording important information about them that would be difficult to show with just words.

Stretch zone: Anteaters feed on ants. Research how the anteater is adapted to eat this food. Draw a picture and label its adaptations.

Computing link: Allow students to work individually or with a partner and allow them access to the internet to research anteaters. They can download photographs to help with their drawings.

> **Possible response:** *Anteaters have an excellent sense of smell, a long sticky tongue, a narrow mouth with no teeth to help the tongue flit in and out, strong acid in the stomach to dissolve ants, strong claws to break into termite nests, and strong tails to help them stand upright to reach further.*

Key idea

Animals are classified into groups by observing their similar and different characteristics.

Summarise the lesson by asking students what they have learned. Let them share their ideas. Ask one student to read out the key idea. Ask students to write down three groups of vertebrates and, for each one, write down one adaptation that helps the animal to live in its habitat.

Workbook activities

Classifying animals (page 18)

This activity supports the investigation on page 18 of the Student Book. Ask students to choose four animals from the list and then use the table to help them classify their chosen animals.

After completing this, students can think about the Stretch zone questions. Ask them which animal in the list is not a vertebrate. They should then name the animal and explain why it is not classed as a vertebrate.

> **Answer:** *Lion or lioness = body covered in fur, four legs. Fur keeps it warm and helps it to be camouflaged. Legs help it to catch prey. Vertebrates – mammal.*
>
> *Snowshoe hare = fur and legs, long ears. Fur keeps it warm and helps it to be camouflaged; good hearing helps it to detect predators. Vertebrates – mammal.*
>
> *Frog = body covered in smooth skin, long back legs, large eyes. Skin helps it to move through water; long legs help it to hop and swim; good eyesight helps it to catch prey. Vertebrates – amphibian.*
>
> *Butterfly = delicate wings, antennae and long tongue. Wings allow flight; antennae sense the environment; the tongue reaches nectar in flowers. Invertebrates – insect.*
>
> *Snake = body covered in hard, dry scales, long flexible body, fangs and poison, forked tongue. Scales protect the snake; flexible body helps it to move; tongue senses the environment; fangs help to catch prey. Vertebrates – reptile.*
>
> *Eagle = large wings, sharp beak, sharp claws and feathers. Wings allow flight; beak and claws help it to catch and eat prey; feathers keep it warm and dry. Vertebrates – bird.*
>
> *Only one animal in the list is not a vertebrate and that is the butterfly. It does not have a backbone or an internal skeleton.*

Different shaped beaks (page 19)

This activity supports the investigation on page 19 of the Student Book. Ask students to try to pick up each of the different seeds. They should use the different tools you have given them. Point out the table and ask them to use it to record the results. They should analyse their results to determine whether the shape of the forceps changes the number of seeds they could pick up. Ask them if the results support or contradict (do not support) the idea that the shape of the beak makes it easier to pick up some seeds more than others.

Computing link: Students can then tackle the Stretch zone. Allow them access to the internet to research the birds listed and plan a short report to tell people how the birds' beaks are adapted to the food they eat.

Review and reflect

Encourage students to reflect on their own learning by pausing for a few moments and thinking about the way they undertook the investigations. Talk about how they managed this and how they worked as a team. Remind them that when they are doing tricky work they are training their brain and this will help their future learning.

Extra activities

1 Students can make card or clay models of a specific animal to show the main adaptations that help it to survive in its habitat but are also used to classify it. You can bring the models together to make a larger animal exhibition.

2 **Computing link:** Ask students to research some of the other beak sizes and shapes found in the Galapagos finches and challenge them to find out information about the islands they come from and the foods they eat.

3 **Maths link:** Students can develop their mathematical skills by repeating the investigations or collating results from other groups and calculating an average for each of the type of seeds picked up by the various forceps. They can present the results of their investigations by drawing bar charts showing the number of seeds picked up by the different tools.

Differentiation

Supporting: Make cards with a photograph of an animal on each one and ask students to sort them into similar groups.

Consolidating: Create a picture wall of animals that are in the same group – for example, a group of reptiles or a group of mammals. Then ask students to make their own group to add to the wall.

Extending: Ask students to research some ways that invertebrates are split into smaller groups and ask them to list some examples.

Differentiated outcomes	
All students	should be able to describe some characteristics used to classify animals into smaller groups
Most students	will be able to classify animals into some major groups through observation of characteristics
Some students	may be able to describe how some invertebrates are classified into smaller groups

Using characteristics to classify plants and microorganisms

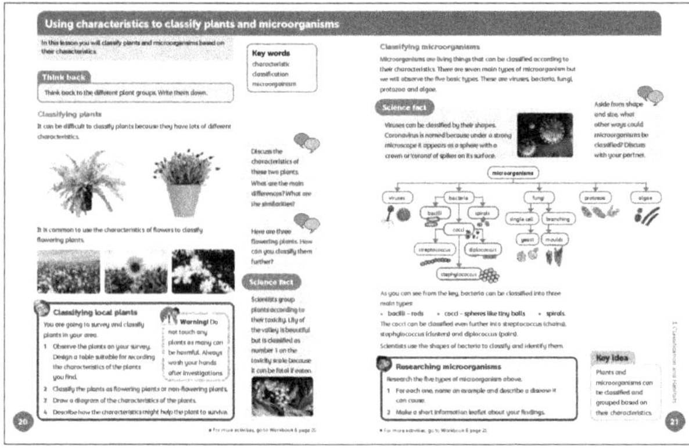

Supporting activities are in Workbook 6 pages 20–21.

Getting started

In this lesson students will study some of the characteristics that can be used to classify plants and microorganisms. They will carry out a survey of local plants to observe and record different characteristics and classify the plants as flowering or non-flowering. They will then explore the classification of microorganisms and use a key to learn about the main groups. Finally, they will carry out a research activity to find out more about groups of microorganisms.

Language support

Students are now familiar with the words 'characteristic' and 'classification' but ask them to tell you what the words mean to review them. Then ask students to try to define the word 'microorganism'. They may have heard of the word before but, if not, give them a clue by asking them to think of other words beginning with the prefix 'micro'. Then tell them it means very small.

Resources

Student Book: access to an outdoor area; writing materials; materials to make leaflets.

Workbook: access to an outdoor area; writing materials.

Key words

characteristic   classification   microorganism

Other words in the lesson

algae   bacteria   cocci   fatal   flowering plant
fungi   non-flowering plant   protozoa   toxicity
virus

Scientific enquiry key words

Plan and/or carry out enquiries to answer questions

Make observations

Record data and results

Analyse data, notice patterns and group or classify things

Report and present findings

Draw conclusions and give explanations

Lesson at a glance

The key teaching points for students in this lesson are:

- plants can be classified as flowering plants and non-flowering plants
- microorganisms can be classified into smaller groups and this is often done based on shape.

In the next lesson, students will learn about the use of keys to classify and identify plants and animals.

Think back: Think back to the different plant groups. Write them down.

Ask students to think about their work from an earlier lesson and then talk about it with a partner. They can make a list and share their list with another group.

> **Answer:** *Plant groups are algae, mosses, ferns, conifers and flowering plants.*

Ask students to look at the photographs and text at the top of page 20. You can also display a variety of potted plants for students to handle and observe. Include some flowering plants and some non-flowering plants such as ferns and mosses.

Discuss the characteristics of these two plants. What are the main differences? What are the similarities?

Students can work in pairs or small groups. Ask them to observe the plants carefully and list some of their characteristics. They can then compare and contrast the plants.

> **Possible response:** *Characteristics include shape, colour and size of leaves, number of leaves, and presence or absence of flowers. Both plants have green leaves but the left plant has no flowers and larger leaves than the right plant.*

Here are three flowering plants. How can you classify them further?

Students can continue to work in pairs or in their small groups. Point out the three flowering plants in the photographs and ask students to look carefully to identify any differences that can be used to split them into separate groups. They can share their ideas with the class.

> **Possible response:** *Characteristics to use to classify the plants further include height of plant, size and shape of leaves, colour of flowers, size of flowers and number of petals.*

Science fact Scientists group plants according to their toxicity. Lily of the valley is beautiful but is classified as number 1 on the toxicity scale because it can be fatal if eaten.

Read out the Science fact and ask students if they have heard of any toxic or poisonous plants. Point out that this fact helps to explain why they should never touch or taste plants without permission. Explain that toxicity isn't the main way plants are classified but it is one more piece of evidence to help scientists.

Investigation: Classifying local plants

Warning! Do not touch any plants as many can be harmful. Always wash your hands after investigations.

Read through the warning with students. Ask them to discuss why they should not touch any plants and why they should wash their hands after an investigation.

Take students out to an area of ground near the school, such as a forest or park, and allow them to work in groups to follow the instructions and complete their survey. Ask them to classify any plants they see as either flowering plants or non-flowering plants. As they record findings, they should draw a diagram of the characteristics of the plants and describe how the characteristics might help the plant to survive.

> **Possible response:** *Students should find a range of flowering and non-flowering plants. Characteristics will be structures that help plants to make food (leaves), obtain water (roots), and prevent animals from eating them (hairs or spines).*

Ask students to read through the text about microorganisms at the top of page 21. Explain that most microorganisms are too small to be seen with the naked eye – a microscope is needed.

Science fact Viruses can be classified by their shapes. Coronavirus is named because under a strong microscope it appears as a sphere with a crown or 'corona' of spikes on its surface.

Read out the Science fact to give students some idea of the shape of one virus. Ask for volunteers to point out the corona of spikes on the surface of the virus.

Aside from shape and size, what other ways could microorganisms be classified? Discuss with your partner.

Students can continue to work in their groups to talk about the ways that microorganisms could be classified. Hint that animals can be classified in part by what they eat – herbivores and carnivores.

> *Possible response: Microorganisms are also classified by what they grow on – such as starchy growth plates or sugary growth plates. They also form different coloured and shaped clumps when they grow and can produce different chemicals.*

Point out the microorganism classification key and ask volunteers to read out the names of the five main groups: viruses, bacteria, fungi, protozoa and algae. Work through the key with students as they trace it with their fingers. Ask them to read the text beneath the key to find out about the shapes of the bacteria. Explain that 'cocci' means 'spheres like small balls'. Point out that diplococcus means 'two cocci', streptococcus means 'chains of cocci' and staphylococcus means 'like a bunch of grapes.'

Explain that larger fungi, such as mushrooms and toadstools, are not classed as microorganisms. The classification of fungi is complicated, but students just need to know that very small fungi, such as yeast and moulds, can be classified as microorganisms.

Investigation: Researching microorganisms

Computing link: Students can work with a partner or in their group to research the five types of microorganisms shown in the key. For each one, ask them to name an example and describe a disease it can cause. They should make a short information leaflet to share their findings.

> *Possible response: Examples of types of microorganisms and a disease include: virus = influenza (flu) and poliovirus (polio); bacteria = Salmonella (food poisoning), Vibrio (cholera) and Clostridium (tetanus); fungi = Microsporum (ringworm) and Trichophyton (athlete's foot); protozoa = Plasmodium (malaria) and Entamoeba (dysentery); algae = blue-green algae (shellfish poisoning).*

Key idea

Plants and microorganisms can be classified and grouped based on their characteristics.

Read through the key idea or ask a volunteer to read it out to the class. Ask students to close their books and write down the names of the five main groups of microorganisms, and then check their answers.

Classifying local plants (page 20)

Warning! Some plants can be toxic or cause irritation. Do not touch any plants. Always wash your hands after investigations.

Read through the warning with students and ask them to discuss why they should not touch any plants and why they should always wash their hands after investigations.

Explain that this activity supports the investigation on page 20 of the Student Book. Allow students to use the worksheet to help them to record their results. They can answer the prompt questions to help them to reflect on their findings.

> *Possible response: Students should find a range of flowering and non-flowering plants. Characteristics will be structures that help plants to make food (leaves), obtain water (roots), and prevent animals from eating them (hairs or spines). They may have found all five groups, but algae may not be seen. They may have placed some plants in either the flowering or the non-flowering group depending on whether plants were in flower or not at the time of the survey.*

Researching microorganisms (page 21)

Ask students to read through the statements and then carry out research on the five plant groups. This will support the research investigation in the Student Book. They should then match each statement to the correct name for the type of microorganism and name one example from each group.

> *Answer: A = algae; B = protozoa; C = bacteria; D = fungi; E = viruses. Examples include: virus = influenza and poliovirus; bacteria = Salmonella, Vibrio, streptococcus, staphylococcus and Clostridium; fungi = Microsporum, yeast (Saccharomyces) and Trichophyton; protozoa = Plasmodium and Amoeba; algae = blue-green algae.*

Review and reflect

Encourage students to reflect on their own learning by pausing for a few moments and thinking about which parts of the work they found tricky. Ask them how they tried to learn the names of the five groups of microorganisms. Talk about how well they managed this.

Conclude the lesson by asking students to look again at question 1 in the 'What have I learned about classification and habitats?' activity on page 40 of the Student Book. They can also complete the second statement on page 40 of the Workbook. (See also the teaching notes in this Teacher's Guide, pages 43–45.)

1 Students could make a 'microorganisms collage' with a large version of the key and labelled drawings of some examples of each organism. Set up a classroom display.

2 Ask students to take leaf rubbings from a range of plants to look at the different characteristics seen on leaves, such as size and position of veins and if they branch out or run parallel along the leaf. They can then look up some common leaf patterns and shapes and try to identify the plants based on the leaves.

Differentiation

Supporting: Use the activities on pages 20 and 21 of the Workbook to provide additional support and structure for the investigations.

Consolidating: Large displays of the key for microorganisms will help students to learn the names and shapes of the five groups.

Extending: Students can study plant identification books to find out more about the detailed features botanists use to classify and identify plants.

Differentiated outcomes	
All students	should be able to state that plants and microorganisms can be classified based on their characteristics
Most students	will be able to identify and name the five groups of microorganisms
Some students	may be able to name specific examples of microorganisms from each group and describe a disease caused by organisms within each group

Using classification keys to group living things

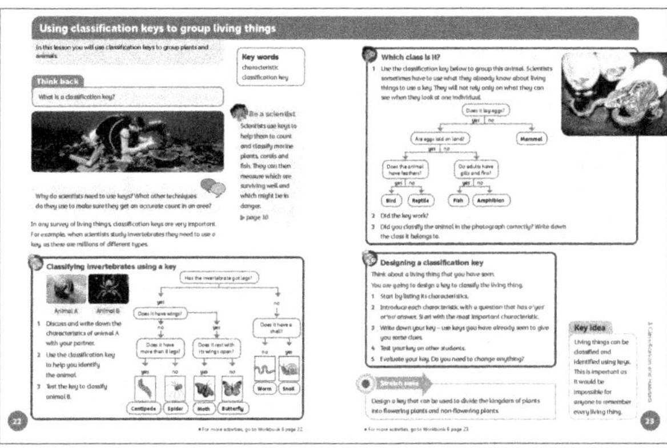

Supporting activities are in Workbook 6 pages 22–23.

Getting started

In this lesson students will study two classification keys used to identify some animals and animal groups. They will learn that many scientists use keys as they cannot learn about every living thing and they will consider some of the characteristics used to classify animals. They will use a simple key to identify which class of vertebrates an animal belongs to and then design their own keys.

Language support

To help develop language skills you could use flashcards with the names of the different animals and groups of animals discussed in the lesson. As you show a card, students have to find the word on the pages in the Student Book and then discuss with a partner any pictures that are associated with the word. You could extend this to include science investigation and enquiry words such as discuss, identify, classify and design.

Resources

Workbook: writing materials; large sheets of paper.

> **Key words**
>
> characteristic   classification key
>
> **Other words in the lesson**
>
> amphibian   bird   butterfly   centipede   fish
> invertebrate   mammal   moth   reptile   spider
> vertebrate

Scientific enquiry key words

Make observations

Record data and results

Analyse data, notice patterns and group or classify things

Report and present findings

Draw conclusions and give explanations

Lesson at a glance

The key teaching points for students in this lesson are:

- it is impossible for anyone to remember every living thing
- living things can be classified and identified using keys
- many keys are based on asking yes/no questions with only one correct answer.

In the next lesson, students will learn about some of the ways we can protect the environment.

Think back: What is a classification key?

Point out the Think back task at the top of page 22. Students can work with a partner to talk about when they have used keys before. They studied and used keys for plants and animals in Year 4.

> **Possible response:** *A classification key is a series of questions that lead to the name of a living thing. It helps us to identify something.*

Why do scientists need to use keys? What other techniques do they use to make sure they get an accurate count in an area?

Ask students to work with a partner to talk about the questions. Suggest they use the photograph at the top of page 22 as a clue. They used keys in Year 4 and in the last lesson, and have previously set up quadrats to make surveys more accurate, so they should be able to share some ideas about this.

> **Answer:** *Scientists need keys as they cannot know about every living thing – there are too many. They also mark out areas to make surveys into squares called quadrats and repeat measurements at different times to make counts more accurate.*

Be a scientist: Scientists use keys to help them to count and classify marine plants, corals and fish. They can then measure which are surviving well and which might be in danger. (Student Book, page 10)

Explain to students that scientists need to monitor habitats regularly and check if animals and plants are dying out, if they are growing well, or if new species are entering the area. Ask them if they have seen TV programmes about divers checking coral reefs. You could show some from the internet.

Investigation: Classifying invertebrates using a key

Allow students to work with a partner or in a small group. Ask them to study the photograph of animal A and then discuss and write down its characteristics. They should then use the classification key to help them identify the animal and test the key to classify animal B. Point out that the more clear-cut an answer is, the easier it is to use the key, so designing the questions correctly is vital when making a key.

> **Answer:** *A = snail; B = spider.*

Investigation: Which class is it?

Students can continue to work with their partner or group. Ask them to use the classification key to group the animal in the photograph into the correct class of vertebrates. They should look at the background as well as the animal for clues. Explain that scientists also use other factors to help them identify an animal – not only what is in the key. Ask students to reflect on how well the key works. Did they find it straightforward to identify the animal? Ask them to write down the class it belongs to.

> **Answer:** *The animal is a reptile. Students may know from prior knowledge that it is a lizard.*

Investigation: Designing a classification key

Explain that students are going to apply their knowledge and experience of using keys to help them to design their own. The worksheet on page 23 of the Workbook supports this investigation.

Allow students to think about a living thing that they have seen and then let them design a key to classify that living thing. They can use the clues in the investigation box, but mainly they should think about the keys they have used and what made them easy to use. Their key should be easy to follow: when they show someone a drawing or photograph of the animal, they should be able to identify it by following the key. Encourage them to let someone test their key and provide feedback so they can think about any improvements. Remind them that this evaluation is an important step in science.

> **Answer:** *Students' own answers: they should produce a simple key describing the characteristics of their animal in yes/no questions.*

 **Stretch zone:** Design a key that can be used to divide the kingdom of plants into flowering plants and non-flowering plants.

Computing link: Allow students to have access to the internet and books to research the characteristics of flowering and non-flowering plants to help them to devise some questions for their key. Ask them to find out about one or two other characteristics, in addition to whether or not they have flowers.

> **Possible response:** *In addition to whether or not a plant has flowers, other differences between flowering and non-flowering plants that could be used in a key are: flowering plants make seeds that are in seed pods or fruit, non-flowering plants produce seeds in cones or have spores; flowering plants also lose leaves in winter and many non-flowering plants do not; flowering plants have tubes (veins) in the leaves and non-flowering plants do not.*

Key idea

Living things can be classified and identified using keys. This is important as it would be impossible for anyone to remember every living thing.

Read through the key idea or ask a volunteer to read it out. This will help students to review the main themes of the lesson. Ask them to list some animals and plants they have identified using keys.

Ask volunteers to tell you the name of a 'kingdom' of living things, and then a 'class' of that kingdom to make sure students have understood the distinction.

Workbook activities

Making your own classification key (page 22)

Explain that students are going to classify a selection of birds and will fill in the question boxes. This is so someone can use the key to identify the birds. Ask students to discuss each question they could ask at each stage. Encourage them to use the drawings of each bird to get some clues about suitable questions. Once they have written the questions in the boxes, they should test them with other students. Ask them to think about which questions worked best and which need to be improved.

> **Possible response:** *First question could be, 'Does it have webbed feet?' The right (no) branch could be, 'Does it have a feathery head?' The 'no' branch leads to the vulture. The 'yes' branch could lead to a question such as, 'Does it have large ears?' 'Yes' would lead to the screech owl and 'no' to the barred owl. The 'yes' branch from the webbed feet question could lead to a question such as, 'Does it have a long neck?' 'Yes' leads to the swan and 'no' leads to the duck.*

 **Designing a classification key (page 23)**

This activity supports the investigation on page 23 of the Student Book. Ask students to follow the instructions to help them in designing their key. They can list possible characteristics in the table and decide on what makes a good classification characteristic and what makes a poor one. The prompt questions lead them onto designing and testing their key.

> **Possible response:** *Examples of characteristics and questions will vary but the characteristics should be ones that do not change and have a definite 'yes' or 'no'. For example, 'does it have legs?' is better than 'length of legs' because animals grow; whether an animal has antennae is better than the colour of animals as this can vary. Size is not a very good characteristic to use in a classification key as animals within a species naturally vary in size and animals grow as they develop, so an adult may be large and their offspring may be small.*

Review and reflect

Encourage students to display their keys and allow them to walk around to see what others have produced. They can have some quiet time to reflect on how they could modify their keys to improve them. Point out that reflecting on their work and learning from others is a very important part of learning.

Conclude the lesson by asking students to complete questions 2 and 3 in the 'What have I learned about classification and habitats?' activity on page 40 of the Student Book and the third statement on page 40 of the Workbook. (See also the teaching notes in this Teacher's Guide, pages 43–45.)

Extra activities

1 Print out some photographs of local animals and plants and ask students to use downloaded classification keys and identification books to identify each one.

2 Give a small group two plants and two animals to classify and ask them to make a small poster to display and show other students their classification system.

3 Show students some film of a duck-billed platypus swimming; ideally show its bill and its webbed feet, and show it laying eggs and feeding its young on milk, if possible. Ask students to make a list of the animal's characteristics. They can then debate which class of vertebrates the duck-billed platypus should be classified into. Point out that some animals are exceptions to the rule.

Differentiation

Supporting: The activity on page 22 of the Workbook will help students to design their own key, as this can be a complex task.

Consolidating: Hand out a range of different keys (you can download them from the internet) related to real living things, or the type that classify shapes, tools or furniture, to give students practice with using keys.

Extending: Allow students to use different types of keys in addition to the ones set out as branches. For example, there are keys where an answer takes them to a different numbered part of the key.

Differentiated outcomes	
All students	should be able to use simple keys to classify some vertebrates and invertebrates
Most students	will be able to design and test their own key
Some students	may be able to describe a range of different styles of key

Looking after our world

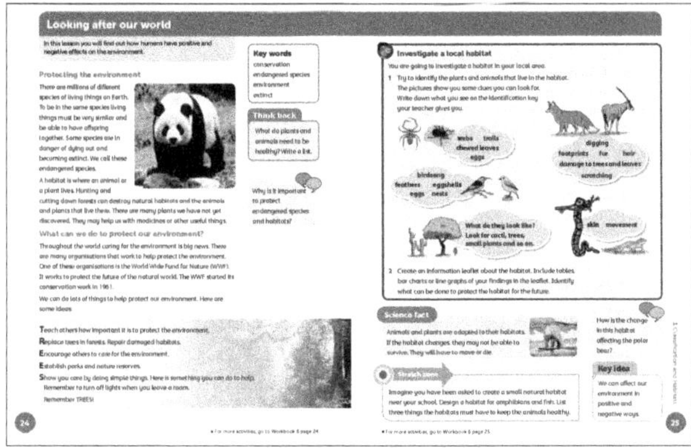

Supporting activities are in Workbook 6 pages 24–25.

Getting started

In this lesson students will study the risk of animals becoming extinct if their habitats are damaged. They will learn about some ways to protect the environment, carry out an investigation on local habitats, and produce a leaflet informing people about the habitats and what can be done to protect the living things that live there.

Language support

To support language development and to extend students' knowledge, create a class display of some extinct and currently endangered species. Ask students to help you by either drawing or bringing in pictures of endangered plants or animals. This type of activity can engage students and also extend their vocabulary as they learn the names of different plants and animals. Put up large labels identifying the 'endangered species' and the 'extinct species' so students learn these phrases.

Resources

Student Book: access to an outdoor area suitable for a plant and animal survey; identification keys for common species found in a local habitat; writing materials; graph paper; rulers; materials to make leaflets.

Workbook: writing materials; access to the internet or local newspapers; materials to make leaflets or booklets.

> ### Key words
> conservation   endangered species   environment   extinct
>
> ### Other words in the lesson
> adapted   affect   habitat   negative   offspring   positive   protect   species

Lesson at a glance

The key teaching points for students in this lesson are:

- humans can affect the environment in many ways
- some of these effects are positive and some are negative
- if habitats are damaged, species can become endangered and eventually extinct.

In the next lesson, students will learn about the negative effects of air pollution.

Think back: What do plants and animals need to be healthy? Write a list.

Allow students to work with a partner to discuss the Think back question and to produce a joint list. Remind them that they have studied plants and animals in earlier years. Ask volunteers to share their lists with others by reading them out.

> *Answer: Animals need food, water, warmth (shelter) and air (oxygen). Plants need water, light, space, nutrients and air (carbon dioxide).*

Ask students to read the text on page 24 about protecting the environment and to study the photograph of the panda. To test understanding, ask, 'What is a species? What do we call animals and plants that are in danger of dying out?' You could use the activity on page 24 of the Workbook to encourage students to interact with the text.

Why is it important to protect endangered species and habitats?

Students can continue to work with their partner or join with another pair to make a larger grouping. Ask them to talk about the need to protect animals and plants and their habitats.

> *Possible response: Animals and plants are part of an interlocking food web, so damaging one part can have an effect on the rest. Many plants are important for food and medicines, so we should look after them.*

After the discussion students can read through the text on page 24 about what we can do to protect the environment. Point out the green box at the bottom of page 24 and ask some students to take it in turns to read out each statement and write the acronym TREES on the board as they do this.

You can test understanding by asking, 'Name an organisation that was set up to protect the environment. When was it formed? What do the R and S stand for in TREES?'

Investigation: Investigate a local habitat

This activity needs to be planned in advance of the lesson.

Carry out a risk assessment: ensure any necessary risk assessments are completed in line with school policy and that parental permission has been given if you are organising field trips away from the school premises.

This investigation is designed to give students the opportunity to look at a local habitat and make informed decisions about how to protect it for future generations. This activity can be done in the school grounds, if there is a suitable area, or as part of a fieldwork exercise.

Provide identification keys to help students identify plants and animals. Ask students to look at the clues given in the Student Book. Instruct them to mark on their identification key if they see an example of the species and how many times they observe it.

They can share their findings by creating an information leaflet about the habitat. Remind them to include tables, bar charts or line graphs of their findings and identify what can be done to protect it for the future.

> *Possible response: Students should find a wide range of habitats in the area. Remind them that habitats can be very small (microhabitats). They may find soil, pond, leaf litter, dead logs, woodland and beach, for example.*

Science fact Animals and plants are adapted to their habitats. If the habitat changes, they may not be able to survive. They will have to move or die.

Read out the Science fact or ask a volunteer to read it out. Point out that because many animals are adapted to a specific habitat, if the habitat is destroyed, it is difficult for the animals to fit in anywhere else.

How is the change in this habitat affecting the polar bear?

Allow students to work with a partner or their small team to talk about the question and reflect on their ideas about the habitat under threat. Ask some groups or pairs to share their ideas with the class and discuss any differences of opinion or interpretation – stressing that this is normal in science.

Stretch zone: Imagine you have been asked to create a small natural habitat near your school. Design a habitat for amphibians and fish. List three things the habitats must have to keep the animals healthy.

Allow students to work in their small team. Point out that for this activity the emphasis is on planning and not constructing. You can ask students to use the internet to find out more about what fish and amphibians need to survive.

Key idea

We can affect our environment in positive and negative ways.

Read out the key idea or ask a volunteer to read it out. Ask students to turn to a partner and tell them one way that they intend to help to protect the environment from now on. Remind them of the acronym from this lesson: TREES.

Workbook activities

Protecting our environment (page 24)

Explain that students should use the words in the box to help them to complete the sentences. You can use this to make the text on page 24 of the Student Book more interactive. Once they have filled in the gaps, students can answer the questions about endangered species and how they can be protected. Finally, they can write down two human activities that destroy natural habitats.

 **Conservation project (page 25)**

Computing link: Ask students to discuss any conservation projects they have heard of before. Encourage them to think about why we need conservation projects and then allow them access to the internet to research a conservation project in their area. Students can also find out about conservation projects by asking people and reading local newspapers. Ask them to choose one project and study this in detail. They could write to the project for more information.

Students share their findings by making an information leaflet or booklet about the project. Make sure their leaflet answers the questions listed on the activity page. Remind them that their leaflet must include drawings of some of the living things protected by this project.

Review and reflect

Encourage students to identify aspects they have not completed correctly and help them to identify improvements. This will help students to develop a positive approach to learning by understanding that learning is a process that will improve with practice and reflection. For example, they can walk round and look at the information leaflets produced by others to identify what they have done well and what they can learn from others as targets to improve their work.

Conclude the lesson by asking students to complete question 4 in the 'What have I learned about classification and habitats?' activity on page 41 of the Student Book and the fourth statement on page 40 of the Workbook. (See also the teaching notes in this Teacher's Guide, pages 43–45.)

Extra activities

1 Allow students to construct the ponds they designed for the Stretch zone task. They can use a corner of the school grounds if one is available. They will need to dig out a circle shape that is 0.4 metres deep at its deepest and line it with sand and then a waterproof liner. The liner can be held in place by rocks around the edge and the pond filled. Plants such as reeds and water lilies can be added.

2 Invite a representative of a local conservation group to meet with your class and talk about their work. Ask them to bring in examples of scientific equipment that they use and also any photographs of the habitats and living things they are working to protect.

Differentiation

Supporting: Ask students to use the activity on page 24 of the Workbook to help them to understand the text about protecting the environment.

Consolidating: Make a large version of the TREES box and display it in the classroom so students can become familiar with it.

Extending: Students can research an animal that has recently become extinct and find out the reasons for its demise.

Differentiated outcomes

All students	should be able to state that damaging habitats can create problems for living things
Most students	will be able to describe what extinct and endangered plants and animals are and give some examples
Some students	may be able to explain some examples of how environmental damage has led to the extinction of a named animal

Air pollution

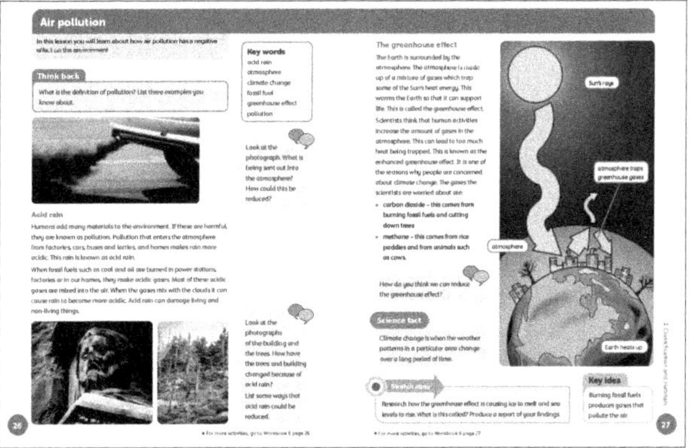

Supporting activities are in Workbook 6 pages 26–27.

Getting started

This lesson considers the negative effects of human activities on the environment and focuses upon air pollution and how this leads to acid rain and the greenhouse effect. Students study examples of air pollution from vehicles and other sources and consider the link between carbon dioxide and methane and the greenhouse effect.

Language support

To support language development refer to this unit's Word wall covering some of the more difficult vocabulary in this unit. You could also ask students to add new words and definitions to their glossaries. A large-scale diagram of the greenhouse effect on Student Book page 27 will help students to place the key words in context.

Resources

Workbook: pieces of dowelling or skewers; clear plastic bags or plastic food wrap; thermometers.

Key words

acid rain   atmosphere   climate change
fossil fuel   greenhouse effect   pollution

Other words in the lesson

carbon dioxide   enhanced greenhouse effect
factory   methane   power station   trapped

Lesson at a glance

The key teaching points for students in this lesson are:

- burning fossil fuels produces gases that pollute the air
- some of these gases cause acid rain and this can damage plants, animals and buildings
- other pollutants in the air are called greenhouse gases and these trap heat and warm up the Earth.

In the next lesson, students will learn how quarrying and deforestation have negative effects on the environment.

Think back: What is the definition of pollution? List three examples you know about.

Ask students to work with a partner to talk about the Think back task. Point out the photograph of the car exhaust at the top of page 26 and suggest they use this as a clue. Volunteers can share their ideas with the class.

> ***Possible response:*** *Pollution is the addition of materials into the environment that are harmful. Examples include pollution from car engines, heating homes, factories, burning wood and disposal of waste such as household objects, sewage and chemicals.*

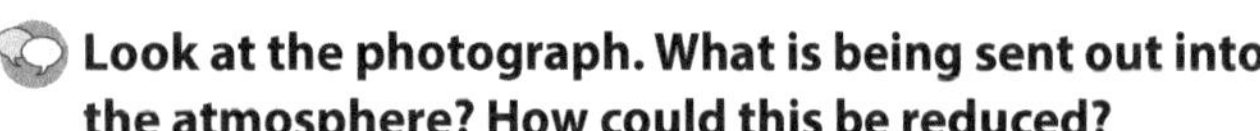 **Look at the photograph. What is being sent out into the atmosphere? How could this be reduced?**

Students can work with their partner to study the photograph and talk about what is coming out of the vehicle's exhaust pipe. They can think about some ways of reducing this emission and share their ideas by moving to join another pair. You can ask volunteer groups to stand up and tell the class about their ideas.

> ***Answer:*** *Car fumes – some students may have heard of carbon monoxide and particles. Car fumes can be reduced by using public transport, walking or cycling, or using electric cars.*

Students can read through the text about acid rain on page 26 and you can check understanding by asking them the questions below. Get them to discuss the questions with their partner: Why do you think rain is important to life? How do you think acid rain is formed? Why do you think that acid rain is a problem?

Look at the photographs of the building and the trees. How have the trees and building changed because of acid rain? List some ways that acid rain could be reduced.

Students can continue to work with their partner to study the photographs and talk about the discussion questions. They can pin their lists up on the wall and students can move around to compare the lists.

> ***Answer:*** *They have been damaged because the acid attacked the wood and stone. Acid rain can be reduced by cutting down on the use of fossil fuels (using less energy and limiting use of cars and aeroplanes).*

Students can read through the text about the greenhouse effect on page 27 and you can ask them to examine the diagram showing the greenhouse effect. Explain that the greenhouse effect is a natural process that keeps the Earth warm enough to support life. Point out that the diagram shows what is called the *enhanced* greenhouse effect: this results from human activities and is leading to climate change.

At this stage you could use the activity on page 27 of the Workbook as this involves labelling a version of the diagram. You can also check understanding by asking the following questions: What do you think we mean by the 'greenhouse effect'? What are the main greenhouse gases? What may happen if human activities continue to increase the amount of greenhouse gases in the atmosphere? What do you think we mean by the 'enhanced greenhouse effect'? What gas is released from burning fossil fuels and cutting down trees?

How do you think we can reduce the greenhouse effect?

Initiate a class discussion about how we can reduce the enhanced greenhouse effect. This can be used to reinforce the idea that we need to be proactive in reducing exhaust fumes and burning fossil fuels and in reducing deforestation (limiting cutting down trees to areas where trees are farmed for timber products). You could talk in simple terms about global agreements, for example the Kyoto Protocol, that aim to reduce global emissions of greenhouse gases.

Science fact Climate change is when the weather patterns in a particular area change over a long period of time.

Read out the Science fact or ask a volunteer to read it out. Ask students to talk about any times they have heard of the term 'climate change' and seen or heard any evidence about it – in magazines or books, on TV or the radio, or near to where they live.

Stretch zone: Research how the greenhouse effect is causing ice to melt and sea levels to rise. What is this called? Produce a report of your findings.

Computing link: Students can work independently or in pairs or small groups. Allow them access to the internet to gather the information needed to produce a report. They can download or print out diagrams and photographs to add to their report.

Answer: The greenhouse effect is causing a warming of the Earth. This is called global warming.

Key idea

Burning fossil fuels produces gases that pollute the air.

Read out the key idea or ask a volunteer to read it out. Ask students to name one of the gases produced when coal, petrol or wood is burned.

Workbook activities

Investigate the greenhouse effect (page 26)

Explain that students are going to make a model to show the greenhouse effect. They make a polythene greenhouse by following the instructions and then investigate to find out if the temperature is warmer inside or outside the greenhouse.

Maths link: Once they have measured and recorded their results five times on day one and five times on day two, ask students if they can see any patterns. They can use the prompt questions to help them to think about this. Ask them why they should take the temperatures at the same time every day.

Point out that there isn't a large greenhouse covering the Earth, but their greenhouse is trapping air inside and this represents carbon dioxide and methane being trapped in the atmosphere.

Answer: Students should find that the temperature inside increases in line with the temperature outside. The temperature inside the greenhouse is higher than the temperature outside. This tells us that trapped gases become warmer and warmer. This is what is happening on Earth.

The greenhouse effect (page 27)

Ask students to label the diagram. This is a copy of the one in the Student Book. Remind them to use only the words in the box. They should then list two gases that help to cause the greenhouse effect and explain how the greenhouse effect can damage habitats and lead to problems for animals and plants. Finally, they describe two ways in which the greenhouse effect could be reduced.

Answer: 1 Labels are (anticlockwise from top-right): Sun's rays; atmosphere traps greenhouse gases; Earth heats up; atmosphere. 2 Two gases that help to cause the greenhouse effect are carbon dioxide and methane. 3 The greenhouse effect warms the Earth and this can dry up areas so habitats do not have enough water; it can melt ice and warm sea water so living things, such as coral, cannot survive. 4 Producing less carbon dioxide can reduce the greenhouse effect. This can be done by using alternatives to fossil fuels and reducing the use of energy in homes, factories and transport. Having fewer animals such as cows on farms can also reduce methane.

Review and reflect

Hold up word cards for the key words in the lesson: acid rain, atmosphere, climate change, fossil fuel, greenhouse effect, pollution. Each time you hold up a word, ask students to turn to a partner and use the word in a sentence. They should write down any of the words they are unsure of and look these up.

Conclude the lesson by asking students to complete question 5 in the 'What have I learned about classification and habitats?' activity on page 41 of the Student Book and revisit the fourth statement on page 40 of the Workbook. (See also the teaching notes in this Teacher's Guide, pages 43–45.)

Extra activities

1 Ask students to create a large colourful poster showing the greenhouse effect. They can use the diagram on page 27 as a basis but can download and add photographs of vehicles, houses, factories and farm animals.

2 **Computing link:** Students can research the effects of global warming on climate change. They can concentrate on one aspect such as melting ice, drought or increased number of storms. They could produce a computer presentation to the class. Make sure they also look at the impact of climate change on living things.

Differentiation

Supporting: Students will benefit from having large-scale posters of the greenhouse effect diagram on display.

Consolidating: Ask students to look around the local area for buildings and statues damaged by pollution in the air. They can also download some examples.

Extending: Students could research predictions about how the world will warm up if greenhouse gases are not reduced.

Differentiated outcomes	
All students	should be able to state that air pollution causes damage to the environment and suggest some ways to reduce this pollution
Most students	will be able to describe how acid rain forms and state that the greenhouse effect warms the Earth's atmosphere
Some students	may be able to describe examples of scientific predictions about global warming

Digging up and cutting down habitats

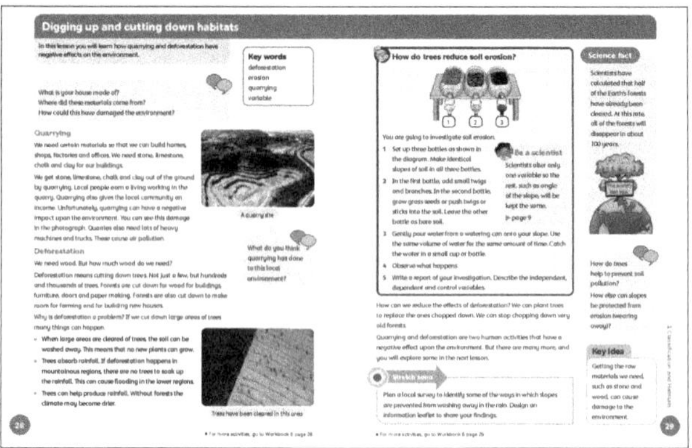

Supporting activities are in Workbook 6 pages 28–29.

Getting started

In this lesson students consider the negative effects of human activities on the environment by looking at the specific activities of quarrying and deforestation. They will use observation skills to study photographs of a quarry and a deforested area to consider the habitats that have been damaged. They then investigate the relationship between deforestation and soil erosion.

Language support

Ensure that students understand that the word deforestation is used in the lesson to mean 'removal of trees'. Explain that the prefix 'de' means 'the opposite of'. Forestation means planting trees, so deforestation means removing trees. Mention other words that use the prefix – such as de-ice; declutter; deskill. Ask students to try to think of or look up other examples to share with a partner.

Resources

Student Book: clear plastic bottles with a section cut out; soil; twigs and sticks; grass seed; measuring jugs; water; watering cans; small cups or bottles.

> **Key words**
>
> deforestation   erosion   quarrying   variable
>
> **Other words in the lesson**
>
> chalk   income   limestone   materials   pollution   raw materials

Scientific enquiry key words

Plan and/or carry out enquiries to answer questions

Make observations

Record data and results

Report and present findings

Draw conclusions and give explanations

Lesson at a glance

The key teaching points for students in this lesson are:

- obtaining raw materials can cause environmental damage
- quarrying removes habitats and creates pollution
- deforestation removes trees that help reduce greenhouse gases. This makes soil more likely to be eroded.

In the next lesson, students will learn about how human activity can cause water pollution.

What is your house made of? Where did these materials come from? How could this have damaged the environment?

The starter discussion activity asks students to think about how their homes are made and to try to identify the types of materials involved in the construction. Ask, 'Where do building materials come from?' Elicit the 'ground' or the 'Earth'.

> *Possible response: Answers will vary but might include bricks, stone, wood, plastics, glass or metal. These are obtained from under the ground by drilling, mining or quarrying or by cutting down trees. This damages the environment by removing plants and soil and by creating noise and air pollution.*

Ask students to read the text on page 28 of the Student Book. This focuses on quarrying, the reasons why quarrying is practised, and what the impact on the environment can be. Ask, 'Why do you think we need to have quarries? Can you name two things we get from quarrying? How can quarrying help the local community?' You could relate the discussions to a local quarry if there is one nearby.

What do you think quarrying has done to this local environment?

Ask students to look closely at the photograph of a quarry, and to answer the question. Ask them to think about what has happened to the trees, soil and living things in the area of the quarry.

> *Possible response: Soil and trees have been removed, rocks have been dug out, and vehicles make noise and air pollution. Roads to the quarry have been constructed.*

Allow students to read the first paragraph of the text about deforestation. This defines deforestation and considers some of the problems caused by cutting down large areas of trees. Ask students to think about the need for wood and the problems deforestation causes. Ask, 'Why do we need wood? Can you give one reason why forests are chopped down?'

Ask students to read through the next section ('Why is deforestation a problem?'). Ask, 'What happens to the soil when we clear large areas of trees? What can happen if we cut down trees on a mountainside? What can happen to the climate if large areas of trees are chopped down?'

Investigation: How do trees reduce soil erosion?

Explain that students are going to investigate soil erosion. Ask them to set up three bottles as shown in the diagram. Ensure they make identical slopes of soil in all three bottles otherwise it won't be a fair test. Point out that the first bottle has small twigs and branches, the second bottle has grass, twigs or sticks, and the third has bare soil.

Demonstrate how to gently pour water from a watering can onto each slope. Stress that students need to use the same volume of water for the same amount of time to make it a fair test. Remind them to catch the water in a small cup or bottle, or use a sink but block the plughole so soil does not block the pipes. Ask students to observe and record what happens. They can share their ideas by writing an investigation report, within which they describe the independent, dependent and control variables.

> *Possible response: The slopes with vegetation should have less soil washed away into the pots or sink. If grass has been grown, the roots are likely to hold the soil very well.*

Be a scientist: Scientists alter only one variable so the rest, such as angle of the slope, will be kept the same. (Student Book, page 9)

Point out the Be a scientist box and ask students to bear it in mind when carrying out their investigation.

Science fact Scientists have calculated that half of the Earth's forests have already been cleared. At this rate, all of the forests will disappear in about 100 years.

Read out the Science fact or ask a volunteer to read it out. Ask students what life without forests would be like for people on Earth and all of the animals. Point out that trees remove a great amount of carbon dioxide and ask students to think back to the greenhouse effect.

 **How do trees help to prevent soil erosion? How else can slopes be protected from erosion (wearing away)?**

Ask students to discuss this with a partner and point out the text at the bottom of page 29. Suggest they will find the answers there if they need a clue. They should also use the results of their investigation. Volunteers can share their ideas with the class.

> *Possible response: The roots and fallen branches help to hold the soil together making it less likely to be washed or blown away. Other methods of reducing soil erosion are to plough around a hill and not up and down (contour ploughing) and to build terraces or retaining walls.*

Stretch zone: Plan a local survey to identify some of the ways in which slopes are prevented from washing away in the rain. Design an information leaflet to share your findings.

Students can work in pairs or in small groups of three or four. Point out that, although they may not be carrying out the survey, the planning will help their scientific skills and their knowledge of soil erosion. The information leaflets can be shared with other groups or put on display as a small library for students to look through so they can learn from each other.

> *Possible response: Students may see walls being used in gardens and alongside roads to hold back soil, or terraces to make slopes less steep. They may see contour ploughing, or stones or wood chips lain over soil that is not going to be planted. They may see temporary crops planted to hold soil in place until the wanted crop is planted. Sometimes people can redirect water into streams to stop it rushing over slopes.*

Key idea

Getting the raw materials we need, such as stone and wood, can cause damage to the environment.

Read out the key idea or ask students to read it themselves. Ask them to work with a partner and, without looking at their books, make a list of ways that getting raw materials can damage the environment.

Workbook activities

Stone survey (page 28)

Point out that stone cut from rocks in the ground has been used for thousands of years. Allow students to survey the local area to find out how stone is being used. You could arrange an educational 'stone tour' for the class or they can observe stone being used on their journey to and from school. Ask them to use the table to record the different examples they see. Finally, ask them to write down why rocks are so important.

> *Possible response: Examples can include stone blocks in buildings, bridges and other structures, rock pieces covering roads and paths, tiles on roofs, and border walls in fields and around gardens.*

Deforestation debate (page 29)

This activity helps students to prepare for a debate about the cutting down of trees. Explain that a debate is where people take it in turns to present their arguments about an issue. They try to persuade other people that their ideas are correct.

Ask students to take the side of one of the people in the drawing. They can use the internet and other sources to find out more about each of the arguments. Point out that a good scientist will also think about possible arguments against their ideas. This is like a sports coach thinking about what the opposition coach might be planning so they can prepare for it. Ask students to plan how they would speak in favour of their argument and be ready to question the other side. They should write down their main points.

You can arrange the class so students can carry out the debate. They can do this in small groups or as a class debate. Split the class into two halves and ask volunteers to present each argument. At the end of the debate, ask students to vote for or against the statement: Cutting down trees should be stopped.

 ## Review and reflect

Ask students to sit down for a few minutes and think about what they have learned from the lesson. They can then think of one idea from the lesson that they would like to investigate and find out more about. Ask them to make a plan about how they would do this. This planning time will allow students to think up some imaginative and original things to do and you could let them carry out these projects. They are highly motivating and reinforce learning.

Conclude the lesson by asking students to revisit the fourth statement on page 40 of the Workbook. (See also the teaching notes in this Teacher's Guide, pages 43–45.)

Extra activities

1 Students can investigate how soil erosion by wind can be reduced. They can use dry sand to represent dry soil and a hair dryer as the wind. You can leave this as an open-ended enquiry or provide more support by explaining that students can modify their soil and water investigation by constructing sand slopes of the same angle and then adding twigs and branches to one slope and comparing this with a bare slope.

2 Ask students to investigate how contour planning or terracing can reduce soil erosion. Again, they can use ideas from their soil erosion and water investigation and use pebbles for retaining walls for their terracing.

Differentiation

Supporting: To explain the impact of quarrying, take students to a flower bed and ask, 'What would happen to the soil, plants and animals if we dig this up to make a car park?'

Consolidating: Leave the soil erosion investigation slopes set up for a few days and then ask volunteers to explain what they were used for and what they found out.

Extending: Students can research where the largest quarries in the world are, what is being removed, and what it is used for.

Differentiated outcomes	
All students	should be able to state that quarrying and deforestation damage the environment
Most students	will be able to describe some reasons why quarrying and deforestation damage the environment
Some students	may be able to discuss the advantages and disadvantages of quarrying and felling/cutting down trees

Water pollution and waste disposal

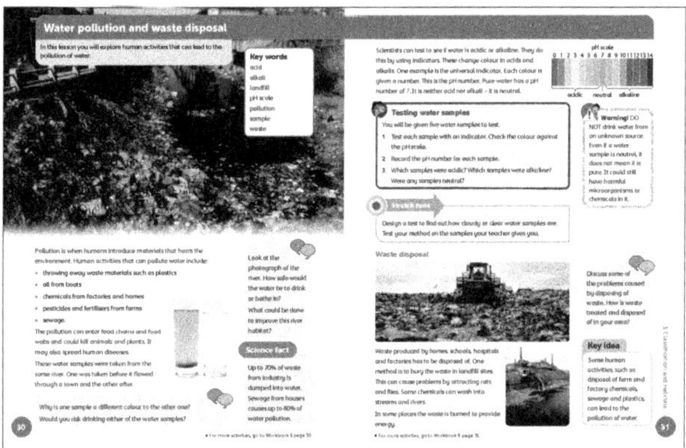

Supporting activities are in Workbook 6 pages 30–31.

Getting started

Students now look at two further types of human activity that can affect the environment – water pollution and waste disposal. In both cases, students are asked to consider what the environmental impacts are. They investigate water samples and discuss the problems associated with disposing of waste.

Language support

Ensure that students are confident with the words and concepts covered in the lesson. Explain to students what is meant by, for example, 'waste disposal' and 'water pollution'. Ask students to identify any words or ideas that they are unfamiliar with and explain what they mean. Ask them to use the photographs to give them clues as to the meaning of the words.

Resources

Student Book: five water samples with different pH values; pH indicator strips.

Workbook: five water samples with varying degrees of turbidity; writing materials; materials to make posters.

Key words

acid   alkali   landfill   pH scale   pollution   sample   waste

Other words in the lesson

chemical   disease   disposal   fertiliser   indicator   neutral   oil   pesticide   plastic   sewage

Scientific enquiry key words

Plan and/or carry out enquiries to answer questions

Recognise and control variables

Make observations

Take measurements, using equipment accurately

Record data and results

Analyse data, notice patterns and group or classify things

Report and present findings

Draw conclusions and give explanations

Identify causal relationships

Lesson at a glance

The key teaching points for students in this lesson are:

- some human activities can lead to water pollution
- examples of human activities include sewage, disposal of plastics, and farm chemicals leaching into waterways.

In the next lesson, students will learn about ways to save electrical energy and how alternative energy sources can help in this.

Ask students to work in pairs to look at the image of the water pollution and describe what they can see. Start a class discussion about the possible impacts of these activities on the environment. During the discussion ensure that you identify enough impacts to enable students to answer the questions.

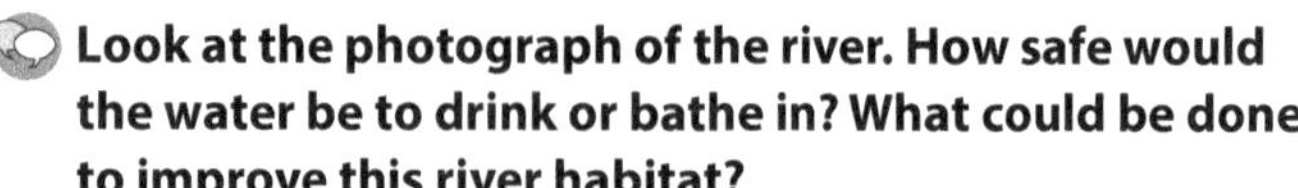 **Look at the photograph of the river. How safe would the water be to drink or bathe in? What could be done to improve this river habitat?**

Students can work in pairs or small groups. Ask them to discuss if they think it is safe to drink the water or swim in it. You can ask them to think about the problems that living things such as fish would have in the water. Once they have thought up some ideas to improve the river allow them to share ideas with the class.

> *Possible response: The river water would not be safe to drink or bathe in. The litter could be removed and people could stop adding substances to the river.*

Ask students to read the text on page 30 and look carefully at the photograph of the glasses of water.

 **Why is one sample a different colour to the other one? Would you risk drinking either of the water samples?**

Ask students which sample looks to be the cleanest. Ask them to share ideas about why one could be clear and the other one cloudy.

> *Possible response: The water in the glass on the right looks the safest to drink, but point out that it could still contain harmful substances and microorganisms that cannot be seen. The one on the left could have soil and other chemicals added.*

Science fact Up to 70% of waste from industry is dumped into water. Sewage from houses causes up to 80% of water pollution.

Read out the Science fact and ask students to think about why dumping waste into water can cause problems for humans and other living things. Ask for volunteers to share their ideas with the class.

Point out the text at the top of page 31, read through it with students, and study the drawing of the pH scale. Explain that this is one way of testing water samples and students are going to use it in their investigation.

Students do not need to know the background science of pH at this stage but tell them that indicators are natural substances that can be extracted from plants. You could add some acidic substances to cabbage juice to illustrate this. Point out that the numbers are used because colours are not easy to describe. Hand out a paint colour guide to make this point – the colours are given names so people can ask for them without having to describe the colour.

Investigation: Testing water samples

Warning! DO NOT drink water from an unknown source. Even if a water sample is neutral, it does not mean it is pure. It could still have harmful microorganisms or chemicals in it.

Read out the warning and ask students to discuss why they should not drink water from an unknown source.

Explain that they will be given five water samples to test. Ask them to test each sample with an indicator. If you cannot obtain commercial universal indicator paper or solution, then you can make your own indicator using red cabbage or other plants with pigments. Students dip indicator paper into the sample or pour a small amount of indicator solution into each sample and check the colour against the pH scale. Remind students to record the pH number for each sample and decide which samples are acidic, alkaline or neutral.

Stretch zone: Design a test to find out how cloudy or clear water samples are. Test your method on the samples your teacher gives you.

Students can work with a partner to plan and carry out their test. Hand out three water samples – one clear, one very cloudy (made by adding soil to water), and one slightly cloudy (made by adding fruit to water). You can use the activity on page 30 of the Workbook to provide more structure and advice for students.

> *Answer: Students should find that using cloudiness (turbidity) is a useful way of finding out if water contains pollutants.*

Show students the photograph and text about waste disposal. Ask them to study it carefully and then talk about the discussion task.

Discuss some of the problems caused by disposing of waste. How is waste treated and disposed of in your area?

Ask students to write down two effects of waste disposal on the environment. They can then research how waste is disposed of in their area. You can provide access to the internet or obtain information from the local government or company that disposes of the waste.

> *Possible response: Problems include attracting rats and other animals and the leaching of chemicals into water, rivers and streams. Landfill sites are also unsightly, they take up land that may have important habitats, and they can smell.*

Key idea

Some human activities, such as disposal of farm and factory chemicals, sewage and plastics, can lead to the pollution of water.

Read out the key idea or ask a volunteer to read it out to the class. Ask students to talk to a partner about any examples of pollution they have seen in their area and to write down a list of ideas about how each could be reduced.

Workbook activities

How cloudy is the water (page 30)

This activity can be used to support the Stretch zone activity in the Student Book. Ask students to read the text about why water can be cloudy (turbid) and then explain that they are going to measure the turbidity of water. Hand out five water samples to test. Ask students to follow the instructions to determine which water sample is the most turbid and which is the least turbid. Ask them to think about why turbidity (cloudiness) is not a completely reliable measure of water pollution.

> *Answer: The turbidity test will not show up any harmful chemicals that are clear (transparent) or microorganisms in the water that are too small to be seen.*

Waste disposal in the local area (page 31)

Ask students to talk to each other about waste. They can use the prompt questions to find out more and then research waste disposal in their local area. Allow them access to the internet and they can interview local people to find out about waste disposal. Students then plan a short report about the information. Remind them to make sure the report answers the questions listed. Finally, students work as a team to make a poster that encourages people to think about reducing waste and recycling. Ensure they copy the symbol onto a piece of paper or card, stick it onto the poster, and add information to persuade people to reduce the amount of waste they produce by

encouraging recycling and reusing of objects. Arrange for them to display the posters around the school so other classes can benefit from them.

Review and reflect

Use the discussion tasks, the research investigation activity and Workbook tasks to encourage students to think about what they understand and what they are finding less straightforward. Discuss the outcomes of each task with students. Encourage them to identify aspects they have not completed correctly and help them to identify improvements. This will help students to develop a positive approach to learning by understanding that learning is a process that will improve with practice and reflection.

Conclude the lesson by asking students to revisit the fourth statement on page 40 of the Workbook. (See also the teaching notes in this Teacher's Guide, pages 43–45.)

Extra activities

1 **Computing link:** Ask students to create an information leaflet about one of the issues covered in the unit so far. You could allocate groups a different topic – such as TREES, acid rain, greenhouse effect, quarrying, deforestation, water pollution, or waste disposal. Let them have access to the internet to find out more information about the topic and to download pictures to make a computer-generated leaflet.

2 **Maths link:** Students could research the types and amounts of pollution that are sent into the ocean, rivers and streams or buried or burned in your area. They can produce tables of the amounts and present these as bar charts or line graphs showing the trends over time. For example, has the amount of waste going to landfill fallen or risen over the last ten years in the local area?

Differentiation

Supporting: Introduce turbidity tests by asking students to write their name on a piece of paper and then try to read it through the different water samples.

Consolidating: Arrange a visit to a local waste disposal site or invite a visitor into the classroom so students can see and hear about waste disposal.

Extending: Students can research some of the most common chemicals that cause pollution in rivers, streams, lakes and the oceans.

Differentiated outcomes	
All students	should be able to state that water becomes polluted when chemicals and waste materials are thrown away into rivers and streams
Most students	will be able to describe how water can be tested for acidity and turbidity
Some students	may be able to list some specific chemicals that pollute water

Caring for the environment

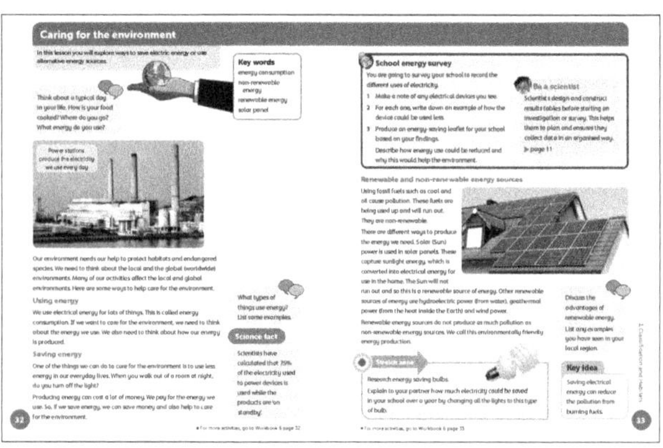

Supporting activities are in Workbook 6 pages 32–33.

Getting started

In this lesson students will study how the energy we use costs money and causes pollution. They will explore ways of saving energy and carry out a school energy survey to record the different uses of electricity. They will consider the differences between renewable and non-renewable energy sources and discuss the advantages of renewable energy sources.

Language support

You can support language development by creating classroom displays of each type of renewable energy source with a simple explanation of how each energy source produces energy.

Resources

Student Book: materials to make leaflets.

Workbook: materials to make leaflets.

> ### Key words
>
> energy consumption   non-renewable energy
> renewable energy   solar panel
>
> ### Other words in the lesson
>
> alternative   environment   fossil fuels   geothermal
> global   hydroelectric   source   wind power

> ### Scientific enquiry key words
>
> Plan and/or carry out enquiries to answer questions
>
> Make predictions
>
> Make observations
>
> Record data and results
>
> Analyse data, notice patterns and group or classify things
>
> Report and present findings

Lesson at a glance

The key teaching points for students in this lesson are:

- many human activities affect local and global environments
- to protect the environment we have to think about energy consumption
- saving energy and using renewable energy sources can reduce pollution.

In the next lesson, students will learn how recycling and reusing materials can help us care for the environment.

Think about a typical day in your life. How is your food cooked? Where do you go? What energy do you use?

Ask students to work with a partner to discuss the questions. They can then share their answers with a nearby pair. Ask the fours to make a list to share with the class.

> ***Possible response:*** *Examples of how food is cooked can include gas, coal, electricity and wood. Students can list examples of transport they use that require energy and also if they watch TV, use laptops or other devices, and visit places with electric lighting such as shops.*

The first section of text on page 32 provides information about the need to protect the environment because human activities can have impacts both locally and globally. Ask students to read the information and then tell them that we need to work together to protect the environment. Explain that the activities in this lesson will focus on energy use, how to reduce energy consumption, and ways to produce energy that don't negatively affect the environment.

Ask students to read the information in the section 'Using energy'. Then ask them to talk about the discussion task.

What types of things use energy? List some examples.

Ask students to work with their group of four. They can talk about the various devices that use electricity and make a list to share with the class.

Possible response: Examples include TVs, laptops, tablets, mobile phones, cookers, heaters and lights.

The section headed 'Saving energy' is designed to encourage students to discuss and think about ways of saving energy. Ask students to read it and then elicit some ideas from them on ways to save energy. Write their suggestions on the board.

Science fact Scientists have calculated that 75% of the electricity used to power devices is used while the products are 'on standby'.

Read out the Science fact and ask for a show of hands to find out how many students realise that devices use electricity when they are on standby. Ask them what they will do in future to save some of this energy.

Investigation: School energy survey

Explain to students that they are going to survey the school to record the different uses of electricity. Ask them to make a note of any electrical devices they see. Suggest they design a table for this purpose and point out the Be a scientist feature to emphasise this. You can use the worksheet on page 33 of the Workbook to support this survey. For each device they find they should write down an example of how the device could be used less. After the survey, ask students to produce an energy-saving leaflet for the school. In it they should describe how energy use could be reduced and why this would help the environment.

Possible response: Examples will vary but students should suggest that devices could be switched off when not in use. Some could be changed for low-energy varieties – and it may be possible to use the devices less.

Be a scientist: Scientists design and construct results tables before starting an investigation or survey. This helps them to plan and ensures they collect data in an organised way. (Student Book, page 11)

Point out the Be a scientist information to students as they plan their investigation survey.

Ask students to read the section of text headed 'Renewable and non-renewable energy sources' and to study the image of solar panels to stimulate thought and discussion about alternative and renewable sources of energy. Encourage students to think about and discuss environmentally-friendly ways of producing electricity. Ask, 'What is the difference between renewable and non-renewable energy sources? List two examples of non-renewable energy sources. List three examples of renewable energy sources.'

Discuss the advantages of renewable energy. List any examples you have seen in your local region.

Ask students to use the information in the text and any prior knowledge to list advantages of renewable energy and list some examples. Volunteers can share their ideas with the class.

Possible response: Advantages are that renewable energy sources will not run out and they cause less pollution than other sources, such as fossil fuels.

Stretch zone: Research energy-saving bulbs. Explain to your partner how much electricity could be saved in your school over a year by changing all the lights to this type of bulb.

Maths link: Allow students to work with a partner and provide access to the internet. Ask students to find out how much electricity an energy-saving bulb uses over an hour and compare this to a more conventional bulb. Students can then plan how to calculate how much electricity could be saved over a year. This is a good opportunity for a cross-curricular link with maths.

Possible response: Students should find various figures depending on the types of bulbs but, for example, a typical energy-saving bulb (12 W) uses 75% less energy than a traditional 100 W bulb. They also last up to 25 times longer. If students find typical cost savings per bulb, they can count the bulbs in the school and calculate the overall saving.

Key idea

Saving electrical energy can reduce the pollution from burning fuels.

Read out the key idea and ask students to turn to a partner and take it in turns to suggest ways to use less electricity.

Workbook activities

Saving energy at home (page 32)

Ask students to study the picture and talk about the questions. They should then choose and find out about one electrical device that is used a lot and draw a table to record how often people use the device and for how long.

Suggest they place the table next to the device for one week and ask people to record every time they use the device and how long they use it for. At the end of the week, students can use the table to calculate the total amount of time the device is used each week.

Computing link: Allow students to use the internet to find out the energy consumption of the device. This is the amount of energy the device uses per hour.

Maths link: Students can then calculate how much energy is being used by the device in their home in a day, a week and a year.

> *Possible response: Examples of heavily used devices in many places include TVs, cookers, washing machines and computers. They can ignore devices that are on all of the time such as freezers and air conditioning.*

School energy survey (page 33)

This activity supports the investigation on page 33 of the Student Book. Explain to students that they are going to survey the school to record the different uses of electricity.

Before they start, ask students to predict where they think the highest use of energy will be found. They then use the table to list the devices they found and how each device could be used less. Next, they can plan an energy-saving leaflet for the school. Point out they should use the checklist to help with their leaflet design.

Review and reflect

Remember to praise the process of learning as much as, if not more than, the outcomes. Encourage students to ask questions and to look back through prior work to review and revise. Use every opportunity to allow students to work together to check understanding and share ideas. After the school energy survey ask students to reflect on how they found out about the devices – and to discuss and list which skills they used.

Conclude the lesson by asking students to complete question 8 in the 'What have I learned about classification and habitats?' activity on page 41 of the Student Book and the fifth statement on page 40 of the Workbook. (See also the teaching notes in this Teacher's Guide, pages 43–45.)

Extra activities

1 Ask students to produce a poster about renewable energy. They can download photographs of different examples and label them to show how they cause less pollution than fossil fuels.

2 Invite a visitor who works for a local company or government body in electricity generation into the classroom. Ask them to talk to students about the sources of energy used to generate the electricity and a little bit about the process. They can also mention the balance between renewable and non-renewable energy planned for the future.

3 Students can research the development of electric cars and the advantages and disadvantages of using them now.

Differentiation

Supporting: Make a collage to display photographs of examples of renewable energy sources on one side of the classroom and non-renewable sources on the other to help students remember these.

Consolidating: Demonstrate the burning of a fossil fuel (a piece of coal or a piece of cotton wool soaked in diesel or oil) to show the pollution that is produced.

Extending: Students can research countries where geothermal energy or hydroelectric power is used a lot. They could also find out why these are not possible in many other countries.

Differentiated outcomes	
All students	should be able to state that using less energy saves money and reduces pollution
Most students	will be able to describe some renewable and non-renewable energy sources
Some students	may be able to explain the advantages of renewable energy sources but that not all are suitable for every country

Recycling and reusing materials

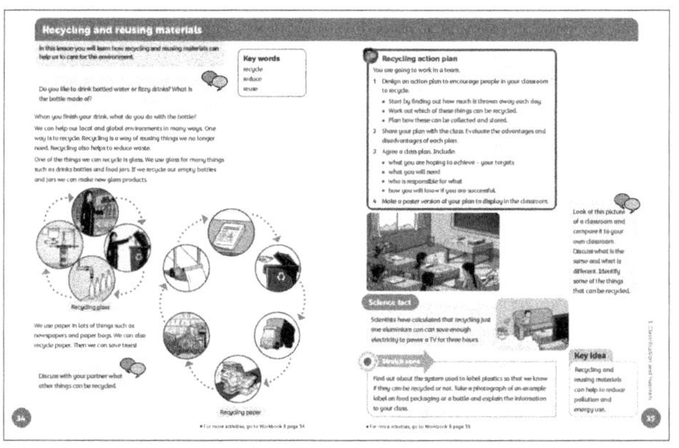

Supporting activities are in Workbook 6 pages 34–35.

Getting started

In this lesson the focus is on recycling and reusing materials as a way to care for the environment. Students are given the opportunity to identify items that can be recycled and items that cannot be recycled.

Language support

To support language development, you could create flashcards of some of the main key words and concepts in the lesson, for example 'recycle', 'environment', 'glass', 'paper', and ask students to tell you what the word is.

Resources

Student Book: materials to make posters.

Workbook: variety of clean waste materials (such as plastic bottles, cardboard, yoghurt pots); string; sticky tape; glue.

Key words

recycle   reduce   reuse

Other words in the lesson

glass   plastic   waste

Scientific enquiry key words

Record data and results

Analyse data, notice patterns and group or classify things

Draw conclusions and give explanations

Lesson at a glance

The key teaching points for students in this lesson are:

- many materials can be recycled and this reduces energy use and pollution
- reusing materials also saves energy and reduces pollution.

In the next lesson, students will learn how managing litter can help us to care for the environment.

Do you like to drink bottled water or fizzy drinks? What is the bottle made of?

The starter discussion activity asks students to think about the water and fizzy drinks they like and what the bottles are made of. Ask students to raise their hands if they drink fizzy drinks or water from bottles. Then ask, 'What is the container made of?'

Ask students to read the text on page 34 and ask them to talk about what they do with plastic bottles when they have finished with them. Then ask students to look at the picture of glass recycling and tell you what they think is happening. Explain in simple terms how glass is recycled using the picture to help with the explanation.

Do the same thing with the picture of recycling paper. Again, ask students to tell you what they think is happening and then provide a simple explanation of the process of recycling paper.

Discuss with your partner what other things can be recycled.

Ask students to think about, in addition to paper and glass, which other materials might be recycled. They might have seen recycling bins near to where they live or they may have taken items to recycling collection points.

Possible response: Metals such as steel and aluminium can be recycled and also wood and cardboard. Objects such as batteries and electrical devices can also be recycled.

Investigation: Recycling action plan

Arrange students so they can work in a team of three or four. Ask them to design an action plan to encourage people in their classroom to recycle. They should start by finding out how much is thrown away each day. They can either research waste in their area by looking on the internet or ask people in the class how much they throw away. Encourage them to list some of the things that are thrown away and decide which of these things can be recycled. Next, they can devise a plan to show how these items can be collected and stored. Ask students to share their plan with the class and evaluate the advantages and disadvantages of each plan. Help them to agree a class plan and make a poster version of this plan to display in the classroom.

Look at this picture of a classroom and compare it to your own classroom. Discuss what is the same and what is different. Identify some of the things that can be recycled.

Ask students to look closely at the picture of a typical classroom and then ask them to look around their classroom. This activity is designed to reinforce the idea that some items can be recycled and some cannot.

Answer: Students should notice the paper, plastic and metal objects that could be recycled.

Science fact Scientists have calculated that recycling just one aluminium can can save enough electricity to power a TV for three hours.

Ask a volunteer to read out the Science fact and ask the class if the saving in electricity is more or less than they expected. They can think back to the last lesson on energy sources and add recycling to their list of ways to save energy.

Stretch zone: Find out about the system used to label plastics so that we know if they can be recycled or not. Take a photograph of an example label on food packaging or a bottle and explain the information to your class.

Computing link: Ask students to work with a partner so they can discuss their ideas. Allow them to have access to the internet to research the labels used to show if something can be recycled. They can also look at some packaging such as clean plastic bottles and plastic wrapping. Let them use a digital camera or a smartphone to take some photographs; they could also download examples.

Key idea

Recycling and reusing materials can help to reduce pollution and energy use.

Read out the key idea or ask a volunteer to read it out. Ask students to tell each other the name of one material that can be recycled and where it should be placed to allow this to happen.

Workbook activities

Make a useful object from waste materials (page 34)

Explain that students are going to look at a variety of waste materials and use their imagination and creativity to make new objects so the material is reused and not thrown away.

Warning! All waste materials should be clean and have no sharp or broken parts.

Read out the warning and ask students to discuss why this is important.

Students can read through the activity to give them ideas and inspiration and then they should work in a small group to design the object. In the left box they should list all the materials they need, and in the right they should sketch out their design. After this stage, they can make the object and evaluate it.

Interpreting recycling data (page 35)

 Ask students to work with a partner and talk about the materials that are recycled in their area.

Maths link: They can then use their maths skills to interpret the line graph. They should talk about and then answer the prompt questions. Remind them that looking for patterns in graphs is a vital skill for scientists. Explain that the units used are percentages. The start and finish percentages are shown in the circles. The lines for each country show the trend or pattern of recycling in terms of percentage of aluminium cans recycled in 2000 up until the latest results were available. You could ask students to follow the line for one country to make this clear. The y (vertical) axis represents percentages but no numbers are given because only the upper part of the graph is shown and the figures in the circles are the key values. The x (horizontal) axis shows years from 2000 to 2012.

Answer: a Brazil; b the USA; c Japan; d Argentina; e In the USA, the recycling fell between 2000 and 2010 and then increased between 2010 and 2012; f In Europe the rate rose from 43% with a steady increase to 66.7% in 2010; g The number of cans that are recycled can be increased by setting up recycling bins and collections and educating people about how important it is.

 Review and reflect

Ask students to sit down for a few minutes and think about what they have learned from the lesson. They can then think of one idea from the lesson that they would like to investigate and find out more about. Ask them to make a plan about how they would do this.

Conclude the lesson by asking students to complete question 6 in the 'What have I learned about classification and habitats?' activity on page 41 of the Student Book and revisit the fifth statement on page 40 of the Workbook. (See also the teaching notes in this Teacher's Guide, pages 43–45.)

Extra activities

1 Ask students to create a poster about recycling. They can investigate other things that can be recycled and create recycling diagrams for these to show how it is done using the paper and glass recycling diagrams on page 34 of the Student Book to guide them. They can then add the recycling diagrams to their poster.

2 Students can design and construct a machine that can sort aluminium cans from steel cans and containers. Allow them to research the properties of aluminium and steel that might help them with their designs. They should test and evaluate their machine. You can hint by handing magnets out to students.

3 Ask students to research how clothing can be recycled and how this can help the environment.

Differentiation

Supporting: Copy the recycling diagrams and cut them into sections so students can reassemble them as a cut-and-paste task.

Consolidating: Show clips from the internet to show the recycling processes for glass and paper so students see the processes in real life.

Extending: Ask students to explain how recycling saves energy and reduces waste but does not eliminate waste completely.

Differentiated outcomes	
All students	should be able to state that recycling and reusing materials reduces pollution
Most students	will be able to describe how paper and glass are recycled
Some students	may be able to explain how recycling reduces energy use and pollution, but that there is still some energy used and pollution produced

Managing litter

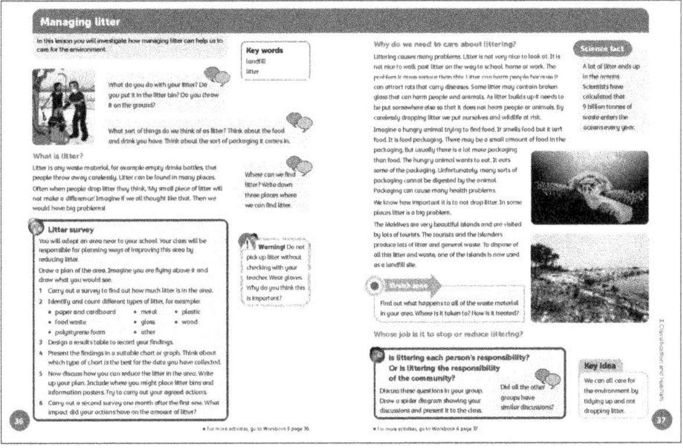

Supporting activities are in Workbook 6 pages 36–37.

Getting started

In this lesson students will study the environmental problems associated with litter. They will learn to define and classify litter and will investigate litter in their area through the planning and carrying out of a litter survey. Students will study some of the effects of litter on wildlife.

Language support

Use this opportunity to clarify the relationship between litter (which is thrown away carelessly) and waste (which may be disposed of correctly but can still cause problems). You could walk around the room and if you drop paper on the floor, students shout 'litter' and, as you put some in the waste basket, they shout 'waste'. Ask them why you should pick up the litter that you dropped.

Resources

Student Book: access to an outdoor area suitable for a litter survey; gloves; graph paper; large sheets of paper.

Workbook: access to an outdoor area suitable for a litter survey; gloves; large plastic bags; materials to make posters.

> **Key words**
>
> landfill   litter
>
> **Other words in the lesson**
>
> packaging   polystyrene   waste

Lesson at a glance

The key teaching points for students in this lesson are:

- litter is waste material that people throw away
  carelessly
- litter can damage habitats and endanger wildlife.

In the next lesson, students will learn about some ways that
people can work together to protect the environment.

What do you do with your litter? Do you put it in the litter bin? Do you throw it on the ground?

Ask students to work with a partner. They should look at
the drawing on page 36 and then talk about how they
handle litter. Remind them that if they put something
in the bin, then it isn't litter – it is waste. To prompt
discussion, as this will set the tone for the lesson, you
could ask questions around a scenario. Tell and ask
students, 'You are walking down the street drinking a
bottle of water. You finish the water. Where do you put the
empty bottle? Do you throw it on the ground, throw it in
a litter bin, take it home to throw away, or take it home for
recycling?' Ask volunteer pairs to share their answers with
the class.

*Possible response: Students should state that they
never drop litter and always put waste in the bin or
take it home for recycling.*

What sort of things do we think of as litter? Think about the food and drink you have. Think about the sort of packaging it comes in.

You can ask pairs to discuss this next question or, for
variety, ask the whole class for suggestions. As they
suggest items and materials that they believe are litter,
you can make a class list on the board.

*Possible response: Elicit that litter includes many
objects and materials such as paper, cardboard,
plastic items, glass containers, wooden objects, waste
food and even metal objects.*

Students can carry out the next discussion with a partner.
Ask them to read the text about 'What is litter?' and then
talk about the discussion question.

Where can we find litter? Write down three places where we can find litter.

Allow students to sit quietly for a few minutes and
imagine walking around their village, town or city and
consider where they are most likely to find litter. They can
then share ideas with their partner and some pairs can
read out ideas to the class.

*Possible response: Students may suggest that litter
gathers near fast-food outlets, where people have
gathered for picnics or barbeques, and where it might
have been blown by wind and caught on fences
and hedges.*

Investigation: Litter survey

Move pairs together to form investigation groups of
three or four. You can use the worksheet on page 36 of
the Workbook to support this investigation.

Warning! Do not pick up litter without checking with
your teacher. Wear gloves. Why do you think this is
important?

Read out the warning and ask students to tell you why
they should not pick up litter without checking with you
and why they should wear gloves. Talk with them about
the need to assess the risks.

Explain that the class is going to adopt an area near to
the school. They will be responsible for planning ways
of improving this area by reducing litter. Start them off
by asking them to draw a plan of the area. Walk around
with them and point out the main features, such as
flowerbeds, paths and fences. Ask them to imagine flying
above it and having an aerial view.

Students can then carry out a survey to find out how
much litter is in the area. They need to identify and
count different types of litter. Point out the examples
of materials they might find. They should then design a
results table and record their findings. They will need to
analyse the results and draw a suitable chart or graph.
Encourage them to discuss how they can reduce the
litter in the area and then they can write up their plan.
Remind them to show where they suggest litter bins
and information posters should be placed. If you have
permission, you could let them carry out their plans and
check how effective it has been by carrying out a second
survey one month after the first one.

In the section headed 'Why do we need to care about
littering?' students explore why littering and waste
disposal is a problem. The photographs show a turtle
eating a plastic bag; and a 'rubbish island' in the Maldives.

Ask students to read the information and look at the turtle. To check understanding you could ask, 'Why do you think that an animal may try to eat litter? What do you think may happen to an animal that eats a plastic bag? Why do you think the Maldives has such a big problem with the disposal of litter?'

Science fact A lot of litter ends up in the oceans. Scientists have calculated that 9 billion tonnes of waste enters the oceans every year.

Ask a volunteer to read out the Science fact. Ask students if they have ever seen litter in the sea. They can share ideas about how it might have got there. Point out the photograph of the turtle and remind them that litter can harm all living things.

Stretch zone: Find out what happens to all of the waste material in your area. Where is it taken to? How is it treated?

You could send off for information from local waste management companies or the government or community department that oversees waste disposal. They should be able to send you details of what is collected, how much is collected, where it is sent and how it is handled. If not, you can ask students to search for this information on the internet and download information sheets about it. Alternatively, students can draw on their own observations of where household waste is taken and what happens to it afterwards.

> *Possible response: Students will discover that the waste is either buried in landfill, composted (only some waste), burned, burned to provide energy, or recycled.*

Investigation: Is littering each person's responsibility? Or is littering the responsibility of the community?

Initiate a discussion within small groups about whose responsibility it is to deal with littering. Ask students to decide whether it is the responsibility of the individual or the community. Then ask students to raise their hands if they think littering is the responsibility of the individual, and again to raise their hands if they think it is the responsibility of the community. It is anticipated that students put up their hands on both occasions. Explain that a spider diagram shows the ideas put forward linked by lines, a bit like a spider web. Encourage groups to have at least four or five ideas in their spider diagram.

> *Possible response: Students may share a variety of ideas and these may include: it is an individual's responsibility as they use the materials; it is the responsibility of the community as there is too much litter for individuals to deal with; it is the individual's responsibility because money will be saved as people would not have to be paid to pick up litter; it is the responsibility of the community as people pay tax for this to be done; it is both the responsibility of the community and the responsibility of the individual because we all have to work together.*

 **Did all the other groups have similar discussions?**

Ask students to pin up their spider diagrams and walk around to look at the ideas generated by other groups. Ask them if they were all having similar discussions. You can walk around and point out any differences.

Key idea

We can all care for the environment by tidying up and not dropping litter.

Read out the key idea or ask a volunteer to read it to the class. Ask students to share any examples of when they have avoided dropping litter and taken an object home to dispose of properly, and any examples of when they have seen people drop litter.

Workbook activities

 **Litter survey (page 36)**

This activity supports the investigation on page 36 of the Student Book. Explain to students that during the litter survey they can use the prompt questions to help them to make decisions and use the table to record their findings.

 **Warning!** Point out the safety warning before allowing students to start their survey. Make sure students check with you before touching anything, and make sure they wear gloves and wash their hands after collecting the litter.

> *Possible response: Students should find a wide range of materials in the litter, with packaging, drinks bottles and cans being the most likely common varieties.*

 After the survey, ask students to imagine that an adult they know wants to encourage people to not drop litter. Encourage them to discuss ways the person could do this.

Design an anti-litter poster (page 37)

Point out that posters are often used to make people think about an issue. Ask students to think about posters they have seen – there may be some in the school about safety, for example.

Allow students to look at the drawing on page 37 of the Workbook to see how an eye-catching image, plus short and accurate information, can have an effect on people. Ask them to design their own poster to encourage people to not drop litter. Explain that the posters will be displayed near litter bins in the school.

Once the posters have been completed, students should evaluate them by asking someone to look at them and asking if it made them think and change their attitude to litter.

Review and reflect

Use the outcomes during the lesson (feedback from discussion tasks, the litter survey investigation activity, the spider diagrams and the poster design task) to encourage students to think about this topic. Encourage students to identify aspects they have not completed correctly and help them to identify improvements. They should also recognise things they are proud of and celebrate this. This will help students to develop a positive approach to learning by understanding that learning is a process which improves with practice and reflection.

Conclude the lesson by asking students to revisit the fifth statement on page 40 of the Workbook. (See also the teaching notes in this Teacher's Guide, pages 43–45.)

Extra activities

1 **Computing link:** Ask students to produce a 'press release' about the problems of littering. They can use the findings from their investigation or any internet research. Students can also include drawings or photographs to help their press release have impact.

2 Allow students to make a 'litter collage' by using washed items of litter to create some artwork. The pieces of litter could be labelled with the type of material and how it could endanger living things.

Differentiation

Supporting: The activity on page 36 of the Workbook will support students in planning and carrying out the litter survey.

Consolidating: Students could carry out a daily check of items placed into the class waste basket and keep a litter diary for a week to show the range and quantity of what is being thrown away.

Extending: Students could evaluate the environmental impact of the different ways that waste is managed.

Differentiated outcomes	
All students	should be able to state that litter can be harmful for living things
Most students	will be able to describe some ways that littering can be prevented and why this is important
Some students	may be able to explain the different ways waste material is handled and what the environmental impact of each is

Protecting the environment

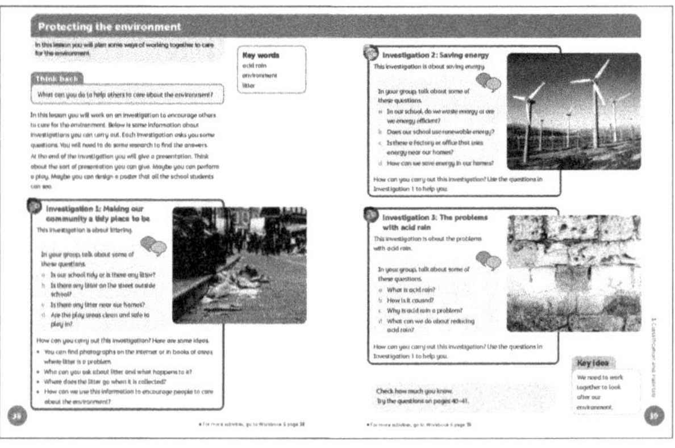

Supporting activities are in Workbook 6 pages 38–39.

Getting started

In this lesson students will carry out an investigation from a choice of investigations to explore the different ways that communities can care for the environment. They will consider the impact of littering in their area and then look at ways the community could save energy. Finally, they will investigate the problems associated with acid rain and what communities can do to reduce this.

Language support

This lesson is the final one in the unit, unless you plan a review lesson (see next section). Encourage students to check their glossaries for completeness and to use the key terms and words in their final presentation. Ask them to look back at the Word cloud on page 14 and make a note of any words they are still unsure of. They can use a dictionary or look through the Student Book to help them to recall any definitions they are unsure of.

Resources

Student Book: pre-prepared packages of information about litter, saving energy, acid rain; additional stimulus materials to capture the imagination of students; access to the internet or books on littering, saving energy, acid rain; access to specialist speakers where possible; writing materials and materials to make leaflets or posters; magazines with suitable images to cut and stick on the leaflets or posters.

Workbook: writing materials; materials to make leaflets or posters; white kitchen towel; tissue paper or pieces of clean fabric; sticky tape.

> **Key words**
>
> acid rain   environment   litter
>
> **Other words in the lesson**
>
> energy   factory   office   renewable

Lesson at a glance

The key teaching points for students in this lesson are:

- communities can work together to care for the environment
- by working together we can reduce problems such as littering, energy waste and acid rain.

Think back: What can you do to help others to care about the environment?

The Think back activity is a rhetorical question as the lesson centres on researching different projects associated with caring for the environment. Tell students that in this lesson they will work in groups to investigate and research one aspect of the environment. They will present their findings to the class.

You can either guide students to choose one of the three projects identified or to suggest a project they are particularly interested in. The projects are:

- Making our community a tidy place to be
- Saving energy
- The problems with acid rain

The main part of the lesson will consist of students working in groups to use their knowledge and understanding to come up with answers to the research questions and to start thinking about creating their leaflets, plays, information posters or other type of presentation. The research questions are merely suggestions to get students thinking about their project and how to present their findings.

Keep the tasks as open-ended as possible but be ready to support students and ensure they have access to the materials they need to conduct their investigations. Make sure that each student is playing an active role in the group. It is suggested that group size be limited to three or four students.

Ideally, the research project will provide an opportunity for all the key words and scientific key words to be included in students' presentations.

Your role throughout the research phase of the project is to facilitate and guide students to find out the relevant information from the resources you have provided and to encourage and stimulate discussions in the groups.

Conducting the research project:

- Ask students to get into groups.
- Ask students to elect a group leader.
- Ask each group to discuss the project proposals given in the Student Book.
- Ask the group leader to make a decision about which project the group is the most interested in based on the interests of the group as a whole.
- Provide a pre-package of information to each group about the project the members have identified as being the one they want to study. This could be in the form of existing leaflets, internet access, access to visiting experts, etc.
- Ask each group leader how the group would prefer to present its findings.
- Provide resources to each group for the production of the end product.
- Ask students to present their findings to the rest of the class. Ensure that every student in the group plays a part in the presentation.

Investigation 1: Making our community a tidy place to be

Explain to students that if they choose this investigation, they are going to be exploring littering. Point out that they will need to think about how they plan to carry out this investigation. Ask them to discuss the questions in the discussion box and study the suggested ideas. Suggest that they find photographs on the internet or in books of areas where litter is a problem and decide on answers to the questions posed in the bullet points. They can then decide how they want to present their ideas and then plan and carry this out.

Investigation 2: Saving energy

Explain to students that if they choose this investigation, they are going to be exploring how to save/conserve energy. Ask them to discuss the questions in the discussion task and then use the internet to prepare for their chosen presentation. Remind them to modify the bullet points in investigation 1 as a prompt.

Investigation 3: The problems with acid rain

Explain to students that if they choose this investigation, they are going to be exploring acid rain. As with the previous investigation, ask them to discuss the questions in the discussion task and then use the internet to prepare for their chosen presentation. Remind them to modify the bullet points in investigation 1 as a prompt.

Once students have finished their research and planned their presentations, you can hold a class conference where each group presents their ideas in turn.

Key idea

We need to work together to look after our environment.

Read out the key idea or ask a volunteer to read it out. Ask students to think about the presentations they have seen and how their ideas need people to work together to care for the environment.

Workbook activities

Air pollution (page 38)

Explain to students that as part of the review of the unit they are going to look back at air pollution. Ask them to look around at all the appliances that use energy and discuss how often these appliances are used.

They should then find out about how electricity is generated in their area and, in particular, which fuels are used to generate electricity. They can then consider the pollution this causes and the impact this has on the environment. Ask them to write a report or information leaflet or make a poster to display their findings.

Next, ask students to carry out an investigation to find out how clean the air in their local area is. They can follow the instructions to make and set up a simple air filter. Allow them to check the filter throughout the day and leave it for more than one day if possible. You could ask students to leave filters on different windows around the school. You can also set up a filter attached to a vacuum cleaner to draw in more air quickly to demonstrate filtering to students. Students should analyse their results and decide if the air is clean or polluted.

Saving energy questionnaire (page 39)

Ask students to use the statements in the table to carry out a survey of people at school or at home. Point out the drawing and remind students how effective a powerful image can be. Ask students to ask as many people as they can and record their responses in the table. They can then use the results of their survey to write a short newspaper report about how well energy is saved in the school and local area.

Review and reflect

Ask students to write down two changes they are going to make personally and two things they are going to do to encourage their community to care for the environment from now on. They can show their points to a partner and work together to agree on the points they are happy to share with the class.

Conclude the lesson by asking students to complete question 7 in the 'What have I learned about classification and habitats?' activity on page 41 of the Student Book and revisit the fifth statement on page 40 of the Workbook. (See also the teaching notes in this Teacher's Guide, pages 43–45.)

Extra activities

1 Ask students to create a mind map covering the unit content including key words and concepts. This will remind students what they have studied and provide the opportunity to revisit words and ideas that they may have found problematic.

2 Students can design and make their own wordsearch or crossword puzzle using some of the key words and ideas from the unit. You can ask them to use one of the free wordsearch or crossword maker websites on the internet, or you can hand out squared paper for them to use. They can test their puzzles on other students and evaluate them.

Differentiation

Supporting: Print out information sheets from the internet to support students in finding information for their presentation.

Consolidating: Film the presentations and show them back after a few weeks to remind students of the concepts.

Extending: Ask students to prepare one presentation that deals with two or three of the themes and not just one.

Differentiated outcomes	
All students	should be able to state that communities working together can help them to care for the environment
Most students	will be able to research a way to help to protect the environment and present their ideas on this
Some students	may be able to research and present ideas on a range of ways to protect the environment

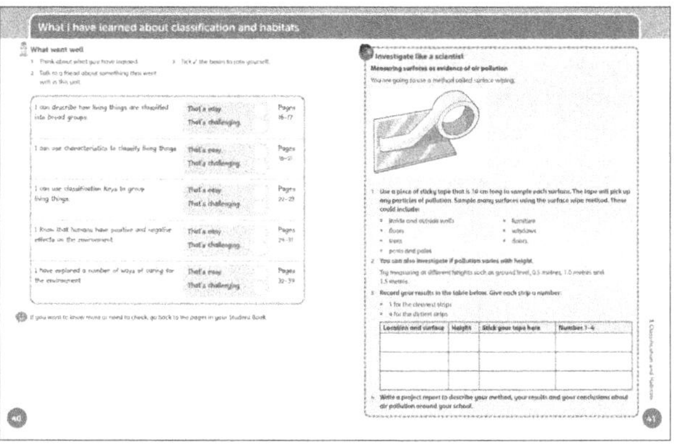

Supporting activities are in Workbook 6 pages 40–41.

Getting started

The aim of this section is to encourage students to review their learning after all of the lessons in the unit. All the lessons in the unit have questions and discussion tasks for students. You will have been using these formatively during lessons to help you assess students' knowledge and understanding of the topics.

On pages 40–41 of the Student Book there are eight questions related to the content of the unit. The questions are arranged in increasing order of conceptual demand and not topic order. Students can tackle these one at a time after individual lessons – specific advice on which questions are most appropriate is given under the 'review and reflect' heading in each relevant lesson. The questions could also be answered as a single summative activity. This could be done by reading out the questions to the class and asking for volunteers to answer them, carrying out the activity as a group work task with students talking about each question, or as an individual written task. Whichever approach is adopted, the questions are designed to give you and the students feedback about progress and to help in identifying targets for development.

On page 40 of the Workbook there is a section containing five 'review and reflect' statements. You can read through the statements and ask students to tick one of the two self-review option boxes to show how confident they are about each topic. You can ask students to reflect on each statement separately after the relevant lesson or carry out the review as an end-of-unit activity. Advice on which specific 'review and reflect' statement is most appropriate is given under the 'review and reflect' heading in each relevant lesson. It would be very useful to do both end-of-lesson and end-of-unit reviews as you will be able to measure students' levels of confidence throughout the unit to respond quickly and also gain a measure at the end of the unit to find out if this has changed for any topics. Time to reflect may make a student more or less

confident and later learning may support understanding of a topic that had been causing problems for some students. As teachers, we know that not all learning takes place at the time the teaching takes place.

What have I learned about classification and habitats? answers

1 Circle the correct phrase or word to finish each sentence.

  a A vertebrate has   no backbone/a backbone

  b The vertebrate classes are fish, amphibians, birds, mammals and   insects/reptiles

  c Prokaryotes is a kingdom of living things that contains   bacteria/trees

Point out that one of the two words or phrases will be correct and students must circle the answer. Stress how important it is to follow question instructions exactly.

Answer: a a backbone; b reptiles; c bacteria

2 Use the key below to identify the plants.

Remind students that to use a key they must answer each question and then follow that branch down to the next question until they reach the answer. Suggest they start with plant A and use its characteristics until they arrive at the name, then they can find the names of the other plants.

Answer: A = bell heather; B = primrose; C = bluebell; D = lesser celandine; E = wood anemone; F = wood sorrel

3 List two observable characteristics that you would use to classify:

  a plant

  b an animal

Remind students that characteristics are what a plant or animal looks like and also remind them that some characteristics are more useful than others.

Answer: a Plant: any two from a wide variety including flowers or no flowers; makes seeds or does not make seeds; shape and size of leaves; position of leaves; shape and size of flower and number of petals. b Animal: any two from a wide variety including wings or no wings; type of body covering; legs or no legs; antennae or no antennae; backbone or no backbone; lays eggs or does not lay eggs.

4 Fill in the missing words in these sentences.

Point out that there are three missing words and that each dash represents a missing letter of the alphabet. Explain that some letters have been added as clues.

5 Label the greenhouse effect diagram by drawing a line from each label below to the correct box on the diagram.

Stress that students should not try to write the words in the boxes as they did in the Workbook activity, but should use a ruler to draw a line from each word or phrase to the correct box. They could tick off each one as they do this.

Answer: Anticlockwise from top: Sun's rays; atmosphere traps greenhouse gases; Earth heats up; atmosphere

6 Arrange the pictures about recycling glass bottles in the correct order. Put the number 1 in the box of the first stage and so on until stage 4.

Explain that the pictures are not in order of the stages for recycling; students should write the numbers 1 to 4 in the empty boxes.

Answer: The order left to right is: 3; 1; 4; 2. This assumes the process starts with buying a bottle, but allow students to start elsewhere as long as the sequence is correct.

 7 Look at the table. It shows what has happened to the fish and coral found on coral reefs between 1990 and 2020.

  a What happened to the number of fish species between 1990 and 2020?

  b What happened to the amount of coral on the coral reefs between 1990 and 2020?

  c Write down two possible reasons for the changes shown in the table.

Point out that students should only use figures from the table to answer parts **a** and **b**. Also mention that the number of species is used and not number of individual fish because it is difficult to count every fish and the variety of species (biodiversity) is a good indication of the health of a habitat.

Answer: a Number of fish species fell from 527 to 390 between 1990 and 2020.
b Percentage of coral cover fell from 66% to 12% between 1990 and 2020.
c Two reasons for this include any two from: global warming is warming seas and making it difficult for coral to grow and this provides less food for fish; pollution such as soil washing into the sea has killed

coral and this affects fish varieties; climate change has resulted in more storms damaging reefs and this limits food for fish; new predators have moved into the area and killed some fish species allowing others to thrive and eat more coral; diseases have affected the coral and the fish; warmer seas have made some fish species move to colder areas and this allows coral-eating ones to thrive.

8 Write down two ways to encourage your friends and family to care for the environment.

Explain to students that they only need to think of two ways from among the many they have studied. Remind them of the litter and energy surveys they carried out and about the causes of air pollution and acid rain.

Answer: Any two from: use posters and information leaflets to tell their friends and family about reducing energy use by using appliances for less time; walking and cycling instead of using cars; switching devices off and not leaving them on standby; not littering; disposing of waste properly; recycling and reusing materials; trying to use renewable energy sources.

Summative assessment

The questions in the 'What have I learned about classification and habitats?' activity on pages 40–41 of the Student Book can be used to consider the progress of each student individually. You can also use the information to create summative reports – such as end-of-term reports – for each student. If you wish to allocate a score or mark for the questions, then the total number of marks you could allocate is 30 (question 1 = 3 marks; question 2 = 6 marks; question 3a = 2 marks; question 3b = 2 marks; question 4 = 3 marks; question 5 = 4 marks; question 6 = 4 marks; question 7a = 1 mark; question 7b = 1 mark; question 7c = 2 marks; question 8 = 2 marks).

You can also keep a record of individual and class responses to the 'review and reflect' statements in the Workbook. This lets you reflect on the overall confidence levels of students to identify areas that may need revision later on.

By reviewing responses to the questions and the self-review of confidence levels you can tailor specific interventions to help students improve. Keep the recording and analysis of the student self-evaluations simple. A general impression of the class's self-evaluation, not individual student records, is all that is required, e.g. 'Fifty per cent of the class were not confident about …'.

Additionally, you can ask students to complete the Quiz Yourself questions for this unit to help check their understanding. These are on pages 142–143 of the Workbook.

The 'Investigate like a scientist' task at the end of the unit on page 41 of the Workbook is designed to encourage students to apply their investigative and creative skills and review key aspects of the content of the unit.

Measuring surfaces as evidence of air pollution

Resources: sticky tape; access to various surfaces, e.g. floors, inside and outside walls, trees, furniture.

Explain that in this investigation students are going to use a method called surface wiping.

Show them the drawing and ask them to use a piece of sticky tape that is 10 cm long to sample each surface. Explain that the tape will pick up any particles of pollution, as it is sticky. Students can then follow the instructions to sample different surfaces using the surface-wipe method. Suggest they also investigate if pollution varies with height.

Explain the scoring system so students can record their results in the table. Point out that the score is an estimate based on how clean or dirty they think the surface is.

After the investigation students can write a project report to describe the method, the results and the conclusions they draw about air pollution around the school.

2 Organs and Systems

In this unit students will:

- learn where the major organs are to be found in the human body
- find out about the functions of the major organs
- identify and name the main parts of the human circulatory system, and describe the functions of the heart, blood vessels and blood
- describe the ways in which nutrients and water are transported within animals, including humans
- know that some diseases can be caused by infection with viruses, bacteria, parasites or fungi that can be passed from one host to another
- describe how good hygiene can control the spread of diseases transmitted in water, food and body fluids, and describe ways to avoid being bitten by insect vectors
- know that humans have defence mechanisms against infectious diseases, including skin, stomach acid and mucus
- recognise the impact of diet, exercise, drugs and lifestyle on the way their bodies function.

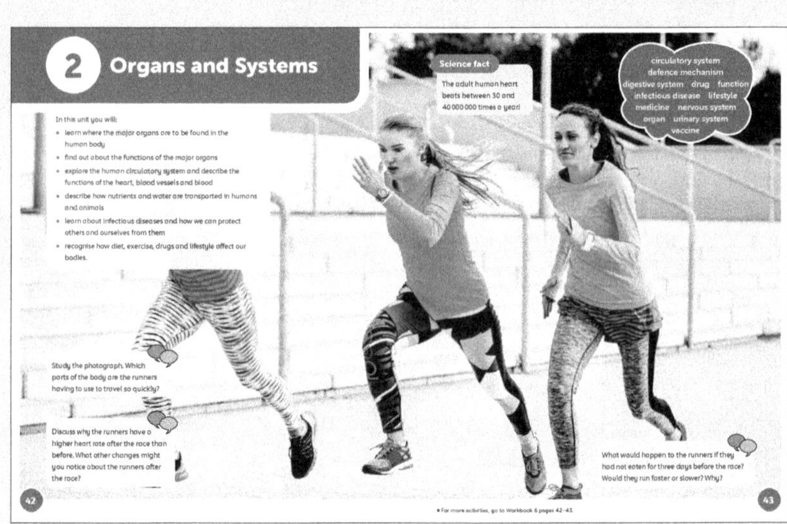

Supporting activities are in Workbook 6 pages 42–43.

Getting started

This unit introduces students to some human anatomy and physiology. The introductory lesson focuses on exercise as a context for reviewing ideas about the heart and nutrition. Subsequent lessons look at the body's major organs and systems, and where they are located and their functions. Students are introduced to the concept of life sustained through food, water and air. This is linked to the functions of the major organs in providing the cells of the body with essential oxygen, nutrients and water. From this, students also learn about another vital function carried out by the major organs in which waste, or poisons, produced by the body are excreted.

Students are encouraged to understand that they have a responsibility to look after their bodies. This includes knowledge of and prevention of the spread of infectious diseases and awareness of the importance of personal hygiene.

Science in context

Place the science in a relevant context for students by allowing them to study and discuss examples of exercise that are familiar to them in their local environment. They can take part in sports in school and visit local sports centres to carry out surveys. Invite local sports people, nutritionists and doctors into school to talk about their work. Also use examples of sports and exercise from around the world to give a global context.

To set the work on food and digestion in context you can discuss the work of nurses, doctors and dentists. Students can survey the produce in local shops and the foods shown in the menus of restaurants to think about the food eaten locally and why it forms part of a healthy diet. They can examine advertisements for foods and drinks to form the basis for discussions about potentially unhealthy foods. If possible, ask health professionals to talk to students about the importance of food and personal hygiene.

Scientific enquiry skills

In this unit students are asked to take measurements of breathing rate and pulse rate. The principle of 'fair testing' is emphasised so that they understand the need to produce evidence that is reliable. They are encouraged to work in a scientific way, to make predictions and to suggest explanations from what they have found out. This is a basic principle of scientific enquiry. There is also the opportunity to revisit the idea of repeating measurements to improve accuracy and organising and interpreting numerical results. Students also analyse and interpret findings and draw conclusions. They present their ideas in a variety of ways.

Student Book: writing materials; large sheets of paper; displays of the major organs posters from last lesson, 30-centimetre rulers; stopwatches or timers; safety glasses; bread; iodine; items to design a computer presentation, leaflet or poster; filter papers or paper towels; filter funnels; beakers or cut-up plastic bottles; soil; water; samples of waste water; plastic bottles; tubing; sticky tape; frames or stands; large flat containers; paperclips; materials for making posters; selection of cans, bottles and packets of food of four types (protein rich, carbohydrate rich, fat rich, and mineral and vitamin rich) showing food labels; thin cardboard; graph paper; items to prepare a presentation or poster that includes drawings or photographs; paperclips.

Workbook: tracing paper; A4 paper; materials for making posters; items to design a presentation (a talk, computer presentation or poster); pieces of cardboard with different sized holes cut in them; box containing large red, small blue, large yellow and small green balloons; a large plastic bag; paperclips; selection of cans, bottles and packets of food of four types (protein rich, carbohydrate rich, fat rich, and mineral and vitamin rich) showing food labels; writing materials; graph paper; large plastic bottles; scissors; elastic bands; sticky tape; straws.

Key words for the unit

Bold words are in the Word cloud and are included in the glossary.

absorption   artery   barrier   bladder   blood
blood vessel   brain   breathing   capillary
cell   circulation   **circulatory system**
defence mechanism   deoxygenated/oxygenated
diabetes   diet   digestion   **digestive system**
drug   enzyme   excretion   exercise
exhalation/inhalation   **function**   health   heart
heart disease   host   hygiene   **infectious disease**
intestines   kidneys   **lifestyle**   liver   lungs
medicine   microorganism   nerves
nervous system   nutrients   obesity   **organ**
plasma   prevention   re-absorption   secretion
smoking   spinal cord   stomach   surface area
system   touch   transmission   **urinary system**
vaccine   vector   vein   water

Scientific enquiry key words

Plan and/or carry out enquiries to answer questions

Make predictions

Recognise and control variables

Make observations

Take measurements, using equipment accurately

Record data and results

Analyse data, notice patterns and group or classify things

Report and present findings

Draw conclusions and give explanations

Identify causal relationships

Language support

This is likely to be the second unit studied by students this year so they will be familiar with the usefulness of starting a unit by reading out the words in the Word cloud on the introductory pages (on page 43 of the Student Book for this unit). Ask students to list each word and tick those they are familiar with. Some of the words may be familiar due to their use outside school, so encourage students to draw on experiences outside the classroom. Remember that students learn about science in many ways – including reading, watching TV and talking to family. Eliciting any prior knowledge of words and concepts is a vital part of learning and teaching.

It is worth creating a Word wall for any unit, but especially this one as it has many complex words and terms. Remind students about the glossary at the back of the Student Book and ask them to add definitions of any key words regularly during lessons. You may have been allowing them to develop this from the last unit; if not, it is important to start now.

To introduce new words, a strategy is to ask students to listen to a word, say it out loud, read it (from a word card, the book or the board) and then write it down. Also encourage them to use the words in context by placing them into a sentence. Remind students that the more they use the words, the better they will understand and learn them. Ask students to also make a list of other, non-scientific words they use in lessons that they are not familiar with. They could write these down in a notebook. Some are highlighted at the start of each lesson in this Teacher's Guide.

The eBook has examples of key words being pronounced. You could use this at the start of the lesson.

If you have started a science library in the classroom, add texts and resources relevant to the unit. If you have a school library, you could ask them to order in texts that will support the unit.

Finally, the first activity in the Workbook (page 42) is linked to the key words for the unit and is designed to support language development.

The key teaching points for students in this unit are:

- to introduce the unit objectives
- to introduce the learning outcomes
- to engage students with the content of the unit
- to review and build on prior learning and understanding of the topics.

In the next lesson, students will learn about the names and positions in the human body of the major organs.

The introductory pages are designed to start students thinking about the human body and how the heart and eating are vital to a healthy lifestyle. Ask students to look at the photograph on the introductory pages. Explain the unit objectives in simple terms: point out that students will be thinking about the organs, digestion, circulation, the nervous system, getting rid of waste, exercise, healthy eating, infectious diseases and personal hygiene. List these on the board and emphasise that they are all linked together and are not separate ideas.

The discussion questions can be undertaken in pairs or small groups of three or four. There is an activity on page 43 of the Workbook that you can use to structure the discussions and help students to take notes.

Study the photograph. Which parts of the body are the runners having to use to travel so quickly?

You can take students outside or into a large inside space and ask them to walk quickly or run and observe each other. They can then talk about which parts of their body they used when running. Students could try running with their arms not moving and held down by their sides. Ask them to refer back to the photograph of the runners and decide which parts of their bodies they are using.

Possible response: Students should note that the runners are using their legs mainly and their ankles and feet also have to be involved. Runners also need to move their arms to help them to run quickly.

Hint to students that they can see the arms and legs moving but inside the body of any runner the different organs are working to carry out vital processes that help them to move and stay alive. Explain that they are going to find out more about these organs and processes during the unit.

Discuss why the runners have a higher heart rate after the race than before. What other changes might you notice about the runners after the race?

Remind students how to take their pulse rate. Explain that they feel a pulse every time the heart beats. Allow them to measure their pulse rate and then predict what will happen to the rate if they exercise. They may have carried out investigations on pulse rate in Years 2 and 3 so these will not be repeated in this unit. However, as an introductory activity you could ask students to exercise gently and compare their resting pulse rate with their pulse rate after exercise as a review. Ask students to then talk about the heart rates of the runners in the photograph. They can also consider what other changes they might notice after the race.

Possible response: The runners have a higher heart or pulse rate after the race because their hearts need to beat faster to take food (glucose) and oxygen to muscles to help them to work better. The breathing rate would also increase, the runners would be hotter and they would perspire.

Science fact The adult human heart beats between 30 000 000 and 40 000 000 times a year!

Ask a student to read out the fact and point out the numbers are 30 million and 40 million.

Maths link: Students could calculate how many times their heart has beaten since they were born – assuming an average of 40 million times per year.

What would happen to the runners if they had not eaten for three days before the race? Would they run faster or slower? Why?

Ask students to think of any times when they have not eaten for many hours. They can share ideas about how they felt. Did they feel lively? Did they feel tired? Then ask them to think back to earlier work on why we need to eat foods – for example in Year 4 – and then decide what would happen to the runners if they did not eat for three days.

Possible response: The runners would not have as much energy if they had not eaten, so would probably run slower. Some students may add that they would also lack protein for growth and repairing any muscles, so might have injuries, and that they would also lack the vitamins and minerals needed to stay healthy.

Key words (page 42)

Explain that this activity will help students to learn some of the key words for the unit. Ask them to find and circle in the wordsearch the six key words given in the box. Then ask students to check the words on this page against the Word cloud on page 43 of the Student Book. They should write down any key words that are on page 43 but missing from the wordsearch.

Answer: Students should find and circle the following words: drug, function, lifestyle, medicine, organ, vaccine. Missing words are: circulatory system; defence mechanism; digestive system; infectious disease; nervous system; urinary system.

Discussion notes (page 43)

This activity can be used to support the discussion tasks on pages 42 and 43 of the Student Book. Allow students to look at the photograph, or they can use the larger colour version in the Student Book. Ask them to discuss the questions with a partner or in a small group. As they talk about the photograph, encourage students to take notes by filling in the boxes in response to the prompt questions. You can demonstrate how to take notes using bullet points to summarise main ideas as this is an important skill for students to develop.

Answer: Students should note that the runners are using their legs mainly and their ankles and feet also have to be involved. Runners also need to move their arms to help them to run quickly.

The runners would not have as much energy if they had not eaten so would probably run slower.

The runners have a higher heart or pulse rate after the race because their hearts needed to beat faster to take food (glucose) and oxygen to muscles to help them to work better.

The breathing rate would also increase and the runners would be hotter and would perspire.

Students should plan to take resting heart and breathing rates, set a time-limited exercise and then take heart and breathing rates immediately after exercise.

Extra activities

1 You could allow students to carry out an investigation to find out changes in breathing and heart rates before and after exercise. This would help them to review prior work and ensure this knowledge is fresh when they study lungs and breathing and the circulatory system.

2 Ask students to download photographs of people carrying out different activities, such as playing sport, walking to work, sitting at a desk, watching TV and sleeping. They can cut out the photographs and make a poster collage showing the people. They can then write down 'low heart rate and breathing rate', 'medium heart rate and breathing rate' and 'high heart rate and breathing rate' on a series of cards and glue them next to the most appropriate picture.

Where are our major organs?

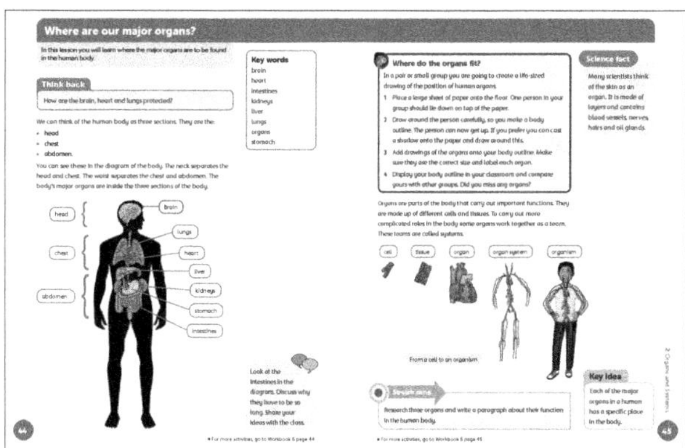

Supporting activities are in Workbook 6 pages 44–45.

Getting started

In this lesson students learn the relationship between cells, tissues, organs, organ systems and an organism. They study the major parts of the body and identify and name the main organs. Finally, they look at a diagram of the human body and label the position of the major organs.

Language support

Students will have heard of the names of the organs in prior work, but make sure that they understand the terms 'cells' and 'tissues'. Explain that the word 'cell' was given to these small parts of the body because early scientists, using newly invented microscopes, thought they looked like small rooms – called *cella* in Latin. Stress that tissue is made up of similar cells that carry out the same job – such as muscle.

Resources

Student Book: writing materials; large sheets of paper.

Workbook: tracing paper; A4 paper; materials for making posters.

Key words

brain   heart   intestines   kidneys   liver   lungs   organs   stomach

Other words in the lesson

abdomen   neck   skin   system

Scientific enquiry key words

Make observations

Record data and results

Report and present findings

The key teaching points for students in this lesson are:

- the human body contains many different organs
- each organ has its own location and function
- organs work together in organ systems.

In the next lesson, students will review their knowledge of the functions of some of the major organs.

Think back: How are the brain, heart and lungs protected?

Ask students to work with a partner to think about the question. Remind students that they studied the human skeleton in Year 3; to help them remember, ask them to locate the brain, heart and lungs on the diagram on page 44 and then feel their own skeleton to see what covers these organs.

> **Answer:** *The skull protects the brain and the ribs protect the heart and lungs. Some students may know that the plate at the front of the ribs (sternum) also protects the heart and lungs.*

Next, ask students to read the text and study the diagram on page 44. They can take it in turns to read out each of the labels with a partner.

Elicit that it is the organs that keep us alive. You may need to prompt to establish that the brain is an organ. Use questions to check understanding. You could ask, 'Do we need the heart to stay alive? What would happen if we did not have a stomach or intestines? Why are the lungs important?' Ask students to stand up and as you say out the parts of the body or names of organs, they have to point to it on their body.

Look at the intestines in the diagram. Discuss why they have to be so long. Share your ideas with the class.

Students can continue to work with a partner. Allow them time to think back to earlier work on digestion, studied in Year 4, and point out that they will be exploring digestion in more detail later in the unit, so it is useful to have prior learning fresh in their minds. Volunteers can share their ideas with the class.

> **Answer:** *The intestines need to be long because the digestive process takes time. Foods are broken down slowly, step by step, and there has to be time for the nutrients to be taken into the body.*

Investigation: Where do the organs fit?

Students can work in their pairs or join together to form a small group. Explain that they are going to create a life-sized drawing of the position of human organs. They can follow the instructions to end up with a large

sheet of paper on the floor with a body outline drawn on it. Two ways of doing this are explained. The casting a shadow method can be used if it is important to avoid any contact between students.

Ask students to draw the organs onto the body outline. Point out that the organs need to be the correct size. Each organ must have a label. Arrange an exhibition to display the body outlines in your classroom and allow students to walk around so they can compare theirs with other groups. Ask them to evaluate their own poster.

Science fact Many scientists think of the skin as an organ. It is made of layers and contains blood vessels, nerves, hairs and oil glands.

Read out the Science fact or ask a volunteer to read it out. Ask students if they think their body outline should have the skin labelled as an organ. Point out that debates like this are common and useful in science. Explain that, at the moment, they are not going to think of skin as an organ, but this may change over the years as they progress through school. This is partly what makes science so interesting. It is always being reviewed and updated.

Ask students to read the text beneath the investigation box and study the diagram showing the development from cell to organism. Point out that the cell is the basic building block for living things like humans. You could explain by using bricks as an example. Bricks (cells) are joined together to make a wall (tissue) and then these join together with window tissue and floor tissue to make a room (organ). The room joins with other rooms of similar types to make all of the rooms in the house (organ system). However, the rooms together do not make a completed house. Other organ systems (the water pipes and electricity cables, for example) are added. Ask students, 'What level of the diagram does the completed house represent?'

Stretch zone: Research three organs and write a paragraph about their function in the human body.

Allow students to work with a partner so they can discuss their ideas. They will need access to the internet and/or science books with sections about the organs. Ask volunteers to take it in turns to read out some of their paragraphs. Point out that they will check their findings, and find out about other organs, in the next lesson.

Key idea

Each of the major organs in a human has a specific place in the body.

Ask students to read the key idea and then close all of their books. They can work with a partner to review their learning from this lesson. On a blank piece of paper they should take it in turns to write down the name of an organ they have learned about in the lesson, so they end up with a list. Remind them that they will need seven organs on their list.

Where are our organs? (page 44)

You can use this activity as an end-of-lesson review, or you could use it at the start to elicit prior learning. Ask students to look at the outline of a human body and then test their knowledge of the organs by drawing them onto the outline. Each organ should be the correct scaled size and in the correct place in relation to each other – not actual size, of course. After completing the drawings of the organs, they should ask a partner to label the organs they have drawn. Pairs can work together to complete all drawings and labels.

Answer: The final labelled diagram should be similar to the one on page 44 of the Student Book.

Tracing the organs (page 45)

This can be an individual or paired activity. Ask students to write the name of each organ in the boxes under the diagrams. Next, they should trace each organ onto a separate piece of paper. Instructions for this are given but you could demonstrate the process. Point out to students the organs are not to scale, so they could draw their own versions. You could also ask them to trace the examples on pages 46–47 of the Student Book so they can see more detail. The traced organs are then cut out and used to make a poster. Remind students to place the organs in the correct place on a body outline and label each one. Encourage students to be creative and colour in their poster so it is eye-catching and then display them in the classroom alongside the other body outline exhibits.

Answer: The final labelled poster should be similar to the one on page 44 of the Student Book.

Review and reflect

Encourage students to reflect on their learning throughout the lesson. Use the discussion tasks, the poster investigation activities and Workbook tasks to encourage students to think about what they understand and what they are finding less straightforward.

Discuss the outcomes of each task with students. Encourage them to think about any questions they have not completed correctly and help them to identify improved answers. When they walk around, they could leave notes containing two comments about what they liked about each display and one suggestion for improvement. They can then go back to their own display and review and reflect on the comments.

Point out that learning is a process and not a one-off event and explain to students that they will learn better if they think about what they have done well and what they need to do to improve. Emphasise that practice and reflection are vital.

Conclude the lesson by asking students to complete question 3 in the 'What have I learned about organs and systems?' activity on page 66 of the Student Book and the first statement on page 66 of the Workbook. (See also the teaching notes in this Teacher's Guide, pages 82–84.)

Extra activities

1 You could obtain plastic models of the body. Show students these and allow them to take them apart and put them back together again. The advantage of models is that they give a 3D view of the organs and how closely packed together they need to be.

2 Ask students to make their own 3D models of the organs of the body. They could use card or modelling clay. Encourage them to make each organ model the correct shape and relative size and, once the body model is assembled, ask them to include label cards.

Differentiation

Supporting: Allow students to refer back to the diagram on page 44 of the Student Book when drawing and labelling the organs.

Consolidating: Keep the body outline displays visible for as long as possible to help students to learn the names and positions of the organs.

Extending: Ask students to draw a diagram of the organisation from cells to organism for another system other than the circulatory system.

Differentiated outcomes	
All students	should be able to name the main body organs
Most students	will be able to locate where the major organs are in the body
Some students	may be able to describe the organisation from cell to organism for a number of systems

What do our major organs do?

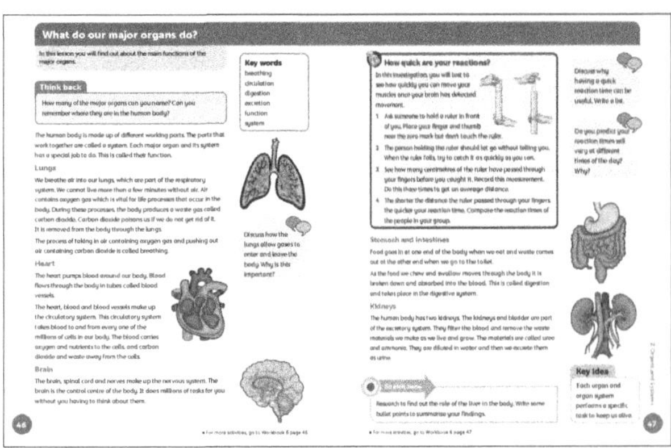

Supporting activities are in Workbook 6 pages 46–47.

Getting started

In this lesson students look at the major organs which are responsible for breathing, taking in food and water, circulating the blood and removing dangerous waste products. The brain controls all of these. They investigate one function of the nervous system – reaction time.

Language support

Write the key words on the board but replace some letters with dashes. Ask students to work out what the words are and then they can write them down. Explain that breathing is a process for getting air into and out of the lungs and the others are systems where organs work together to carry out important functions.

Resources

Student Book: displays of the major organs posters from last lesson, 30-centimetre rulers.

Key words

breathing   circulation   digestion   excretion   function   system

Other words in the lesson

absorbed   ammonia   blood vessels   carbon dioxide   circulatory system   digestive system   excretory system   filter   nervous system   oxygen   reaction   respiratory system   urea

Scientific enquiry key words

Make observations

Take measurements, using equipment accurately

Record data and results

Report and present findings

Lesson at a glance

The key teaching points for students in this lesson are:

- organ systems carry out vital tasks in the body
- vital systems include the circulatory, nervous, digestive, and excretory systems
- the brain is the control centre of the body.

In the next lesson, students will learn more about the lungs and breathing.

Think back: How many of the major organs can you name? Can you remember where they are in the human body?

Support students as they recap the names of the human organs. They can work with a partner to make a list and after five minutes you could allow them to look back in the Student Book to fill any gaps. Regular reviews like this are very important to learning and aiding recall. You will help to develop the skill of regular review and help students to realise that this is a better way of learning than last-minute and extensive revision of concepts they have not thought about for a long time. 'Little and often' is a useful phrase.

Answer: Students may recall: brain, lungs, heart, liver, kidneys, stomach, intestines.

Lungs

Students can read the text on page 46 about the lungs and study the diagram showing lung structure. Point out that the lungs work 24 hours a day throughout a person's lifetime. To check understanding you can ask, 'What do the lungs breathe in? What do the lungs breathe out?' Elicit that the lungs work to keep us alive.

 **Discuss how the lungs allow gases to enter and leave the body. Why is this important?**

Ask students to work with a partner. Suggest they use the text but also the details in the diagram to help them to decide on their answers. They can also draw on prior learning about exercise and health from Year 3. Volunteers can share their ideas with the class.

Possible response: The lungs are open to the air and they have many branches inside that have a good blood supply. This helps gases to move from the lungs to the blood and from the blood to the lungs. This is important because the gas oxygen is needed for life and carbon dioxide is a waste product that has to be removed from the body.

Heart

Students can then read the text about the heart and look carefully at the heart diagram. Point out the different chambers and the vessels leaving the heart. You can ask students some questions about the text such as, 'What does the heart pump around the body? What is the circulatory system? What is taken to and away from cells in the body?'

Brain

Ask a volunteer to read out the text about the brain and point out the diagram of the brain. Ask students to think about why the brain is so important for the body and ask them to name five things they do that are controlled by the brain. They can work in groups to list the activities, which should include heart beating, breathing, moving, seeing, hearing, feeling, smelling, tasting, thinking, eating, talking, etc. Share activities with the class. Lead students to realise that every single part in their body is under the control of the brain. Use the analogy of a giant master computer that fits into a tiny skull.

Investigation: How quick are your reactions?

Arrange students into small groups. Explain that in this investigation, students will test to see how quickly they can move their muscles once their brain has detected movement. Ask them to follow the instructions. You can demonstrate how the test is carried out and allow some practice as it takes a few times to get the coordination correct. Ask students to record the measurements and discuss why they should do this three times to calculate an average distance. Students can then compare the reaction times of the people in their group.

Possible response: Students will find a slight variation in reaction times. You could discuss issues such as reaction time when someone is tired through lack of sleep or exercise.

Discuss why having a quick reaction time can be useful. Write a list.

Students can discuss this with their investigation group immediately after the investigation. If they need any support, suggest they imagine people playing sports, driving vehicles or flying aircraft.

Possible response: People need quick reaction times to catch or kick cricket balls, hockey balls or footballs. Fast reactions can also help to prevent accidents between runners, vehicles and aircraft.

Do you predict your reaction times will vary at different times of the day? Why?

Allow students to continue with their investigation group to think about whether reaction times will vary during the day. Ask them to think about how they feel just after they have woken up or late at night. You could also hint that if they haven't eaten or consumed water for while this may have an effect.

Possible response: Reaction times do vary during the day. Reaction time increases (gets worse) at night and is at its slowest in the early morning. It is also slower when people are hungry or thirsty.

Stomach and intestines

Students can next read the text about the stomach and intestines silently. Remind them that they studied the digestive system in Year 4. Elicit that the digestive system absorbs food, which contains energy that keeps us alive. This is an opportunity to reinforce energy as the fuel of life. A good analogy is petrol or diesel in cars.

Kidneys

Students can then read the text about the kidneys and look carefully at the diagram of part of the excretory system. This is also called the urinary system. The excretory system also includes waste leaving the large intestine, materials leaving the skin and gases leaving the lungs. Remind students of earlier work in Year 3 when they filtered muddy water to clean it. If possible, demonstrate this.

Stretch zone: Research to find out the role of the liver in the body. Write some bullet points to summarise your findings.

This could be an individual or paired challenge. Students will need access to the internet or biology books. Stress that the liver has many functions and students only need to produce a bullet point list, not details of each function.

Answer: The liver makes bile to help with digestion of fats; it removes some waste products; it stores sugar in the form of glycogen; it stores vitamins and minerals; it helps to make blood plasma; it purifies blood by removing harmful substances.

Key idea

Each organ and organ system performs a specific task to keep us alive.

Ask students to read the key idea. They then work with a partner to list the organs and write one function next to each one. They can self-assess by looking back at their Student Book.

Workbook activities

Our heart, lungs and brain (page 46)

This can be an individual or paired activity. Ask students to look at the diagrams of the organs and write about the function of each organ. Next, ask them to think about and write down which part of the skeleton protects the organ in the head and which part of the skeleton protects the organs in the chest.

> **Answer:** *Heart – pumps blood around the body; lungs – allow gases (oxygen) to enter the body and waste gases (carbon dioxide) to leave the body; brain – control centre of the body. The skull protects the brain, and the ribs and sternum protect the heart and lungs.*

Summary table (page 47)

Explain to students that this is an opportunity to create a summary table of what each organ does. Ask them to complete the table and remind them to only use words from the box. After completing this, ask them to work with a partner. They should take it in turns to read out the name of an organ from the table and point to where in their body that organ is found.

> **Answer:** *The table should be completed as: brain; stomach and intestines; heart; kidneys; lungs.*

Review and reflect

Ask students to sit down for a few minutes and think about what they have learned from the lesson. They can then think of one idea from the lesson that they would like to investigate and find out more about. For example, they may wish to test the reaction times of people at home. Ask them to write a plan outlining how they would do this. This planning time will allow students to think up some imaginative and original things to do and you could let them carry out these projects to motivate and reinforce learning.

Extra activities

1 Ask students to plan and carry out an investigation to test the hypothesis that reaction times vary throughout the day. They can discuss their plans with you and then test reaction times, using the technique in the lesson, at different times in the day. Try to coordinate groups so they test at the same times. This will allow you to work out class averages (means) based on more data. Students could produce a reaction times scientific article.

2 Tell students that whales and penguins are mammals like us and have lungs. Ask students to research how long these mammals can stay under water before coming up for air. (Emperor penguins can dive hundreds of metres searching for food before returning to the surface! Show a photograph or a film of whales surfacing and blowing water out of their air holes after diving and swallowing hundreds of kilograms of krill – plankton and small crustacea – on each dive.)

Differentiation

Supporting: Students could complete the activities on pages 46–47 of the Workbook as they read through the text in the Student Book to help make the content more interactive.

Consolidating: At regular intervals across a few days, say the name of an organ and ask volunteers to point to where it is in the body and describe its functions.

Extending: Ask students to think about how the structure of each organ allows it to carry out its functions.

Differentiated outcomes	
All students	should be able to state the functions of the heart, lungs and brain
Most students	will be able to describe the functions of the major organs
Some students	may be able to link the structure of an organ to its ability to carry out its function

Lungs and breathing

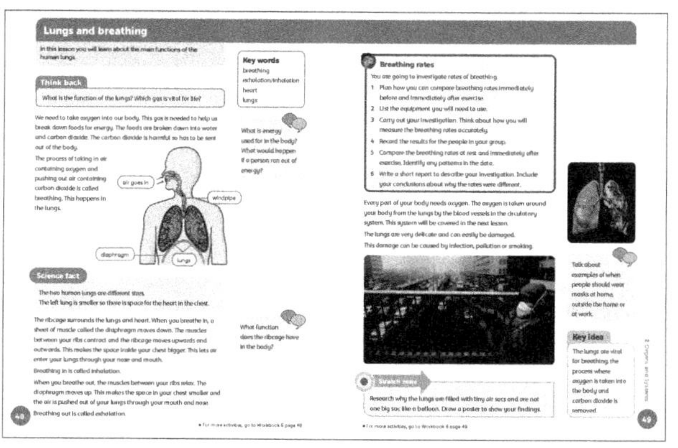

Supporting activities are in Workbook 6 pages 48–49.

Getting started

In this lesson students will study the position and structure of the lungs. They will learn about the process of breathing and the exchange of oxygen and carbon dioxide in the lungs. They will investigate breathing rates before and after exercise and link this to the body's increased need for oxygen to supply muscles. Finally, they will consider the importance of protecting the lungs.

Language support

Ensure that students understand that the word 'exhalation' means when someone is breathing out. Explain that the prefix 'ex' means 'out'. Ask them if they have seen signs that say 'exit'. Point out that the word 'inhalation' means when someone is breathing in. Ask students to stand up and when you say 'inhalation' they breathe in very slowly. When you say 'exhalation' they breathe out very slowly. Repeat this a few times.

Resources

Student Book: stopwatches or timers.

Workbook: items to design a presentation (a talk, computer presentation or poster).

Key words

breathing   exhalation/inhalation   heart   lungs

Other words in the lesson

blood vessel   carbon dioxide   circulatory system
diaphragm   energy   food   oxygen   ribcage

Scientific enquiry key words

Plan and/or carry out enquiries to answer questions

Recognise and control variables

Make observations

Take measurements, using equipment accurately

Record data and results

Analyse data, notice patterns and group or classify things

Report and present findings

Draw conclusions and give explanations

Identify causal relationships

Lesson at a glance

The key teaching points for students in this lesson are:

- oxygen needed by the body is taken in via the lungs
- carbon dioxide, a waste gas, leaves the body through the lungs
- the diaphragm, ribs and muscles are needed for the breathing mechanism.

In the next lesson, students will learn about the human circulatory system.

Think back: What is the function of the lungs? Which gas is vital for life?

Allow students to work with a partner to review the work from the last lesson. Ask them to write down the function of the lungs and the name of the gas and pass their answers to another pair for checking.

> *Answer: The lungs allow gases to enter and leave the body. The gas vital for life is oxygen.*

Ask students to read the text at the top of page 48 and study the diagram of the respiratory system. To check understanding you could ask, 'What is oxygen used for in the body? Which waste gas is produced and has to be sent out through the lungs? What is breathing?'

 **What is energy used for in the body? What would happen if a person ran out of energy?**

Pairs can continue to work together to talk about what energy is used for in the body. Remind them of the work they have done in earlier years on food and energy, for example in Year 4, and their ideas from the introductory lesson of this unit. Confident students can share their ideas with the class.

> *Possible response: Energy is used to help muscles to work (contract), to keep us warm, and for growth and repair.*

Science fact The two human lungs are different sizes. The left lung is smaller so there is space for the heart in the chest.

Read out the Science fact and ask students if they are surprised that the lungs are different sizes. Ask, 'How else would the heart fit?' They can look back at the diagram on page 44 of the Student Book to see the relationship between the heart and lungs.

Allow students to read through the text about the breathing mechanism at the bottom of page 48 and ask them to keep looking at the diagram of the lungs, ribcage and diaphragm as they do this. They can then talk about the discussion task that follows.

What function does the ribcage have in the body?

Students can read the text again and also hold their ribs as they breathe in and out. Ask them to try to breathe in and out without their ribs moving. Ask some students to share their ideas with the class.

Answer: The ribcage expands and contracts to make space for the lungs and it also protects the lungs.

Investigation: Breathing rates

Explain that students are going to investigate rates of breathing. The worksheet on page 49 of the Workbook supports this investigation.

Leave this open-ended as students have explored breathing rates in Year 3 at a simpler level and so should be able to build on these ideas. Ask them to plan and carry out an investigation to compare breathing rates immediately before and immediately after exercise. They will need to list the equipment needed and consider how they will measure breathing rates accurately and record the results. Encourage them to think carefully about any risks and how to keep everyone safe.

Students should set a measurable time for the exercise and it must be the same every time they carry out the test. The resting breathing rate should only be taken after the person has rested and the breathing rate after exercise should be taken immediately the exercise has finished. They should count breathing rate over the same amount of time. For example, they count the number of breaths in 30 seconds and double it to get the rate per minute.

To allow students to reflect on their ideas and share them with others, ask them to write a short report and include their conclusions. Encourage them to think about fair testing, accuracy and how they can analyse the data.

Ask a volunteer to read out the text beneath the investigation and encourage students to study the photograph of the person wearing the mask. You can also ask students to look carefully at the picture showing the healthy and unhealthy lungs. Point out that this is made up from more than one picture as in real life both lungs are likely to be healthy or both unhealthy. Ask students to decide which lung is healthy and which

is unhealthy and talk to a partner about the clues they used. Elicit that the dark lung is the unhealthy one because of the dark patches and lack of red colour (blood). Ask students to suggest causes for the unhealthy lung. Link this to the photograph of the person wearing the mask. Stress how delicate the lungs are and how important it is to look after them.

Talk about examples of when people should wear masks at home, outside the home or at work.

Encourage students to think about the importance of masks. You can ask them about jobs that people do where they have to wear masks. Use evidence from the photograph to help students think about polluted air and to recall that infectious diseases can travel through the air from person to person.

Possible response: Dusty jobs require people to wear masks and it is also useful to wear masks in polluted areas, such as near traffic. Doctors, nurses and other medical staff and scientists also wear masks to prevent microorganisms from spreading.

Stretch zone: Research why the lungs are filled with tiny air sacs and are not one big sac like a balloon. Draw a poster to show your findings.

Ask students to look carefully at the diagrams of the lungs on pages 44, 46 and 48 to see if they can see any tiny sacs. Point out that in real life the sacs, or alveoli, would be too small to see (they are 0.2 mm in size) but they have been made bigger for the diagrams. Allow them access to the internet to find out more about alveoli.

Possible response: Alveoli increase the surface area of the lungs so the gases can be in contact with more surface to bring them closer to blood vessels.

Key idea

The lungs are vital for breathing, the process where oxygen is taken into the body and carbon dioxide is removed.

Read out the key idea or ask a volunteer to read it out. Ask students to discuss which vital gas enters the lungs and which waste gas leaves the lungs. They can talk to each other about the breathing mechanism and why breathing rates increase during exercise.

Workbook activities

How we breathe (page 48)

You can use this activity to help students to learn the names of the parts of the respiratory system and to check understanding of the text on page 48 of the Student Book. Ask them to label the diagram and remind them to only use words from the box. Next, they should describe what happens to the ribs and diaphragm as a person breathes and in what happens to the ribs and diaphragm as a person breathes out.

> *Answer: Labels top left and anticlockwise are: nose; mouth; diaphragm; lungs; windpipe. As the person breathes in the diaphragm moves down and the ribcage moves upwards and outwards. As the person breathes out the diaphragm moves up and the ribcage moves down and inwards.*

Breathing rates (page 49)

This activity supports the investigation on page 49 of the Student Book. Explain that the workbook page provides prompts and spaces to write responses to structure the investigation. Students write down their plan and record how they will measure the differences between people. They then think about how they will record results and reflect on safety issues. There is a table for students to record their findings. Finally, they can tick which type of presentation they have decided to do.

> *Possible response: Responses will vary but students should include a set, measurable time for the exercise that is the same every time they carry out the test. The resting breathing rate should only be taken after the person has rested and the breathing rate after exercise should be taken immediately after the exercise has finished. They should count breathing rate over the same amount of time. For example, they count the number of breaths in 30 seconds and double it to get the rate per minute. For safety, they should make sure that people do not exercise for too long, that they stop if they feel ill, and that they do not trip over anything.*

Review and reflect

Encourage students to reflect on their own learning by looking through the activity pages they have completed and the report they produced, and thinking about their stand-up talk, computer presentation or poster. Talk about how they managed this work and ask them to think of two learning targets to improve what they did.

Extra activities

1   You could carry out a class demonstration of breathing. Ask students to sit up straight in their seats or to stand up and tilt their heads forward so that they can see their chest, and then take a deep breath in and out. They should be able to see that their chest moves up and down as they expand and contract the chest when breathing. This could be done in pairs. The breathing exercise is to demonstrate the position of the lungs in the chest cavity.

2   Ask students to work in a small group to make a large poster-sized version of the diagram on page 48 of the Student Book. They can then add information boxes to their poster explaining the breathing process and why breathing is so important.

Differentiation

Supporting: Use the activity on page 48 of the Workbook to help students to learn the parts of the respiratory system and to support them in understanding the text about breathing.

Consolidating: Display a poster of the respiratory system and show an arrow with 'oxygen' written on it entering the lungs and an arrow with 'carbon dioxide' written on it leaving the lungs.

Extending: Ask students to research the composition of inhaled and exhaled air.

Differentiated outcomes	
All students	should be able to label the parts of the respiratory system
Most students	will be able to describe the process of breathing
Some students	may be able to relate the structure of the lung to its function of transferring oxygen and carbon dioxide to and from the blood

The human circulatory system

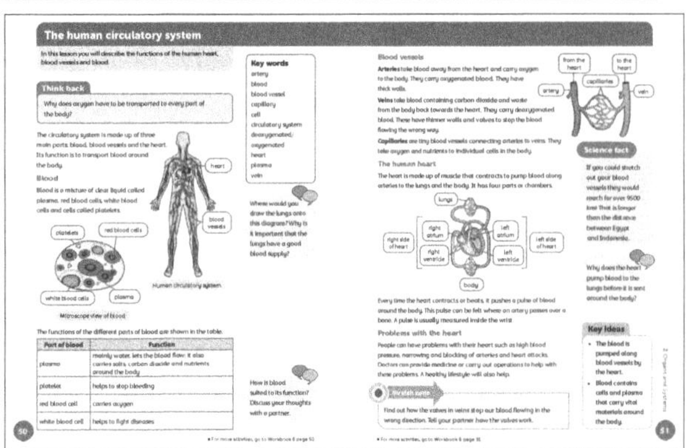

Supporting activities are in Workbook 6 pages 50–51.

Getting started

In this lesson students will learn that the heart and blood vessels are called the circulatory system. They study the composition of blood and the functions of each component. They will learn the names of different blood vessels and their functions. Students then consider the structure of the human heart in more detail and consider some potential problems with the heart.

Language support

There are a number of new words for students in this lesson, though they may have heard of some outside school. For example, they may have heard of artery, vein and plasma. Check prior knowledge by reading out all of the key words and asking students to hold up a number 1 to 5 based on their level of knowledge of the words: 1 = 'never heard of it' and 5 = 'can define it and use it in a sentence'. Students can return to their scores at the end of the lesson to see if confidence levels have risen. It is worth explaining that 'oxygenated' means to have oxygen added. 'Deoxygenated' means oxygen is removed. Remind them that 'de' added to a word means the opposite – they may have heard of the words de-ice or deconstruct. Write the key words on the board or on word cards, and you can also use the eBook so students can hear them pronounced.

Key words

artery blood blood vessel capillary cell
circulatory system deoxygenated/oxygenated
heart plasma vein

Other words in the lesson

chamber function mixture muscle platelet
pulse salts transport valve

Scientific enquiry key words

Make observations

Draw conclusions and give explanations

Identify causal relationships

Lesson at a glance

The key teaching points for students in this lesson are:

- blood is made up of liquid plasma, platelets, red blood cells and white blood cells
- blood vessels take blood from the heart to the body and back to the heart
- the heart is made of muscle; it contracts to pump blood.

In the next lesson, students will explore how nutrients are broken down in the digestive system.

Think back: Why does oxygen have to be transported to every part of the body?

Ask students to work with a partner to review their prior work on oxygen. Once they have arrived at an answer, ask volunteers to share their answer with the class.

> *Possible response: Oxygen is needed for humans to obtain energy from foods. Every part of the body needs to make energy so every part will need oxygen.*

Allow students time to read the text on page 50 and study the diagram of the circulatory system. Ask them to make bullet point notes of the main points – this is an important skill to develop. You can check understanding by asking, 'What are the three main parts of the circulatory system? What is the function of the circulatory system?'

 **Where would you draw the lungs onto this diagram? Why is it important that the lungs have a good blood supply?**

Students can work with a partner to study the circulatory system diagram and think back to their work on the lungs in the last lesson. They can point to where the lungs would be. They can then think about why the lungs should have a good blood supply. Remind them of the function of the lungs.

> *Possible response: The lungs are either side of the heart. They need a good blood supply because gases have to enter and leave the blood at the lungs.*

Blood

The section on blood is complex so you may prefer to read through it with students or ask volunteers to read out a sentence at a time and talk to the class about the ideas covered. Stress that blood is a mixture of liquid and cells and each component has a different function. Show them the table on page 50 and allow students to study it for three minutes. They then close their books and discuss these questions with a partner: Which part of the blood carries oxygen? What is the liquid part of the blood called? What might happen if you did not have white blood cells? What do platelets do?

How is blood suited to its function? Discuss your thoughts with a partner.

After discussing the text allow students to think about the discussion question. They will need to bring together the different roles carried out by the cells and you can point out that if these were not in a liquid (plasma), the blood would not flow. After the discussion you could ask some students to suggest answers to the class.

Possible response: Blood is well suited to its function as it is mainly liquid – so it can flow through vessels – and it has a mixture of cells that carry out the different functions. The plasma also has more than one function.

Blood vessels

Point out the text and the diagram at the top of page 51. Write the words 'artery', 'vein' and 'capillary' on the board or on a sheet of paper on a wall and set students the challenge of finding out what each one is and what its function is. You can allow five minutes for them to read the text and study the diagram and then ask volunteers to give a simple definition of each of the words. Suggest they remember the difference between arteries and veins by underlining the 'i' and 'n' at the end of the word ve<u>in</u> and say these vessels take blood <u>IN</u> to the heart; or they could underline the 'a' at the start of <u>a</u>rtery as a reminder that arteries take blood AWAY from the heart.

Science fact If you could stretch out your blood vessels they would reach for over 9500 km! That is longer than the distance between Egypt and Indonesia.

Read out the Science fact or ask students to read it themselves. Ask them if the fact surprises them. Point out that it gives a good clue as to how small many of the blood vessels are – some are only wide enough to let single red blood cells through (0.005 millimetres!).

The human heart

Ask students to study the text and the diagram about the heart on page 51. There is a larger version of the diagram on page 50 of the Workbook that you could use to accompany this text. Explain to students that the circulatory system is like a racing track. Tell them to place their finger at the start line – in the left ventricle. Tell them that the ventricles and atria (plural of atrium) are medical words for chambers or rooms. Students may be interested to know that an atrium was a central courtyard in a house in Ancient Rome.

When you say 'Start', they can follow the race track along the small arrows to the body, through the body, and back to the heart. Ask them which chamber of the heart they have arrived in. Next, they race through the valve to the right ventricle, on through the lungs, and then back to the left atrium. Finally, they arrive back in the left ventricle after finishing a lap. Before students talk about the discussion question, you could give them some rules to help them to understand the circulation of blood:

- Blood leaves the heart from ventricles.
- Blood enters the heart through atria.
- The tiny valves make sure blood flows through the heart in the correct direction.
- Oxygenated blood is shown as red and deoxygenated blood is shown as blue.

Students may be interested to know that only one artery carries deoxygenated blood – the pulmonary artery to the lungs. Only one vein carries oxygenated blood – the pulmonary vein from the lungs. Otherwise, arteries carry oxygenated blood away from the heart and veins return deoxygenated blood into the heart from the body.

Why does the heart pump blood to the lungs before it is sent around the body?

Ask students to talk about the question with a partner. Point out the two colours on the diagram and, if they need a clue, tell them that the blue represents deoxygenated blood and the red represents oxygenated blood. Students can share their ideas with a nearby pair.

Answer: The blood is pumped to the heart so it can pick up oxygen (oxygenated blood) before it is sent around the body.

Problems with the heart

Students can read through the section of text describing some problems that can occur with the heart. Ask them to list three examples of heart problems. They should pick out high blood pressure, narrowing and blocking of arteries, and heart attacks. These will be mentioned in later lessons in the unit that are more directly linked to health, so make sure students are aware of them here.

Stretch zone: Find out how the valves in veins stop our blood flowing in the wrong direction. Tell your partner how the valves work.

Computing link: Allow students access to the internet to find out more about valves. They can demonstrate the action of valves using their hands.

Possible response: The valves are flaps that point in one direction, like an open door. When blood is flowing in the correct direction, the doors are pushed open. If the blood tries to flow in the opposite direction, it presses against the valves and closes them.

Key ideas

- *The blood is pumped along blood vessels by the heart.*
- *Blood contains cells and plasma that carry vital materials around the body.*

Ask a volunteer to read out the key ideas and then ask students to close their books and take out a blank piece of paper. Ask them to list the components of blood and the names of the three different blood vessels. They can then open their books and check their answers.

Workbook activities

Label the parts of the circulatory system (page 50)

You can use this activity to support students as they work through the text on page 51 of the Student Book or as an end-of-lesson review. Ask students to label the diagram. Remind them to only use the words in the box. Warn them that they will have to think carefully to decide which vessel is the vein and which is the artery. You can point out that the artery will leave the heart from a ventricle (bottom chamber) and the vein will enter the heart at an atrium (top chamber).

Once the diagram is labelled, students can complete the Stretch zone. Ask them to look carefully at the diagram, and they can look back to the one on page 51 of the Student Book to see the colours, and explain why the artery that transports blood from the heart to the lungs is the only artery that transports blood that is not rich in oxygen.

> *Answer: Labels are (top right and clockwise): lungs; vein; capillaries; heart; artery. The (pulmonary) artery to the lungs leaves the heart – so it is an artery and not a vein – but it carries blood that has had most of the oxygen taken out of it. It must take the blood to the lungs to be oxygenated.*

Exercise, pulse rates and fitness (page 51)

Ask students to study each of the drawings and then draw a line from each activity to the correct likely pulse rate. They should use a ruler and a pencil to keep the lines neat and tidy. Remind them that bpm means beats per minute. Suggest they start with the drawing showing the most exercise and assign it the highest pulse rate, and then assign the lowest pulse rate to the drawing showing the least exercise. Once students have completed the matching activity, ask them to think about when they exercise most and what happens to their pulse rate when they exercise.

> *Answer: Students should draw lines to match the following: sprinting 140 bpm; reading 78 bpm; jogging 120 bpm; sleeping 67 bpm; walking quickly 105 bpm; walking slowly 95 bpm. Students may suggest different times when they exercise most, but they should all state that their pulse rate increases.*

 ## Review and reflect

Hold up word cards for the key words in the lesson: artery, blood, blood vessel, capillary, cell, circulatory system, deoxygenated, oxygenated, heart, plasma, vein. Each time you hold up a word ask students to turn to a partner and use the word in a sentence. They should write down any of the words they are unsure of.

Ask students to reflect on the discussions and activities they have carried out and think about how well they worked with another person and how well they communicated with each other. Ask them to think of one way of working that they are proud of and one thing they could improve on. Tell them this is their target. They can see if they meet their target in the next few lessons.

Conclude the lesson by asking students to complete questions 5 and 7 in the 'What have I learned about organs and systems?' activity on page 67 of the Student Book. (See also the teaching notes in this Teacher's Guide, pages 82–84.)

Extra activities

1. **Computing link:** Students could research how the valves in the heart work and find out how heart problems can be helped with surgery and mechanical devices such as pacemakers and even mechanical valves.

2. Ask students to make a large circulation game. You could mark out the circulatory system similar to the one on page 51 of the Student Book in chalk in a large open space. Most students can act as red blood cells moving around the system. Some students can act as the heart and gently push the 'red blood cells' in the correct direction. Other students can be the lungs and hand pieces of paper with 'oxygen' written on to the red blood cells and other students can act as the body and take 'oxygen' cards from the red blood cells.

Differentiation

Supporting: Demonstrate the pumping of the heart by allowing students to place their hands in a bowl full of water and squeeze them together to feel the water squeeze out.

Consolidating: Ask students to use materials such as tubing, a balloon, strands of wool and plastic bottles to make a labelled 3D model of the circulatory system.

Extending: Students can research why the walls of arteries are thick and muscular and the walls of veins are thin.

Differentiated outcomes

All students	should be able to state that the circulatory system comprises the heart, blood and blood vessels
Most students	will be able to name and describe the main vessels of the circulatory system and the components of blood
Some students	may be able to link the composition of blood to its various functions

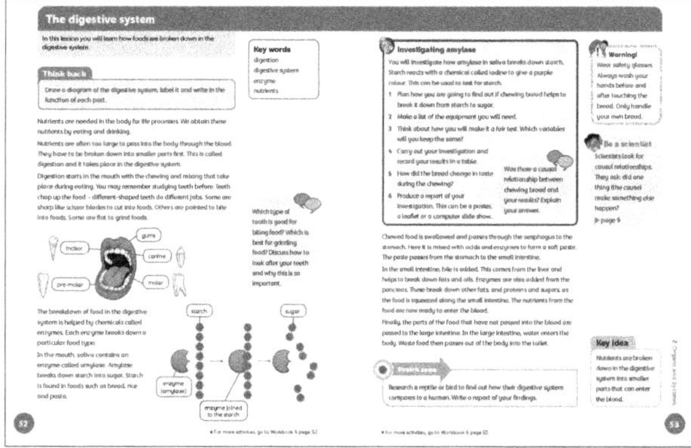

Supporting activities are in Workbook 6 pages 52–53.

Getting started

In this lesson students will review their knowledge of the digestive system from Year 4. They will extend their knowledge of the importance of enzymes and consider the important role of teeth in the breaking down of food. During the stages of digestion they will look at the roles of chewing, mixing and enzymes and will investigate the role of amylase.

Language support

To help develop language skills and review prior knowledge you could use flashcards with the names of the different parts of the digestive system and ask students to point to them on a diagram. You could extend this to include words such as enzyme, teeth, chewing and mixing. To help students pronounce 'amylase' spell it out phonetically as 'AM-uh-laze'.

Resources

Student Book: safety glasses; bread; iodine; items to design a computer presentation, leaflet or poster.

> ### Key words
>
> digestion   digestive system   enzyme   nutrients
>
> **Other words in the lesson**
>
> amylase   bile   chewing   intestine   iodine
> mixing   oesophagus   pancreas   saliva   starch

> ### Scientific enquiry key words
>
> Plan and/or carry out enquiries to answer questions
>
> Recognise and control variables
>
> Make observations

Record data and results

Report and present findings

Identify causal relationships

Lesson at a glance

The key teaching points for students in this lesson are:

- nutrients in food are broken down step by step in the digestive system
- the breakdown of foods is helped by enzymes
- chewing and mixing also help the breakdown of nutrients.

In the next lesson, students will learn how nutrients and water are absorbed into the body from the digestive system.

Think back: Draw a diagram of the digestive system, label it and write in the function of each part.

Hand out a blank piece of paper to students and ask them to sit quietly and think back to their work in Year 4. They can do this as an individual challenge or work with a partner to share ideas. You could use the activity on page 52 of the Workbook if you want to provide more structure. Once they have completed as much of their diagram as they can you can show them one on the screen or from a book and they can fill in any gaps.

Allow students to read the text and study the diagram of the teeth on page 52. To check understanding you can ask, 'How do we obtain nutrients? Why do nutrients have to be broken down? Where does digestion start?'

Which type of tooth is good for biting food? Which is best for grinding food? Discuss how to look after your teeth and why this is so important.

Ask students to work with a partner to discuss the questions. Point out that they will need to use the text and diagram to help them. They can make their own table to summarise their ideas about the functions of the teeth and list any ways they can look after their teeth. They can share their ideas with a nearby pair.

> *Possible response: Incisors bite food and canine teeth grip and tear food. Molars are best for grinding food. Ways to look after teeth include brushing regularly, avoiding acid and sugary foods, and having regular check-ups at a dentist.*

Read through the text at the bottom of page 52 explaining the way that enzymes work and point out the diagrams that show the process. Explain that the enzyme is shown in blue and it has a shape that matches the nutrient to be broken down. Each enzyme fits a different nutrient.

Investigation: Investigating amylase

Explain that students are going to investigate how one enzyme, amylase, works to break down starch.

Warning! Wear safety glasses. Always wash your hands before and after touching the bread. Only handle your own bread.

Start by reading out the safety warning and ask students why these warnings are so important. Allow students to use the information in the investigation to help them to plan and carry out the investigation and only support them when needed. This will help them to develop independent learning skills and will encourage them to persevere and puzzle things out. If you need to give advice, tell them that they will need to compare unchewed and chewed food by testing with iodine – which will show a purple colour if amylase has broken the starch in the bread down to sugar (maltose). Hint that they could have controls by just adding water to some bread and testing it.

Allow students to display their leaflets/posters or if they have designed a computer presentation this can be presented to the class.

> *Answer: Students should plan to test unchewed bread with iodine; they then chew another piece of the same bread and place this into a bowl to test with iodine – it should turn purple. The variables are: independent = presence of amylase (chewing or not chewing); dependent = colour change due to iodine; control = amount of iodine used, amount of bread used, type of bread used.*

Be a scientist: Scientists look for causal relationships. They ask: did one thing (the cause) make something else happen? (Student Book, page 9)

Ask students to read the Be a scientist information and remind them that a causal relationship is only valid if nothing else could have caused the change. For example, in their investigation the amylase could not be said to cause the change if some bread soaked in water or just left for two hours in the sunlight also broke down and gave the purple colour.

Was there a causal relationship between chewing bread and your results? Explain your answer.

Ask students if they noticed any causal relationships in their investigation. Allow them to think about what they did and whether their results were caused by amylase and no other reason.

> *Answer: Students should note that the breakdown of starchy bread to sugar was caused by the amylase in their saliva as they chewed.*

Stretch zone: Research a reptile or bird to find out how their digestive system compares to a human. Write a report of your findings.

Computing link: Ask students to work with a partner so they can share ideas and allow them access to the internet. They can either study a reptile or a bird, or you could allocate reptile to some pairs and bird to others. The report can be word processed and include downloaded photographs and diagrams.

Possible response: Bird and reptile digestive systems are similar to a human's but with some differences:

Most birds have a crop – a place in the oesophagus where food is softened. After storage in the crop, the food passes into the first half of the stomach where acid, mucus and enzymes are added. The food passes to the bottom half of the stomach – the gizzard. This muscular organ acts like human teeth to grind the food mix into smaller pieces. The food is then a paste and it passes into the small intestine where the food nutrients are digested and absorbed, as in humans.

Reptiles have longer tongues to sense the environment and some have salivary glands adapted to produce venom. Otherwise the digestive system has a similar pattern to a human's – but the oesophagus can be expanded in some to allow large prey to be swallowed and the stomach may contain stones to help with the grinding of food.

Key idea

Nutrients are broken down in the digestive system into smaller parts that can enter the blood.

Ask a volunteer to read out the key idea and ask students to look again at the drawing they did at the start of the lesson to see if they can add anything to it. Ask them to add the labels 'enzyme' and 'teeth' to their diagram.

Workbook activities

Parts of the digestive system (page 52)

This activity can be used to help students to review their prior knowledge of the digestive system at the start of the lesson, or it can act as a useful review at the end. Ask students to label the diagram. Remind them to use the words in the box.

Answer: Labels are (top left then anticlockwise): mouth; oesophagus; liver; pancreas; small intestine; large intestine; stomach.

Once the labelling is complete you can check the answers and ask students to try the Stretch zone task.

Answer: The common name for the oesophagus is food pipe or gullet.

Breaking down food (page 53)

Ask students to study the diagram and then identify the different teeth. Explain that they should identify which number matches each description.

Students can then think about the Stretch zone task. Ask them to recall or find out which types of teeth are called incisors, canines, pre-molars and molars. They can use a mirror to find them in their mouth and then draw a picture of each. They can use the diagram at the top of the page to help. Remind them to label each one.

Finally, ask students to research and write down two ways that they can look after their teeth.

Answer: 1 = 2 (canine); 2 = 1 (incisor); 3 = 3 and 4 (pre-molar and molar). Students should draw an incisor so it looks like drawing 1; a canine like drawing 2, a pre-molar like drawing 3; a molar like drawing 4. Ways of looking after teeth include brushing with a toothbrush regularly, using toothpaste regularly, avoiding sugary foods and drinks, avoiding acidic foods and drinks and visiting a dentist regularly.

Review and reflect

Remember to praise the process of learning as much as, if not more than, the outcomes. The amylase investigation, especially if carried out with little support from you, is complex and students will have done very well to complete it without intervention. Encourage students to ask questions and to look back through prior work to review and revise. Use every opportunity to allow students to work together to check understanding and share ideas. After the research investigation, ask students to reflect on how easy they found the planning and the identification of the variables.

Extra activities

1 Ask students to research and describe some enzymes that break down other carbohydrates and proteins. They could make a table to share the names, the function and the location in the digestive system of these enzymes.

2 Students can make a 3D model of the digestive system. Many different materials can be used – for example dry pasta, plastic tubes and bottles, card, string and balloons. You can also ask them to make their own digestive system lab coat or overall. Hand out white cloth or aprons and ask them to use coloured marker pens to draw the digestive system so that when they wear the apron the system is visible at the front of their body.

Differentiation

Supporting: Hand out the activity on page 52 of the Workbook at the start of the lesson to help students review prior work on the digestive system.

Consolidating: Display students' work on the digestive system as soon as it is completed and leave it on display for a few weeks to help students to learn the main parts and functions.

Extending: Ask students to research the lock-and-key theory of enzyme action.

Differentiated outcomes	
All students	should be able to state that teeth and enzymes help in the breakdown of nutrients
Most students	will be able to describe an investigation to show that amylase breaks down starch to sugars
Some students	may be able to explain the action of enzymes on nutrients

Absorbing nutrients and water

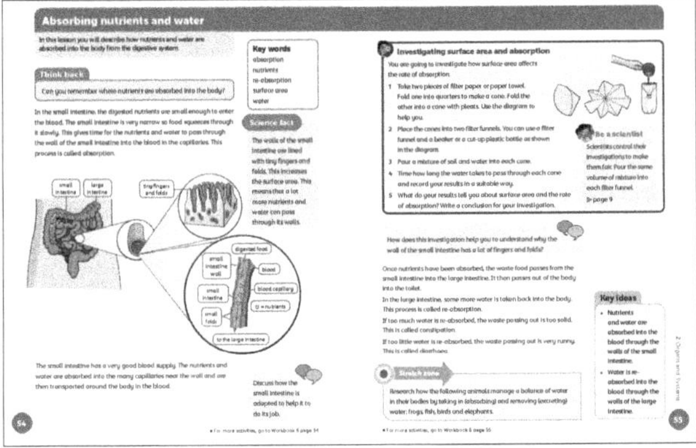

Supporting activities are in Workbook 6 pages 54–55.

Getting started

In this lesson students will continue their study of the digestive system. They will learn how the digested nutrients enter the body through the walls of the small intestine and into the blood. They will also learn how water is absorbed into the body through the walls of the large intestine.

Language support

Students will be familiar with the words 'blood' and 'surface' but you need to explain the key word 'absorption'. Point out the spelling and let students practise writing down the word. Explain that absorption means 'taking in'. In the digestive system this means the taking in of nutrients into the body. Students may have heard of a sponge absorbing water or a person absorbing knowledge.

Resources

Student Book: filter paper or paper towels; filter funnels; beakers or cut-up plastic bottles; soil; water; timers or stopwatches.

Workbook: A4 paper.

> **Key words**
>
> absorption   nutrients   re-absorption
> surface area   water
>
> **Other words in the lesson**
>
> blood   capillaries   constipation   diarrhoea
> intestine

Scientific enquiry key words

Plan and/or carry out enquiries to answer questions

Make observations

Take measurements, using equipment accurately

Record data and results

Analyse data, notice patterns and group or classify things

Report and present findings

Draw conclusions and give explanations

Identify causal relationships

Lesson at a glance

The key teaching points for students in this lesson are:

- nutrients are absorbed into the body through the walls of the small intestine
- the small intestine is adapted to have many folds and a good blood supply.

In the next lesson, students will learn about water transport and the urinary system.

Think back: Can you remember where nutrients are absorbed into the body?

Ask students to sit quietly for a few minutes to think back to their earlier work on the digestive system. Ask for volunteers to suggest answers.

Answer: Through the small intestine.

Ask students to read the text on page 54 and study the diagrams. These show the structure of the small intestine with an increasing magnification. Ask students to decide which picture shows the highest magnification. Point out the fingers and folds and the way that the blood capillary is very close to the wall of the small intestine.

Science fact The walls of the small intestine are lined with tiny fingers and folds. This increases the surface area. This means that a lot more nutrients and water can pass through its walls.

Read out the Science fact. This explains why the wall of the small intestine is so folded. Explain to students that they will be investigating why this is useful later in the lesson.

Discuss how the small intestine is adapted to help it to do its job.

Ask students to work in pairs or in groups of three or four. They should discuss the question and draw on the information in the text and diagrams on page 54 to arrive at some suggestions. Volunteers can stand up and share their ideas with the class.

Answer: The small intestine is long and narrow, has thin walls that have many fingers and folds, and has a good blood supply.

Investigation: Investigating surface area and absorption

Arrange students into groups of three of four. Explain that they are going to investigate how surface area affects the rate of absorption. Remind them that surface area is a measure of how much surface something has. A stamp has a small surface area and a football pitch has a large surface area. Ask students which desk or table in the room has the largest surface area.

Be a scientist: Scientists control their investigations to make them fair. Pour the same volume of mixture into each filter funnel. (Student Book, page 9)

Point out the Be a scientist information and ask students to think about this, and other ways to make their investigation a fair test, as they plan and carry out their filtering.

Show students how to take two pieces of filter paper or paper towel and fold one into quarters to make a cone. Demonstrate how to fold the other into a cone with pleats. Students can then follow the instructions to complete the investigation. After analysing their results, ask them to reflect on what the results show about surface area and the rate of absorption. They can then write a conclusion for their investigation. Point out that, although they are looking at filtering and soil drainage, this is very similar to absorption.

Possible response: The investigation shows that the water passes through the filter paper with the largest surface area quicker than the other filter paper. This is like the folds in the small intestine.

How does this investigation help you to understand why the wall of the small intestine has a lot of fingers and folds?

Ask students to talk about their results from the investigation and look back at the diagrams of the small intestine on page 54. Some students can share their ideas with the class.

Possible response: It shows that a folded surface has a larger surface area and this lets more water/liquid pass through it.

Ask volunteers to read out a sentence each of the text at the bottom of page 55. To check students in the class understand the ideas, you can ask them, 'What happens to the waste food after nutrients have been absorbed? What happens if too much water is re-absorbed in the large intestine?'

● **Stretch zone:** Research how the following animals manage a balance of water in their bodies by taking in (absorbing) and removing (excreting) water: frogs, fish, birds and elephants.

● **Computing link:** Allow students access to the internet and other sources of information such as biology books and science encyclopaedias. You could allocate one animal to each pair to save some time or ask students to research all of them. They can share ideas in a variety of ways such as a computer slide show, leaflets, a science article or poster.

> *Possible response: Students should find that birds and elephants drink water and elephants excrete it as urine. Birds pass out excess water with waste food. They do not have a bladder. Frogs can absorb water through their skin and pass out urine. They have a bladder. Fish take in a lot of water through their gills and can excrete materials such as salts back out through their gills. Fish have a urinary pore that takes waste from their kidneys and sends it outside. Some types of fish have a small bladder.*

Key ideas

- *Nutrients and water are absorbed into the blood through the walls of the small intestine.*
- *Water is re-absorbed into the blood through the walls of the large intestine.*

Read out the key ideas and ask students to talk to a partner about how the small intestine is adapted to carry out its function. They can write down their ideas or tell another group.

Workbook activities

How nutrients enter the blood (page 54)

You can use this activity to help students to review their knowledge of the digestive system. Ask them to link the part of the digestive system to its function. They can then discuss why water is easily absorbed into the body and use their creativity and imagination to write a short story to describe the journey that food takes after it enters the mouth. Remind them to explain how the food is broken down and where this happens. They should include the words: nutrients, teeth, enzymes and absorption.

> *Answer: The links are: liver = bottom statement; small intestine = 3rd statement; large intestine = 2nd statement; stomach = 4th statement; mouth = top statement; oesophagus = 5th statement. Water is easily absorbed into the body because the particles are small and pass through the walls of the small intestine easily. Also, the small intestine walls are adapted to be thin and have a large surface area.*

Investigating surface area and absorption (page 55)

Explain that this activity supports the investigation on page 55 of the Student Book. Point out that it will demonstrate how folding can increase the surface area without something having to take up more space.

● **Maths link:** Remind students how to calculate area and then ask them to find the area of each of the three pieces of paper. If you have given out identical pieces of paper, then this should be the same for each piece.

Encourage students to follow the instructions in the diagram to make the folded versions of the papers but you may need to demonstrate how to fold the three pieces of paper into a concertina. Ask students to think about what the surface area of each folded piece of paper is and then they can lay the folded pieces of paper over the flat piece. Check that they are not overlapping the pieces. Students can calculate the total area of the three folded pieces of paper. They can then discuss how folding the paper has helped them to increase the surface area of the paper.

> *Answer: The total area of the three folded pieces of paper will be three times the area of the flat piece – even though they are not taking up more space on the desk.*

Review and reflect

Encourage students to identify aspects they have not completed correctly and help them to identify improvements. This will help them to develop a positive approach to learning by understanding that learning is a process that will improve with practice and reflection. For example, they can look at the completed activity pages to identify what they have done well and what they can learn from others as targets to improve their work. Ask them how they coped with the mathematics in the activity on page 55 of the Workbook.

Extra activities

1 Ask students to compare the ways that surface area is increased in the small intestine and in the lungs. Ask them to make a small poster that shows the similarities and differences.

2 Students could draw a large version of the diagram on page 54 of the Student Book. They can colour it in and add labels and then use a digital camera or a smartphone to make an instruction video that explains how nutrients are absorbed into the body.

Differentiation

Supporting: Allow students to start the lesson with the activity on page 54 of the Workbook to review the digestive system.

Consolidating: Poster displays are an excellent way to consolidate learning, so buy or encourage students to make large poster versions of the digestive system and the small intestine diagrams.

Extending: Students could research how water is transported in other animals.

Differentiated outcomes	
All students	should be able to state that nutrients are absorbed into the body in the small intestine
Most students	will be able to describe how the small intestine is adapted to increase the efficiency of absorption
Some students	may be able to compare how water is transported in different animals

Water transport and the urinary system

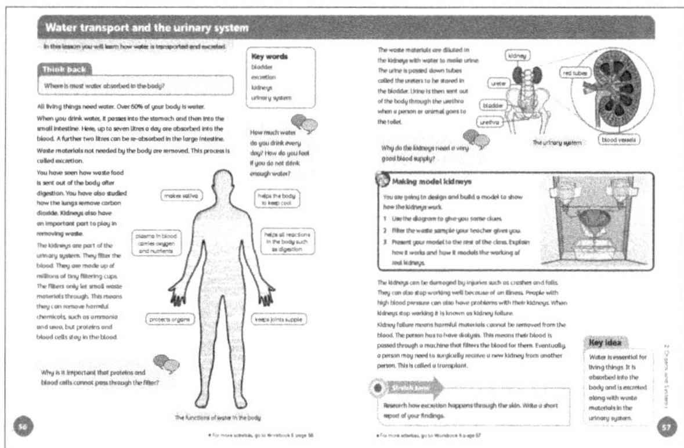

Supporting activities are in Workbook 6 pages 56–57.

Getting started

In this lesson students will learn where water is absorbed into the body and why water is important. They will look at the roles water plays in the body and then study in detail how water is regulated, including through excretions from the urinary system.

Language support

Some of the key words in the lesson – excretion, urinary system – are not simple and they are not easy to spell. A strategy you can use is to ask students to close all books and then write the first letter of each word on the board, ask students to decide what the next letter should be and so on until the words are fully displayed. Students can copy the words and then share examples of when they have heard the words before so you can elicit prior understanding.

Resources

Student Book: samples of waste water; funnels; filter paper; plastic bottles; tubing; sticky tape; frames or stands; large flat containers.

Workbook: pieces of cardboard with different sized holes cut in them; box containing large red, small blue, large yellow and small green balloons; a large plastic bag.

> **Key words**
>
> bladder   excretion   kidneys   urinary system
>
> **Other words in the lesson**
>
> dialysis   filter   re-absorbed   ureter   urethra   urine   water

Scientific enquiry key words

Plan and/or carry out enquiries to answer questions

Make observations

Take measurements, using equipment accurately

Record data and results

Report and present findings

Draw conclusions and give explanations

Lesson at a glance

The key teaching points for students in this lesson are:

- water is absorbed into the body mainly through the walls of the small intestine
- excess water is excreted with waste materials through the urinary system.

In the next lesson, students will learn how the nervous system functions.

Think back: Where is most water absorbed in the body?

Ask students to work with a partner to talk about the work from the last lesson and decide on an answer to share with the class.

> **Answer:** *The small intestine – some students may say the large intestine, but point out that this is reabsorption and is not the main way water enters the body.*

 **How much water do you drink every day? How do you feel if you do not drink enough water?**

Students can continue to work with their partner. After comparing notes about how much water they drink each day, they should share experiences of when they have not drunk enough water; some could share their experiences with the class.

> **Possible response:** *Volumes of water per day will vary but should be around 1.5 to 2.0 litres per day. Students may report feeling thirsty and having a headache.*

Allow students to read the text on page 56 or ask volunteers to read a sentence each. To check understanding you could ask, 'How much of your body is water? How many litres per day of water can be passed into the blood? What is excretion? Which organ system are the kidneys in? What is their function?' Let students study the diagram to learn more about the reasons why water is so important to the body. The diagram shows the main uses of water in the body. Ask students to make a list of the six functions and, for each one, ask students to research what would happen to the body if a lack of water meant the function stopped.

 **Why is it important that proteins and blood cells cannot pass through the filter?**

Students can continue to work with a partner to talk about the discussion question. Ask them to imagine what might happen if the kidneys removed everything from the blood. They can share their ideas with the class.

> **Answer:** *It is important that proteins and blood cells are not lost as these substances are vital for health and are difficult to re-make.*

Students can read the information at the top of page 57 and look at the diagram of the urinary system. Explain that the large picture of the kidney shows a blue tube where water leaves the kidney and red tubes that represent the blood entering the kidneys. The outer layer, with all of the small blood vessels, is where the filtering takes place. You could use the activity on page 57 of the Workbook to support this.

 **Why do the kidneys need a very good blood supply?**

Ask students to use the text and the diagram to help them to answer the discussion question. Remind them of the function of the kidneys. If they need another clue, ask them how they would have been able to filter the soil and water in the last lesson if they didn't have any soil and water.

> **Answer:** *A good blood supply is needed so the kidneys can filter all of the blood in the body in a short amount of time.*

Investigation: Making model kidneys

Arrange students into groups of three or four. Explain that they are going to design and build a model to show how the kidneys work. Point out that they can use the diagram in the investigation box. Make up a waste sample for students to filter – this can be any insoluble substance mixed with water, such as sand and water, iron filings and water, sawdust and water or chalk powder and water.

Students should use cotton wool, filter paper and/or sand in the objects used to represent the kidneys (for example the tops of plastic bottles). They pour the mixture through the 'kidneys' and let the filtered water run into the bladder (the funnel) and then on down the urethra.

Set students the challenge of filtering the sample using their model urinary system. Remind them that filtering takes place in the kidneys, as they might think this is taking place in the bladder in the diagram within the investigation box. Allow students to present their model to the rest of the class. They should explain how it works and how it models the working of real kidneys. They should be able to label the parts of their model to match the urinary system.

Students can then read the information at the bottom of page 57 about problems of the urinary system. You can check understanding by asking, 'How can the kidneys be damaged? What is kidney failure? Why do some people need dialysis? How does dialysis work? What is a transplant?'

Stretch zone: Research how excretion happens through the skin. Write a short report of your findings.

Computing link: Students can work individually or with a partner. Allow them access to the internet and/or books and science encyclopaedias so they can find out about the skin. They can draw or download a cross-section diagram of the skin and label where substances can be excreted and what these substances are.

Possible response: Excretion through the skin occurs at sweat glands. Sweat is mainly water but also contains ammonia, urea and salts.

Key idea

Water is essential for living things. It is absorbed into the body and is excreted along with waste materials in the urinary system.

Ask a volunteer to read out the key idea. Ask students to close their books and then work with a partner to list three reasons why water is important to the body. They can then try to draw from memory the urinary system and open their books to check their answers.

Workbook activities

Model the function of the kidney (page 56)

This is a whole-class activity to model the urinary system. You will need a large space such as a hall or outside space and you need to set up the system as shown in the diagram. You should also blow up coloured balloons to different sizes according to the key under the diagram. Cut various-sized holes out of a rigid sheet of cardboard so that some of the balloons (blue and green) fit through but the red and yellow do not. Set up the path of chairs and draw a chalk line or fix tape to show the path of the blood. Put the balloons in a box and allocate roles to students. Once you have positioned the filter (students with the card) and the rest of the 'kidney' students on the chairs, ask students modelling blood to pick up a balloon and walk around the blood vessel until they reach the cardboard filter.

If their balloon fits through the holes, they pass it through and it is moved down to the plastic bag (bladder).

If the balloon does not fit through the holes, the student takes it back to the box and picks up another balloon of any colour. After a few minutes you can empty the plastic bag and ask students to analyse the contents and think about what this tells them about how the kidneys filter blood and form urine.

Answer: The bag (bladder) will contain water and urea to make urine, and the other substances (sugar and proteins) will stay in the box (blood).

How the urinary system works (page 57)

This activity will support students in understanding the text and diagram on page 57 of the Student Book. Ask them to study the diagram and label it using words from the top box. Once this is completed, you can ask them to read through the paragraph and use words in the bottom box to fill in the gaps. Explain that they can use each word more than once. Allow them to complete any gaps they cannot fill by turning back to the Student Book.

Answer: Labels are (top to bottom): kidney; ureter; bladder; urethra. Missing words in order are: kidneys; urine; kidneys; blood; poisons; poisons; urine; bladder; kidneys; poisons; blood; dialysis.

Review and reflect

Make a large copy of the diagram of the urinary system on page 57 of the Student Book and make some word cards with the parts of the system: kidney; bladder; ureter; urethra. Also make up some cards that do not fit: lungs; capillaries; teeth; enzymes; small intestine; platelets; right atrium; liver; large intestine; heart; oesophagus; plasma; left ventricle; stomach. Place the cards face down on a desk at the front of class and ask individual students to come up and pick up a card. They then try to either stick the card onto the diagram in the correct place or, if the word card isn't part of the urinary system, they place it back on the desk face down without letting anyone else see it. Do this until all students have tried the activity and the diagram is complete.

Conclude the lesson by asking students to complete question 4 in the 'What have I learned about organs and systems?' activity on page 66 of the Student Book. (See also the teaching notes in this Teacher's Guide, pages 82–84.)

Extra activities

 1 **Computing link:** Ask students to research how a dialysis machine works. They can download photographs and diagrams and make a small poster.

2 Students could draw around a person or the shadow of a person as they did when drawing the organs. This time they make a poster of the importance of water to the body by colouring the outline blue and adding information boxes showing the uses of water in the body. They could also research and add information about the symptoms a person has if they have a lack of water.

Differentiation

Supporting: Set up your own version of the model kidney so students can get clues and inspiration for their own designs.

Consolidating: Use plastic model kidneys to show the structure of the kidney; you can point out the blood vessels and where water leaves through the ureter to move to the bladder.

Extending: Ask students to research how water enters the tubes of the kidney from the small blood vessels.

Differentiated outcomes

All students	should be able to state that water and waste substances are excreted through the urinary system
Most students	will be able to name the parts of the urinary system and describe the functions of each
Some students	may be able to link the structure of the kidneys to its function as a filter

The brain and the nervous system

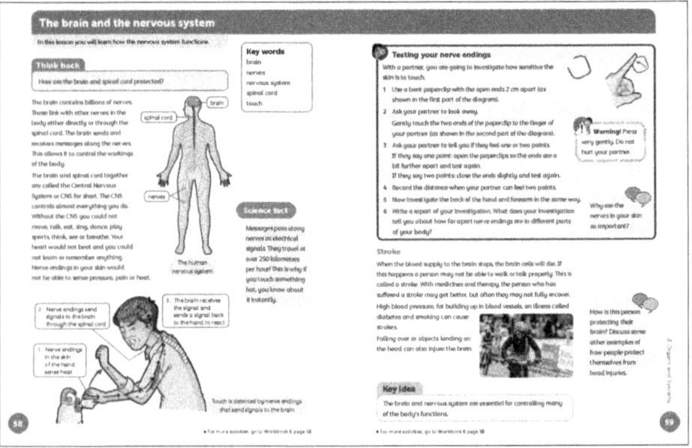

Supporting activities are in Workbook 6 pages 58–69.

Getting started

In this lesson students will study the nervous system and learn that nerves run all over the body. These send and receive messages to and from the brain. The brain is the control centre of the nervous system. Students will learn that the brain and spinal cord are called the central nervous system and they will investigate nerve endings.

Language support

Review the key words. They should be familiar to students from prior work, but check that they know how to spell the words and how to pronounce them. Read out a word and ask for volunteers to define it and use it in a sentence.

Resources

Student Book: paperclips.

Workbook: paperclips.

> ### Key words
>
> brain   nerves   nervous system   spinal cord   touch
>
> **Other words in the lesson**
>
> Central Nervous System (CNS)   messages   nerve endings   stroke   touch

Scientific enquiry key words

Plan and/or carry out enquiries to answer questions

Make observations

Take measurements, using equipment accurately

Take repeated readings where necessary.

Record data and results

Analyse data, notice patterns and group or classify things

Report and present findings

Draw conclusions and give explanations

Identify causal relationships

Lesson at a glance

The key teaching points for students in this lesson are:

- the brain controls many functions in the body by sending and receiving messages along nerves
- the brain and spine are known as the Central Nervous System (CNS)
- nerve endings in the skin help us to feel.

In the next lesson, students will learn about infectious diseases.

Think back: How are the brain and spinal cord protected?

Ask students to reflect on their earlier work on the skeleton from Year 3 and their discussions from earlier in this unit. They can feel their head and backbone to investigate this.

> **Answer:** *The brain is protected by the skull and the spinal cord is protected by the backbone.*

Ask students to read the text on page 58 and study the diagram of the nervous system. You could ask volunteers to read out a sentence at a time. To check understanding, ask, 'How does the brain send and receive messages to and from the body? Where is the spinal cord? What is the CNS? Why is the CNS so important?'

Science fact Messages pass along nerves as electrical signals. They travel at over 250 kilometres per hour! This is why if you touch something hot, you know about it instantly.

Read out the Science fact and ask students if they realised that nerve messages (impulses) travelled so quickly.

To illustrate the importance of having nerves and a brain to detect and interpret messages, show students the diagram at the bottom of page 58. Ask them if they have ever accidentally touched something hot or sharp and allow them to discuss examples.

 **Investigation: Testing your nerve endings**

Explain that students are going to investigate the sensitivity of the skin. You could use the worksheet on page 59 of the Workbook to support this investigation.

 **Warning!** Press very gently. Do not hurt your partner.

Start by reading out the safety warning and asking students why the rule is so important.

Demonstrate how students can make a sensitivity detector by bending a paperclip to leave open ends that are 2 cm apart. They can then follow the instructions to use this device to test the sensitivity of the skin on the fingers. The person being tested reports if they feel one or two points. Explain that if they say one point, the tester should open the paperclip, so the ends are a bit further apart, and test again.

Ask students to record the distance when their partner can feel two points. Then they should investigate the back of the hand and the forearm in the same way. Students should take it in turns to be the tester and the person being tested and then write a report of their investigation.

> **Answer:** *Students should notice that the nerve endings are closer together on the fingers where sensitivity is most important.*

 **Why are the nerves in your skin so important?**

Students can discuss this question with their group after they have completed the investigation. Ask them to imagine what life would be like if they could not feel anything with their skin.

> **Possible response:** *Students should suggest that if their skin has no nerves, they would not be able to feel hot and sharp objects and this could be dangerous. We also use nerves in the skin to know how hard to grip when we are picking up objects, using tools or playing sports.*

Stroke

Students can then read the information about strokes. Explain that, although the problem is caused by blocked blood vessels, it leads to the brain having problems controlling nerves. This is why a person who has had a stroke may have difficulty walking and talking. Point out to students that they will be learning more about health issues such as diabetes and high blood pressure in later lessons in this unit.

How is this person protecting their brain? Discuss some other examples of how people protect themselves from head injuries.

Ask students to look at the photograph of the cyclist and discuss the question with a partner. You can ask them to share examples of times when they have worn a helmet and to write a list of when people should wear a helmet.

Possible response: Students could share times they have ridden a cycle or been on a motorbike. People who should wear helmets include construction workers, motorcyclists, racing drivers, canoeists, climbers and cavers. Students may also say that keeping fit is helping to protect the brain because physical activity has been shown to make people feel happier and healthier.

Key idea

The brain and nervous system are essential for controlling many of the body's functions.

Ask a student to read out the key idea. Ask students to make a list of three body functions controlled by the brain.

Workbook activities

The brain and nerves (page 58)

You can use this activity to support students when they are working through the text on page 58 of the Student Book or you can use it as a review task. Ask students to label the diagram by using the words in the box. They should then work individually or with a partner to answer the questions.

Answer: 1 labels are (top to bottom): brain; spinal cord; nerves; 2 Central Nervous System (CNS); 3 any three, for example controlling muscles for eating, singing, moving, seeing, sensing smell, touching objects, detecting heat, pressure and sharp objects and heart beating; 4 any two from high blood pressure, fat in the blood vessels, diabetes, smoking, head injures due to falls, etc.

How sensitive is your skin? (page 59)

Explain that this activity supports the investigation on page 59 of the Student Book. Ask students to follow the instructions, including those in diagrammatic form, to make the sensitivity tester and then test their partner. They should record findings in the table provided and answer the questions.

Answer: Students should discover that the nerve endings on the fingers are closer together than on the forearm and back of the hand. They should suggest that is it important to have very sensitive fingers as we hold and touch things with our fingers.

Review and reflect

Ask students to think of one idea from the lesson that they would like to investigate and find out more about. Ask them to make a plan about how they would do this. This planning time will allow students to think up some imaginative and original things to do and you could let them carry out these projects.

Conclude the lesson by asking students to complete the second statement in the 'What I have learned about organs and systems' activity on page 66 of the Workbook. (See also the teaching notes in this Teacher's Guide, pages 82–84.)

Extra activities

1 **Computing link:** Ask students to research a cross-section diagram of skin to find out about the nerve endings. They can download a diagram and label the nerve endings. If they produced a skin cross-section to find out about excretion from sweat glands in an earlier lesson, they can modify this diagram.

2 Students can research the reflex action or reflex arc. Explain that when the body has to move quickly, the nervous system has got a short cut. If a person puts their hand on a pin the nerve endings send a message along a nerve and this goes to the spinal cord. Here it goes in a quick loop and back to the arm and hand to allow them to move very quickly. The message does not have to go to the brain, so this saves time. They could find a diagram of the reflex arc and print it off for their notes.

Differentiation

Supporting: Allow students to use the activity on page 58 of the Workbook to support them as they read through the text in the Student Book.

Consolidating: The investigation instructions on page 59 of the Workbook can be used to consolidate learning about nerve endings.

Extending: Ask students to research what the peripheral nervous system is.

Differentiated outcomes	
All students	should be able to state that the nervous system contains the brain, the spinal cord and nerves
Most students	will be able to describe examples of the functions of the nervous system
Some students	may be able to explain what the Central Nervous System and peripheral nervous system are

Infectious diseases and their prevention

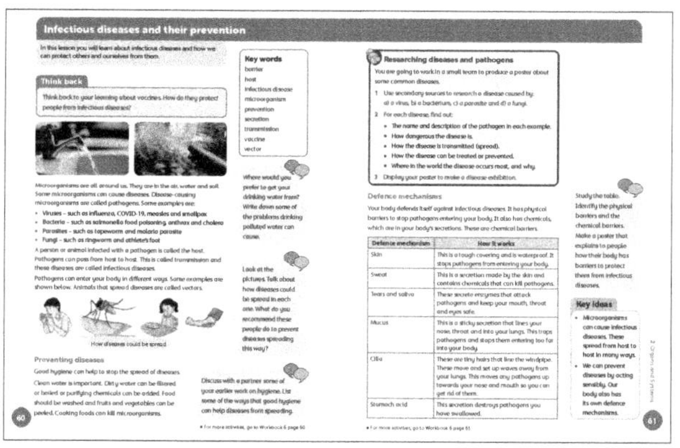

Supporting activities are in Workbook 6 pages 60–61.

Getting started

In this lesson students will review the main groups of microorganisms and learn how some can cause infectious diseases. They will research some examples of pathogens and the diseases they cause and consider some of the ways that infectious diseases can be prevented. They will then study the body's defence mechanisms.

Language support

Students will be familiar with the words microorganism, vaccine, prevention and disease, but you need to explain the keys words barrier, host, infectious, secretion, transmission and vector. Point out the spelling of each word and let students practise writing down the words. They can hear them pronounced in the eBook. Don't define the new words at the start of the lesson but tell students that they will learn about them as they study the topic. When you do reach a key word in context allow them time to write down a definition.

Resources

Student Book: materials for making posters.

Workbook: materials for making posters.

Key words

barrier   host   infectious disease   microorganism
prevention   secretion   transmission   vaccine
vector

Other words in the lesson

bacteria   cilia   chemical barrier
defence mechanism   fungi   hygiene   mucus
parasites   physical barrier   viruses

Scientific enquiry key words

Plan and/or carry out enquiries to answer questions

Record data and results

Report and present findings

Lesson at a glance

The key teaching points for students in this lesson are:

- infectious diseases are diseases that spread from host to host in many ways
- infectious diseases are caused by microorganisms
- the body has its own defence mechanisms to help us to fight infectious diseases.

In the next lesson, students will learn about the impact of diet on health.

Think back: Think back to your learning about vaccines. How do they protect people from infectious diseases?

Remind students that they studied vaccines in Year 3. They may also have heard of vaccines relating to the Covid 19 pandemic and may have had vaccinations to help protect them against other diseases. If they need support, allow them to spend time researching vaccines on the internet or in science books and encyclopaedias. You could produce an information page based on page 109 of Student Book 3 or by downloading information from the internet.

Possible response: Vaccines protect people because parts of the pathogen causing the disease (or a weakened or dead version) is injected into a person and the person's normal immune response is triggered. They may recall that this results in antibodies. These attack the pathogen if it attacks the body in the future.

Where would you prefer to get your drinking water from? Write down some of the problems drinking polluted water can cause.

Point out the photographs at the top of page 60. Allow students to work together to study them and think about the consequences of drinking water from both sources. They can share their ideas with the class.

Answer: The photograph on the right shows dirty and polluted water so the tap water should be the chosen source.

Ask students to read the text on page 60. Remind them that they learned about some microorganisms when they were studying classification in the last unit. Ask, 'Which type of microorganism causes measles? Which type causes cholera? What is a host? List two vectors.' You can also use the activity on page 60 of the Workbook to help students to study the text.

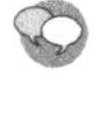 **Look at the pictures. Talk about how diseases could be spread in each one. What do you recommend these people do to prevent diseases spreading this way?**

Students can continue to work with their partner. Allow them to look at the drawings and discuss the questions. They can share their experiences of seeing these and other potential ways that disease-causing microorganisms (pathogens) can be spread. Ask them to recommend ways that the people could prevent the spread or transmission of pathogens. Ask volunteer pairs to share their ideas with the class.

> *Possible response: Diseases could spread by the following (left to right):*
>
> *droplets from sneezing and coughing spread in the air and can be transmitted via surfaces or directly by other people breathing them – prevented by using a tissue, wearing a mask and cleaning surfaces regularly*
>
> *biting insects can spread blood from person to person and the blood can have pathogens – prevented by limiting the number of insects indoors, covering the body with clothes, having insect nets and using insect repellent*
>
> *microorganisms growing on food can be eaten and cause disease in the person eating it, and flies attracted to the food can spread microorganisms with their feet and when they are eating – prevented by storing food in a cold place, cooking it so it gets very hot, peeling or washing fruits and vegetables, and covering food so it is not left out in the open*
>
> *microorganisms can enter the body through cuts – prevented by washing cuts with clean water and covering with a bandage or plaster.*

Preventing diseases

Allow students to read the text about preventing diseases with good hygiene to help them with the discussion task alongside it.

 **Discuss with a partner some of your earlier work on hygiene. List some of the ways that good hygiene can help diseases from spreading.**

Allow students time to reflect on earlier work but encourage them to draw on their own experiences and other sources of information, such as advice from parents and family, health leaflets, TV advertisements and public health films and posters.

> *Answer: Students should list some actions for good hygiene from the following: washing their body, washing their hair, brushing teeth, washing clothes, brushing hair, clipping nails, washing hands regularly and always after visiting the toilet or before handling food.*

 ### Investigation: Researching diseases and pathogens

Arrange students into groups of three or four. Explain that they are going to work in a team to produce a poster about some common diseases. The worksheet on page 61 of the Workbook provides a structure for the research. Remind students that secondary sources are ones they have looked up rather than information that they have found out themselves through investigations. Ask them to read through all of the instructions and then carry out their research. Point out that they must use the checklist of issues for each pathogen. Once they have designed and completed their posters you can display them to make a disease exhibition.

> *Possible response: Examples will vary but may include: virus = COVID 19, measles or influenza; bacterium = cholera, anthrax or salmonella food poisoning; parasite = malaria or tapeworm; fungi = athlete's foot. Check they have mentioned the mode of transmission, treatment/prevention and global location. Many will be throughout the world but some can be more localised.*

Defence mechanisms

 **Study the table. Identify the physical barriers and the chemical barriers. Make a poster that explains to people how their body has barriers to protect them from infectious diseases.**

Point out the text and table on page 61 and ask students to use it to help them to consider the physical and chemical barriers that protect us from infections. Ask them to work together to plan a poster to explain the body's barriers. These can be displayed next to their disease and pathogen posters.

Explain to students that a physical barrier is something solid that they can touch – such as the skin and the tiny hairs called cilia. Chemical barriers are made up of chemical substances such as acids and enzymes. You can emphasise the point by using weeding as an example. A physical barrier to weeds might be a plastic sheet on the ground, gravel or even a fork with which to lift weeds. A chemical barrier to weeds would be to spray the weeds with a herbicide chemical.

> *Answer: Posters should contain all of the barriers – skin, sweat, tears, saliva, mucus, cilia and stomach acid.*

Key ideas

- *Microorganisms can cause infectious diseases. These spread from host to host in many ways.*
- *We can prevent diseases by acting sensibly. Our body also has its own defence mechanisms.*

Ask a student to read out the key ideas. Ask students to turn to a partner and take it in turns to give an example of how a pathogen can be spread and how this spread can be prevented.

Workbook activities

Learning about microorganisms (page 60)

This is an example of a DART activity – so called because of the acronym <u>D</u>irected <u>A</u>ctivity <u>R</u>elated to <u>T</u>ext. It is a useful way of making reading more active and engaging and it also allows you and students to check understanding. (Other DART activities you will have used include: questions at the end of text to check comprehension; asking verbal questions during the reading of text; reading to the class and then stopping so students have to tell you what should be in the gaps; and presenting text with gaps in for students to complete.) In this example, ask students to follow the instructions and answer the questions. Make sure they are clear when to circle and when to underline words. They then use the text to answer the questions.

Answer: Students should circle the word 'pathogens'. They should underline 'host'. Two examples of diseases caused by viruses are any two from influenza, COVID 19, measles or smallpox. A disease caused by bacteria can be any one from salmonella food poisoning, anthrax or cholera. A type of fungi causes athlete's foot. A parasite is a living thing that lives in or on another living thing.

Researching diseases and pathogens (page 61)

This activity supports the investigation on page 61 of the Student Book. Remind students that they are going to use secondary sources to research diseases. They should start by defining what a secondary source is and then during their research they can use the table to record their research findings. This will help them to organise their findings and check that they are not missing anything out. Once the table is complete, they can use it to help them to design their poster.

Answer: Secondary sources are sources of information that a person has not found out for himself or herself. Examples are books, newspapers, magazines, encyclopaedias, websites and scientific journals and films.

Review and reflect

Use the discussion tasks, the research investigation poster and the Workbook tasks to encourage students to think about what they understand and what they are finding less straightforward. Discuss the outcomes of each task with students. Encourage them to identify aspects they have not completed correctly and help them to identify improvements. This will help

students to develop a positive approach to learning by understanding that learning is a process that will improve with practice and reflection.

Conclude the lesson by asking students to complete question 2 in the 'What have I learned about organs and systems?' activity on page 66 of the Student Book and the third statement on page 66 of the Workbook. (See also the teaching notes in this Teacher's Guide, pages 82–84.)

Extra activities

1   Students could make their own yoghurts to help them practise clean food handling hygiene, but also to show that many microorganisms are harmless and in fact helpful to people. They will need clean yoghurt pots, a pan, a thermos flask or a heavy pot to keep the yoghurt warm, milk and live plain yoghurt. They simply place two tablespoons of live yoghurt into a thermos flask or heavy pot. The milk is heated in a pan so it is just bubbling. You could provide pre-heated milk if you wish. Leave the milk to cool so it is still hot but can be touched with a finger (46°C if you want to give students practice in measuring temperature). The warm milk is added to the thermos or heavy pan and stirred gently to mix the milk and yoghurt. Put a lid on and leave it for at least eight hours without it being moved. The yoghurt can be placed into smaller pots and fruit can be added if you wish.

2   Ask students to work in groups of five or six to create a short play about a person who has an infectious disease. They can act out the person, family, nurses and doctors to demonstrate how the disease was caught, how it could have been prevented, and how it can be treated. You can arrange a 'show lesson' and the plays can be put on and possibly filmed.

Differentiation

Supporting: Hand out the activity on page 60 of the Workbook to support students in their learning from the Student Book.

Consolidating: Allow students to walk through the poster exhibition to review their understanding of infectious diseases and defence mechanisms.

Extending: Students can research some of ways that insect vectors are controlled and the potential environmental impact of this.

Differentiated outcomes	
All students	should be able to name some infectious diseases and the microorganisms that cause them
Most students	will be able to describe the body's defence barriers and how these work to help prevent infectious diseases
Some students	may be able to explain how insect control helps to prevent infectious diseases but can have an environmental impact

A healthy diet

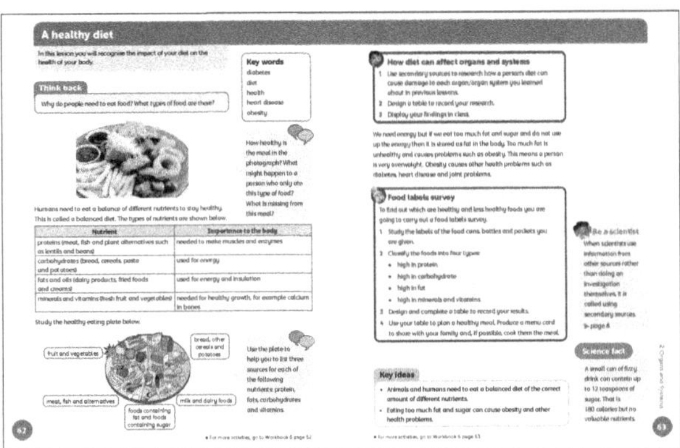

Supporting activities are in Workbook 6 pages 62–63.

Getting started

In this lesson students will explore further the idea that a healthy and balanced diet contains the correct types and amounts of nutrients. They will review how foods with high sugar or fat content can damage health, for example causing obesity and diseases linked to being overweight – such as diabetes, heart disease, strokes and some types of cancer.

As with all topics linked to health, you will need to be sensitive to issues such as students' body self-image. They may be upset if they feel they or family members are overweight or underweight. This is also true if they or family members suffer from any of the diseases mentioned in the lesson.

Language support

Review the words proteins, carbohydrates, fats, oils, vitamins and minerals and remind students that they studied these food groups in Year 4. Write these words on word cards and hold them up in turn. Ask students to volunteer examples of foods rich in each food group. Read out the key words for the lesson and ask students to turn to a partner and discuss each one to review their understanding. Remind them that diabetes is a disease that stops sugar being controlled in the blood, and obesity describes the situation when a person is a great deal overweight.

Resources

Student Book: writing materials; large sheets of paper; selection of cans, bottles and packets of food of four types (protein rich, carbohydrate rich, fat rich, and mineral and vitamin rich) showing food labels; thin cardboard.

Workbook: selection of cans, bottles and packets of food of four types (protein rich, carbohydrate rich, fat rich, and mineral and vitamin rich) showing food labels.

Key words

diabetes   diet   health   heart disease   obesity

Other words in the lesson

carbohydrate   fat   minerals   oils   protein   vitamins

Scientific enquiry key words

Plan and/or carry out enquiries to answer questions

Make observations

Record data and results

Analyse data, notice patterns and group or classify things

Report and present findings

Draw conclusions and give explanations

Lesson at a glance

The key teaching points for students in this lesson are:

- a healthy diet has a balance of the correct amount of nutrients
- eating too much fat and sugar can cause health problems such as obesity and diabetes.

In the next lesson, students will learn about healthy life choices

Think back: Why do people need to eat food? What types of food are there?

Allow students to work with a partner so they can share ideas. Encourage them to think back to earlier work on diet and food groups and ask them to make a list to share with the class.

Answer: Food is needed for energy and raw materials for growth and movement. Students may recall proteins, carbohydrates, fats, minerals and vitamins.

How healthy is the meal in the photograph? What might happen to a person who only ate this type of food? What is missing from this meal?

Ask students to look carefully at the plate of food at the top of page 62. Ask them to identify any of the foods that might be less healthy than others on the plate. Volunteers can share their answers with the class.

Answer: The meal is not very healthy. It has a lot of fried and fatty foods. If a person ate this type of food in excess, they would have problems with their weight and health. There are no vegetables or fruits.

Point out the text and table on page 62 and ask students to discuss the nutrients. Ask, 'Which types of foods contain a lot of protein? Why do we need to eat carbohydrates? If you were short of minerals and vitamins, which foods would you eat?'

Then encourage students to study the healthy eating plate. Tell them they will need to look at this carefully to help them to complete the discussion task.

Use the plate to help you to list three sources for each of the following nutrients: protein, fats, carbohydrates and vitamins.

Encourage students to think about the difference between the different nutrients – proteins, carbohydrates, fats, minerals and vitamins – and the different sources of food that are rich in these – meats, pulses, bread, pasta, butter, cooking oils, vegetables, fruit, etc. They can then look at the food plate to make the link between the nutrient and a food rich in it. Point out that no food has only one type of nutrient. For example, vegetables will have a small amount of carbohydrate, fat and protein but much, much less than beans.

> *Answer: Protein = meat, fish and alternatives (pulses and dairy products for example); fats = fatty meats, dairy products and biscuits/cakes; carbohydrates = bread, cereals and potatoes; vitamins = any of the fruits and vegetables.*

Investigation: How diet can affect organs and systems

Allow students to work in groups of three or four. They will need access to the internet to find the secondary sources to research how a person's diet can cause damage to each organ/organ system mentioned in previous lessons. You can ask each group to study all of the organs or allocate a different organ to each group. Once they have found out about the health risks due to diet, you can ask them to design a table to record their findings and these can be displayed in the classroom.

> *Possible response: There will be a wide range of health risks but check that students include: circulatory system = fat deposited in blood vessels leads to heart attacks and high blood pressure; lungs = overweight people can find it difficult to breathe and take exercise and too much red meat and fatty foods makes breathing less efficient; digestive system = too many fatty foods can cause digestive problems and cancer; kidneys = some foods, especially processed foods, salty foods and too much meat can cause kidney disease; brain = fatty and sugary foods can effect memory and learning and lead to diseases such as dementia.*

Read through the text on page 63 with students and then explain that they are going to survey some food labels to see which foods might contain potentially unhealthy levels of substances such as fats and sugars.

 **Investigation: Food labels survey**

The worksheet on page 62 of the Workbook supports this investigation.

Hand out a range of food packaging or download and print out food labels from the internet. Select examples of healthy and less healthy foods. Ask students to work in a small group. Ask them to classify the foods into four types: high in protein; high in carbohydrate; high in fat; high in minerals and vitamins. Students then design and complete a table to record their results. They should then use their table to plan a healthy meal and communicate this by producing a menu card.

Remind students that the healthy eating plate on page 62 shows the ratios of each type of food – this will help them to avoid having too much fat or sugar and also make sure they add enough vegetables or carbohydrates.

Suggest that they share this with their family and, if possible, cook them the meal. Point out that some fatty and sugary treats taken in small amounts will not cause health problems, but eating far too much sugar and fat will be unhealthy. You can also point out that many people prefer protein sources that do not come from animals and there are many of these to ensure a healthy diet.

The healthy meal should contain a balance of proteins, carbohydrates, fats, vitamins and minerals in the ratio shown in the healthy eating plate.

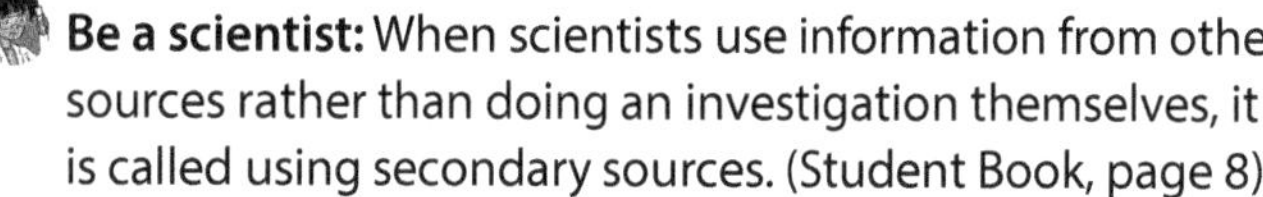 **Be a scientist:** When scientists use information from other sources rather than doing an investigation themselves, it is called using secondary sources. (Student Book, page 8)

Ask students to read the Be a scientist information to remind them that they often use secondary sources. You can also remind them that some secondary sources are not reliable, so they should always check where the information comes from and try to use a number of sources, not just one that might be inaccurate.

Science fact A small can of fizzy drink can contain up to 12 teaspoons of sugar. That is 180 calories but no valuable nutrients.

Read out the Science fact to students and ask them to think about how much fizzy drink they consume every week. If they drink a lot of fizzy drinks, did they realise they were also having so much sugar? Tell them that if a food or drink has a lot of sugar but no other nutrients, it is sometimes called consuming 'empty calories'.

Key ideas
- *Animals and humans need to eat a balanced diet of the correct amount of different nutrients.*
- *Eating too much fat and sugar can cause obesity and other health problems.*

Ask volunteers to read out the key ideas and ask students to close their books and write down the names of the main nutrients. They can then tell a partner one example of a health problem caused by an unhealthy diet.

 **Food labels survey (page 62)**

This activity supports the food labels survey investigation on page 63 of the Student Book.

Show students the examples of food labels and point out the units and the places where nutrients are listed. Different countries have different regulations about food labelling so, if you do not have this type of food label, then use local examples. If these do not show the nutrients and amounts, then you can download some from the internet for the investigation.

Students can complete the table to record their results. Remind them to add the units each time, as the amounts are very important. They then plan their healthy meal and start to plan their menu card.

> *Possible response: 2 Values will vary according to the food labels you have handed out. 4 The healthy meal should contain a balance of proteins, carbohydrates, fats, vitamins and minerals in the ratio shown in the healthy eating plate.*

Make your own healthy eating plate (page 63)

Ask students to think back to what they have eaten in the last 24 hours. Suggest they make a list. They then add the foods to the correct part of the blank healthy eating plate. They can write the names of the foods or suggest that they make their plate more eye-catching by drawing them. Next, they should study their plate and decide if it shows a healthy balanced diet. This is called evaluation.

Ask them to write down anything they should consider adding to their diet and anything they should consider eating less of.

> *Possible response: Plates will vary but students should identify if they are eating too much sugary or fatty foods and not enough fruit, vegetables and protein.*

 ## Review and reflect

Students can draw two large speech bubbles. In one they can finish the sentence, 'One important thing I have learned about healthy diet is …' and in the other they can finish the sentence, 'One thing I am most proud of in this lesson is …'. Pin these to a wall so students can look at each other's answers and achievements.

Conclude the lesson by asking students to complete question 6 in the 'What have I learned about organs and systems?' activity on page 67 of the Student Book. (See also the teaching notes in this Teacher's Guide, pages 82–84.)

Extra activities

1 Students could study the menus and choices available for meals provided in school. If the school does not have a cafeteria or dining hall, then they can study menus from local restaurants. These can be collected or found on the internet. They can select a three-course meal that would be very unhealthy and a three-course meal that would be very healthy, and give reasons for their choices.

2 Arrange a class picnic. Ask students to bring what they decide are healthy snacks and drinks and take them outside to eat the food and discuss their choices. Alternatively, you could ask local food outlets and shops to donate some foods and you and the students could make the snacks before the picnic.

Differentiation

Supporting: Allow students extra time to review their prior knowledge of nutrient groups. Display a large version of the healthy eating plate to remind them of work in earlier years.

Consolidating: Hand out old food magazines and supermarket leaflets and ask students to cut out examples of foods that are rich in each of the nutrient groups.

Extending: Students can research some diseases caused by nutrient shortages. For example: scurvy (vitamin C); rickets (vitamin D) and kwashiorkor (protein).

Differentiated outcomes	
All students	should be able to list the nutrient groups and describe their roles in the body
Most students	will be able to describe some health problems linked to an unhealthy diet
Some students	may be able to describe some health problems caused by nutrient deficiency

Healthy life choices

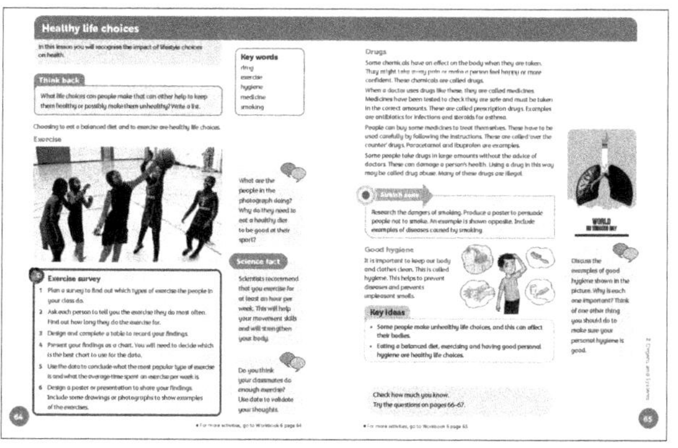

Supporting activities are in Workbook 6 pages 64–65.

Getting started

In this lesson students will learn about the importance of exercise and good hygiene in contributing to a healthy lifestyle. They will carry out an exercise survey and present their ideas to the class. Students will also learn the distinction between medicines that are prescribed by a doctor and those that can be purchased in shops. They will also learn about other substances that are harmful to the body and how taking them is called 'abuse'.

Language support

Students will be familiar with the words exercise, hygiene and medicine. However, drug and smoking are words that need careful interpretation. Explain the distinction between a medicine that is tested and taken to make us feel better and a drug that can be harmful and often illegal. Use your discretion when dealing with smoking and drugs and check what is acceptable within school and local guidelines. This is a sensitive area.

Resources

Student Book: writing materials; graph paper; items to prepare a presentation or poster that includes drawings or photographs.

Workbook: writing materials; graph paper.

Key words

drug   exercise   hygiene   medicine   smoking

Other words in the lesson

asthma   chemicals   diet   ibuprofen   illegal
pain   paracetamol   prescription   steroids
tobacco

Scientific enquiry key words

Plan and/or carry out enquiries to answer questions

Record data and results

Analyse data, notice patterns and group or classify things

Report and present findings

Draw conclusions and give explanations

Lesson at a glance

The key teaching points for students in this lesson are:

- exercise, a balanced diet and good personal hygiene are healthy lifestyle choices
- unhealthy lifestyle choices can have a detrimental impact on human health.

In the next lesson, students review their understanding of the content of the unit.

Think back: What life choices can people make that can either help to keep them healthy or possibly make them unhealthy? Write a list.

Ask students to think about their prior learning from school and outside school and encourage them to make a list of life choices. They can share their list with a partner and make a joint list. Ask volunteer pairs to share their lists with the class.

> ***Possible response:*** *Lists may include: eating a healthy diet; good hygiene; exercise; plenty of sleep; taking care crossing roads; wearing a helmet if cycling; drinking enough water.*

Exercise

What are the people in the photograph doing? Why do they need to eat a healthy diet to be good at their sport?

Students can continue to work with their partner. Point out the photograph on page 64 and ask them to talk about the activity they can see taking place. Ask them to talk about when they have done similar activities – this will help to prepare them for the survey to follow. To link back to work on diet, ask students to consider why a healthy diet is needed if doing a lot of exercise.

> ***Answer:*** *The people are playing basketball. They need a healthy diet to give them enough energy and the raw materials for growing and repairing muscles.*

Science fact Scientists recommend that you exercise for at least an hour per day across the week. This will help your movement skills and will strengthen your body.

Ask a volunteer to read out the Science fact. Explain that the World Health Organisation (WHO) guidelines state that children and adolescents aged 5–17 years should:

- do at least an average of 60 minutes per day of moderate-to-vigorous intensity, mostly aerobic, physical activity, across the week
- incorporate vigorous-intensity aerobic activities, as well as those that strengthen muscle and bone, at least 3 days a week
- limit the amount of time spent being sedentary, particularly the amount of recreational screen time.

Ask students to reflect on whether they do an hour of exercise per day across the week and follow the WHO advice.

Investigation: Exercise survey

Explain that students are going to plan and carry out a survey to find out which types of exercise people in the class do. Allow them to ask each person which exercise they do most often and how long they do the exercise for.

Students can design and complete a table to record their findings, or you can use the activity on page 64 of the Workbook as this provides a blank table. Encourage students to present their findings as a chart – the Workbook activity supports this by providing graph paper with the axes. Once they have decided which type of graph to use, they can interpret the data to conclude what is the most popular type of exercise and the average time spent on exercise per week. Ask students to share their ideas by designing a poster or presentation. Allow them to take photographs of students modelling the exercise or download pictures from the internet to add to the posters.

Do you think your classmates do enough exercise? Use data to validate your thoughts.

Encourage students to analyse the data to draw conclusions about the amount of exercise taken by people in the class. Remind them of the Science fact. Ask students to share their ideas and discuss with the class to find out if everyone agrees. If the exercise levels are low, you could consider discussing an exercise plan for your class.

> **Possible response:** *If people are exercising less than one hour per day, it is possible they are not getting enough exercise. The exercise can include walking to school, for example.*

Drugs

The next section of text on page 65 relates to the distinctions between prescription medicines, over the counter medicines and drug abuse. Some of this content may be sensitive. If you have decided to consider these issues in the lesson, read through the text with students and write down the terms 'medicine' and 'drug abuse' on the board and clarify that one is useful and the other is dangerous. Ask students if they have ever taken painkillers or other medicines on the advice of a doctor or a parent. All medicines can be harmful and should only be taken in the correct doses. They should only be taken for the correct reasons. Ask, 'If you have a younger brother or sister, why is it important that medicines are locked away or placed on a high shelf? Why are medicines often kept in bottles with hard-to-open tops?'

Stretch zone: Research the dangers of smoking. Produce a poster to persuade people not to smoke. An example is shown opposite. Include examples of diseases caused by smoking.

Point out the diagram of the lungs on page 65 and ask students to work with a partner to research the dangers of smoking. Remind students of their earlier learning about the structure and functions of the lungs and ask them to turn back to page 49 of the Student Book to look again at the picture of the lungs there. Laws about smoking vary from country to country but it is likely to be illegal for children under the age of 16, so stress that not only is smoking very bad for a person's health, but young people who smoke are also breaking the law. Encourage students to make their posters bright and eye-catching.

> **Possible response:** *Students should find many diseases caused by smoking including: cancer, heart disease, stroke, lung diseases, diabetes, emphysema, bronchitis, eye diseases and rheumatoid arthritis.*

Good hygiene

Discuss the examples of good hygiene shown in the picture. Why is each one important? Think of one other thing you should do to make sure your personal hygiene is good.

Ask students to read the text and study the picture showing aspects of personal hygiene. These do not cover all aspects as students are being encouraged to apply earlier learning. Students can compile a list and share this with the class. If you wish, they can do drawings for each one, similar to the ones on page 65.

> **Possible response:** *Students should suggest that showering and washing hands removes microorganisms and helps to keep them clean and healthy. Brushing teeth removes bacteria and stops teeth from being damaged. Combing or brushing hair helps to keep it clean and removes things from it. They could add the following: clipping nails, washing clothes, changing clothes regularly, using a tissue if coughing or sneezing, wearing a face mask, and staying indoors away from others if you feel unwell to try to avoid disease transmission.*

Key ideas

- *Some people make unhealthy life choices, and this can affect their bodies.*
- *Eating a balanced diet, exercising and having good personal hygiene are healthy life choices.*

Ask a volunteer to read out the key ideas. Ask students to imagine they had to write a 'tweet' to someone explaining what they had learned in the lesson. Ask them to compose a tweet using a maximum of 280 characters – a character being a letter of the alphabet, a punctuation mark and a space between words. For example, *Personal hygiene is important to a healthy lifestyle* is 52 characters long.

Workbook activities

Exercise survey (page 64)

This activity supports the investigation on page 64 of the Student Book. Explain to students that they should ask six people to tell them the exercise they do most often and for how long. If they gather data from more than six people, they will need to write the table into their books to give them more space.

Maths link: Students should complete the table and then present their findings as a bar chart. A bar chart is used as the *x*-axis contains separate categories (exercise type). Point out that a graph has already been started for them. They then analyse the findings and determine which was the most popular type of exercise and the average time spent on exercise per week. To calculate mean (average) time spent exercising, students will need to add up the total time every member of the class spends exercising and then divide this by the number of students in the class. They can then design their poster or presentation.

> **Possible response:** *Values will vary but students should produce a bar chart with the bars along the x (horizontal) axis being each type of exercise and the number of people who do that exercise up the y (vertical) axis.*

Personal hygiene (page 65)

This can be an individual or paired activity. Ask students to study the pictures and complete the table by naming each object and explaining why it is important for personal hygiene. Encourage them to use the term 'microorganism' in their explanations to make the link to how good personal hygiene helps to prevent disease transmission.

> **Answer:** *A toilet paper helps to keep microorganisms off hands when using the toilet (should still wash hands afterwards though); B toothbrush and paste cleans teeth and helps to stop tooth decay;*

> *C soap and towel used to wash hands to remove harmful microorganisms; D deodorant or anti-perspirant to reduce under-arm sweating (some students may see this as sun cream to protect from harmful sunlight); E tissues to trap microorganisms if coughing or sneezing; F shower to keep the body clean and remove dry sweat; G antiseptic spray/wipes to clean surfaces and kill some microorganisms; H hair brush to keep hair clean and remove objects.*

 Review and reflect

This is the final lesson of the unit, unless you include a review lesson. Ask students to look back through their notes and any models and posters they have made to select two pieces of work they are keen to learn more about. They can set this as an individual learning target.

Allow students a few moments of quiet reflection so they can enjoy the feeling of satisfaction from learning so many new things and working so well with others.

Conclude the lesson by asking students to complete question 1 in the 'What have I learned about organs and systems?' activity on page 66 of the Student Book and the fourth statement on page 66 of the Workbook. (See also the teaching notes in this Teacher's Guide, pages 82–84.)

If you wish, you can ask students to revisit and discuss all of the questions on pages 66–67 of the Student Book and all of the statements on page 66 of the Workbook.

Extra activities

1 **Computing link:** Students can research different sports to find out which use up the most energy per hour. They can make a top-ten list and display this with the energy used. Allow students to download pictures of people doing the sports and activities. Encourage discussions about the differences between sports that take many hours (professional cycling in the Tour de France and mountain climbing), take an hour or so (football, hockey and basketball) and those that only take seconds or minutes (100-metre and 400-metre sprints). You could also ask students to think about activities that take a long time but do not use up as much energy as climbing or long-distance cycling, such as hill walking and golf.

2 Ask students to keep an exercise diary to log when they are active. They can keep this for a week and could also add times when they are awake but sitting down – such as their time in school, when watching TV or if they play computer games. They can make a personal decision about whether or not they are taking enough exercise.

Differentiation

Supporting: Use the activity on page 64 of the Workbook as this has been designed to provide extra support for students.

Consolidating: Show a film of people taking part in sports to promote thought and discussion about exercise.

Extending: Students could research some over the counter medicines and find out about their uses and some of their side effects.

Differentiated outcomes	
All students	should be able to list some positive life choices such as exercise, healthy diet and good personal hygiene
Most students	will be able to describe how to investigate the types and amount of exercise people take
Some students	may be able to describe some uses of over the counter medicines and their side effects

What have I learned about organs and systems?

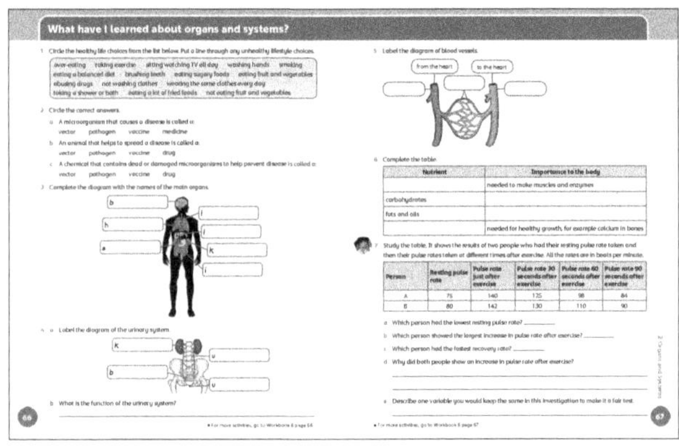

Supporting activities are in Workbook 6 pages 66–67.

Getting started

The aim of this section is to encourage students to review their learning after all of the lessons in the unit. As you will know from the last unit, all lessons have questions and discussion tasks for students to answer and these can be used formatively during lessons. You can also use responses to these to help you to assess students' knowledge and understanding of the topic.

If you are working through the units in the order of the book, then you will already be familiar with the ways to use the review questions in the Student Book and the self-reflection statements in the Workbook. However, if you have covered Unit 2 first, you may find the following information useful.

On pages 66–67 of the Student Book there are seven questions related to the content of the unit. The questions are arranged in increasing order of conceptual demand and not topic order. Students can tackle these one at a time after each relevant lesson – specific advice on which questions are most appropriate is given under the 'review and reflect' heading in each relevant lesson. The questions could also be answered as a single summative activity. This could be done by reading out the questions to the class and asking for volunteers to answer them, carrying out the activity as a group work task with students talking about each question, or as an individual written task. Whichever approach is adopted, the questions are designed to give you and students feedback about progress and to help in identifying targets for development.

On page 66 of the Workbook there is a section containing four 'review and reflect' statements. You can read through the statements and ask students to tick one of the two self-review option boxes to show how confident they are about each topic. You can ask students to reflect on each statement separately after the relevant lesson or carry out the review as an end-of-unit activity. Advice on which specific review and reflect statement is most appropriate is given under the 'review and reflect' heading in the relevant lesson. It would be very useful to do both approaches, as you will be able to measure student levels of confidence throughout the unit to respond quickly and also gain a measure at the end of the unit to find out if this has changed for any topics. Time to reflect may make a student more or less confident and later learning may support understanding of a topic that had been causing problems for some students. As teachers, we know that not all learning takes place at the time the teaching takes place.

It is important students feel confident that their reflections will be listened to and dealt with sensitively so that they do not hesitate to report areas they are not confident with. This information is vital for you in providing support strategies in the end-of-unit summative assessment.

What have I learned about organs and systems? answers

1 Circle the healthy life choices from the list below. Put a line through any unhealthy lifestyle choices.

Make sure students are not daunted by the large selection of words in the word box. Encourage them to work through the phrases methodically so they only concentrate on one at a time.

Answer: Students should circle: taking exercise; washing hands; eating a balanced diet; brushing teeth; eating fruit and vegetables; taking a shower or bath. The rest should have a line through them.

2 Circle the correct answers.

a A microorganism that causes a disease is called a: vector/pathogen/vaccine/medicine

b An animal that helps to spread a disease is called a: vector/pathogen/vaccine/drug

c A chemical that contains dead or damaged microorganisms to help prevent disease is called a: vector/pathogen/vaccine/drug

Remind students that with a multiple-choice question of this type they only need to circle one answer from the four choices. They could start by eliminating any obviously wrong answers to limit the choices.

Answer: a pathogen; b vector; c vaccine

3 Complete the diagram with the names of the main organs.

Explain that in this question there is not a word box to select from, but the first letter of each word is given as a clue. Point out that two of the words begin with 'l' so students need to look carefully to identify the organ.

Answer: b = brain; h = heart; s = stomach; l = lungs; l = liver; k = kidney; i = intestine

4 a Label the diagram of the urinary system.

Give students the same instructions as you gave for question 3. Remind them not to mix up the two tubes that have names beginning with the letter 'u'.

Answer: k = kidney; b = bladder; top u = ureter; bottom u = urethra

b What is the function of the urinary system?

Point out that when answering a question in a test or quiz students should look at the amount of space they are given for the answer. This will help them to work out how much to write. In this case, they only have space for a short sentence.

Answer: To remove harmful substances and excess water from the body.

5 Label the diagram of blood vessels.

Suggest to students that they start by looking at the clues showing which way the blood is travelling in the blood vessels. Remind them that the word 'vein' ends with the word 'in'. The colours of the vessels also give a clue.

Answer: Left label = artery; middle label = capillary; right label = vein.

6 Complete the table.

Remind students that this is the same table they studied in the lesson on healthy diet. They should add two nutrients and two examples of their importance to the body.

Answer: Top nutrient = protein; bottom nutrient = vitamins and minerals; top missing importance = used for energy; lower missing importance = used for energy and insulation.

7 Study the table. It shows the results of two people who had their resting pulse rate taken and then their pulse rates taken at different times after exercise. All the rates are in beats per minute.

a Which person had the lowest resting pulse rate?

b Which person showed the largest increase in pulse rate after exercise?

c Which person had the fastest recovery rate?

d Why did both people show an increase in pulse rate after exercise?

e Describe one variable you would keep the same in this investigation to make it a fair test.

Point out that the answers to parts a to c will be found in the table. Students need to analyse the data. This will involve some mathematics, as they will have to compare the numbers by using subtractions. Part d will require prior knowledge of exercise and part e will need them to apply their prior knowledge of investigations.

Answer: a A; b A; c A; d pulse rate increases as the heart beats more often to send blood around the body to supply oxygen and sugar for exercise; e any one from: time of exercise; type of exercise; time of day the resting pulse rate was taken; resting in the same way.

Summative assessment

The questions in the 'What have I learned about organs and systems?' activity on pages 66–67 of the Student Book can be used to consider the progress of each student individually. You can also use the information to create summative reports – such as end-of-term reports – for each student.

If you wish to allocate marks to the questions on pages 66–67 of the Student Book, then, assuming each answer has one mark, you could allocate as follows: question 1 = 6 (for every circled answer); question 2 = 3; question 3 = 7; question 4a = 4; question 4b = 1; question 5 = 3; question 6 = 4; question 7 = 5; this gives a total of 33 marks.

You can also keep a record of individual and class responses to the review and reflect statements in the Workbook. This lets you reflect on the overall confidence levels of students to identify areas that may need revision later on.

By reviewing responses to the questions and the self-review statements, you can tailor specific interventions to help students improve. Keep the recording and analysis of the student self-evaluations simple. A general impression of the class's self-evaluation is all that is required, e.g, 'Fifty per cent of the class were not confident about …'.

Additionally, you can ask students to complete the Quiz Yourself questions for this unit to help check their understanding. These are on page 144 of the Workbook.

Investigate like a scientist

The 'Investigate like a scientist' task at the end of the unit on page 67 of the Workbook is designed to encourage students to apply their investigative and creative skills and review key aspects of the content of the unit.

Investigation: Making a model lung

Resources: large plastic bottles; scissors; balloons; elastic bands; sticky tape; straws.

Explain to students that they are going to design and construct a model lung. Show them the equipment you have provided, especially if this is different from the examples shown on page 67. Encourage students to use the diagram shown as step 2 to help them construct the lung. The key steps to point out or demonstrate are cutting the top off the bottle to make the 'ribcage', fixing the straw through the bottle neck so it is airtight, making sure the straw sticks into the balloon and making sure the balloon representing the diaphragm is fixed tightly across the bottom of the bottle.

Once the model lungs are constructed, students can try to get the lung to inflate. You can show them how to pull down and push up on the model diaphragm. They can then feel their own diaphragm as they breathe.

Answer: 1a To inflate the lungs students will have pulled down on the balloon (the diaphragm) so it makes the space in the bottle (chest cavity) larger. The lungs can expand into it. 1b To deflate the lungs students will have pushed up on the balloon (diaphragm) so it moved up into the bottle (chest). 2 Students should be able to feel their ribcage rising and falling as they breathe in and out.

Finally, ask students to think about the Stretch zone challenge.

Answer: Students should discover that the diaphragm is at the bottom of the ribs where they have been feeling. It is a sheet of muscle that separates the chest from the rest of the body. The diaphragm contracts and moves downwards to help the lungs inflate and relaxes and moves upwards to help the lungs deflate during inhalation and exhalation.

3 The Way We See Things

In this unit students will:

- recognise that light appears to travel in straight lines and can be represented by ray diagrams
- revise that we see things because light travels from light sources to our eyes or from light sources to objects and then into our eyes
- explore how light rays change direction (can be reflected) from surfaces, including mirrors, into the eye
- describe how a ray of light changes direction when it travels through different mediums and know that this is called refraction
- explore how we see colour
- investigate shadows to understand their shape and size
- discover how light is measured.

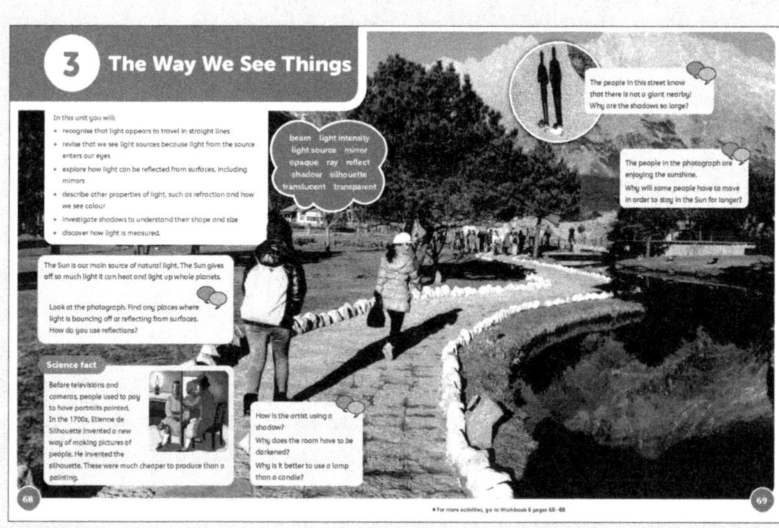

Supporting activities are in Workbook 6 pages 68–69.

Getting started

This unit allows students to review that we see things because light enters our eyes from a light source. Students become aware of the properties of light such as reflection and refraction and how a beam or ray of light travels in a straight line. They will explore how light can be reflected from surfaces including mirrors. Students investigate and measure light. They also investigate how light travels through some materials better than others but is blocked by opaque materials. Students investigate how shadows are cast and how they can change the shape and size of a shadow. The content of this unit can be observed and experienced in everyday life. The Workbook provides additional activities and alternatives, which are all supported in this Teacher's Guide.

Science in context

Use the lessons in this unit to demonstrate how we use different light sources. The properties of light allow scientists and engineers to select the most appropriate material to use for a specific purpose. For example, students can observe local buildings to see how light can travel through some materials to get light into dark spaces, and apply this to building designs. Provide opportunities for students to observe how light travels and how the intensity of sources can be measured. Students need to understand that light can be reflected from surfaces, but these are not sources of light, so show them examples of mirrors being used in buildings and on vehicles. Provide opportunities for students to observe how shadows are formed naturally with light from the Sun. They can observe shadows being cast and know how the size and shape of them change by changing an object's position in front of the source of light. Take students out to see examples of sundials to show how ways of telling the time have changed over time and you can also let them survey examples of people creating shade – such as with umbrellas and awnings. You can also encourage students to consider ways that science helps people who are not able to see.

Scientific enquiry skills

Throughout this unit, students will try to answer questions by collecting evidence through observation, contributing to discussions, and making predictions about the properties of light. They will follow scientific enquiry methods and will collect and make observations to test their predictions. They will make comparisons and decide whether their predictions are supported or refuted by their results. Students will collaborate with others to problem solve and communicate their ideas through tables of results and presentations.

You can use the Investigation master sheet on pages 4–5 of this Teacher's Guide to support investigative work. This provides prompts and structure to support students in planning and carrying out fair tests and recording and drawing conclusions about their findings.

Student Book: pencils; rulers or tape measures; paper cups; paperclips; pieces of cardboard; writing materials; materials to make posters; torches; small mirrors; large mirrors; small cardboard boxes; pieces of thin card; scissors; sticky tape; foil; dark card; materials to make leaflets; rulers; large sheets of paper; ray boxes made previously; protractors or other angle measurers; circular pieces of card; pieces of dowelling or pencils; samples of materials that are opaque, transparent and translucent; selection of small objects; cardboard or other material to use as a screen; materials to make shadow puppets; sheets or blinds (optional); coloured filters; large screens or a blank wall; small opaque objects such as building blocks; access to outdoors at various sunny times; drawing materials; 1-metre sticks; compasses; chalk; bright light sources; solar-powered calculators; tissue paper; access to the internet or books with information on the brightness of stars.

Workbook: range of food samples; blindfolds; walking sticks; sets of sticky notes; writing materials; smartphones or digital cameras; torches; small mirrors; large mirrors; small cardboard boxes; pieces of thin card; scissors; sticky tape; shoe boxes or other narrow boxes made of thin card; magazines showing interiors with mirrors (or access to the internet); tissue paper; glue or sticky tape; lamps or candles; flexible mirrors or mirrors on a roll; pieces of thick card; a hole punch or pencil; modelling clay or reusable adhesive; torches or lit bulbs; a small area of blank wall or large pieces of card covered with paper; coins, cups; water; coloured objects; black object; coloured acetate or filter; selection of objects that are opaque, transparent or translucent; large sheets of white paper or a white wall; sharp pencils; torches or lamps; pieces of black card, shadow puppets made previously; access to outdoors at sunny times; paper plates; pencils; bright light sources; solar-powered calculators; tissue paper; access to the internet or books with information on measuring light intensity; graph paper; large cardboard boxes; glue; small toys; measuring tapes long enough to measure an outside area; materials to create posters.

Key words for unit

Bold words are in the Word cloud and are included in the glossary.

angle of incidence   angle of reflection   **beam**

blind spot   colour   direction   eyes   horizon

incident ray   **light intensity**   **light source**

lumen   lux   midday   **mirror**   normal   **opaque**

optical illusion   periscope   **ray**   ray diagram

reflect   reflected ray   refraction   sense

shadow   shadow stick   sight   **silhouette**

sundial   **translucent**   **transparent**

Scientific enquiry key words

Plan and/or carry out enquiries to answer questions

Make predictions

Recognise and control variables

Make observations

Take measurements, using equipment accurately

Record data and results

Analyse data, notice patterns and group or classify things

Report and present findings

Draw conclusions and give explanations

Identify causal relationships

Language support

Introduce the key words to students at the start of the unit. Read all of the words out loud to allow students to hear their pronunciation. The eBook also has a useful resource that gives the correct pronunciation of key words.

Students could create a miniature Word wall in the back of their books. They can add pictures and simple definitions that are meaningful to them as they become more confident with each of the words. Discuss each word and explain that students will learn and use them throughout the learning of the unit. Encourage students to say the words out loud to support them with pronunciation. Do not correct their pronunciation but model the correct ones by repeating back what they say. For example, 'So you are saying that reflection means . . .'.

Remind students that there is a glossary in the back of their Student Book where definitions of words on the introductory pages can be added to when they have a good working understanding of them. The completed glossary can be used as a revision tool and in later years as a reminder of prior learning.

At the start of the learning in this unit, try to add resources to a classroom library including books, magazines and leaflets that include content about light, and in particular shadows, reflection and refraction.

Unit at a glance

The key teaching points for students in this unit are:

- to introduce the unit objectives
- to introduce the learning outcomes
- to engage students with the content of the unit
- to review and build on prior learning and understanding of the topics.

In the next lesson, students will learn more about sight.

The purpose of this introductory lesson is for students to start thinking about different sources of light and how shadows are cast. The introductory pages show a range of different ways that light is used and the images are used as a starting point. Ask students to look closely at the photographs on the pages and to read the text about how the Sun is our main source of natural light and heat. Then ask them to engage with the discussion questions.

Look at the photograph. Find any places where light is bouncing off or reflecting from surfaces. How do you use reflections?

Ask students to discuss this with a partner then encourage them to share their ideas with the class. Explain that light bounces off some shiny surfaces, like the water, into our eyes.

> *Possible response: Students should suggest that light is being reflected from the water, the white stones alongside the path, the snow on the mountain, the light parts of the person's dark jacket and the windows in the distant building. Some will realise that in order to see an object light must bounce or reflect from it so every object could be listed. They may suggest they use reflections when they use a mirror and also to see any objects.*

Science fact Before televisions and cameras, people used to pay to have portraits painted. In the 1700s, Étienne de Silhouette invented a new way of making pictures of people. He invented the silhouette. These were much cheaper to produce than a painting.

Read out the Science fact or ask a volunteer to read it out. Ask students to imagine making a portrait of someone using this technique. Ask them how they would take a picture today. They will probably suggest using a camera or a smartphone. Point out that these devices need light to enter them in order to take a photograph but they work in a different way to the silhouette.

How is the artist using a shadow? Why does the room have to be darkened? Why is it better to use a lamp than a candle?

Students can work with their partner again. Point out the drawing of the person creating a silhouette and ask them to work through the questions in turn. Ask volunteers to tell you what they discussed about each question and ask the rest of the class to evaluate and come to a conclusion about each question.

> *Possible response: Students should suggest that the artist is using light to cast a shadow of the person onto the canvas or paper. The room has to be darkened so that the source of light casts a good shadow when it is blocked by the person being drawn. A lamp would make more light and is less likely to flicker so the shadow would be darker and clearer to draw around.*

Point out the photograph in the circle at the top of page 69. Ask students to look at the photograph of the people and their shadows on the ground, and to discuss the question with their partner.

The people in this street know that there is not a giant nearby! Why are the shadows so large?

Encourage students to think about why the shadows look much larger than the people who are casting them. Remind them that they studied shadows in Year 5 and it is useful to review prior work before moving onto new ideas.

> *Possible response: Students should suggest that the light from the Sun is being blocked by the people. This is casting a large shadow on the ground. Elicit that the shadows are longer than the people because the light source (probably the Sun in this case) is low in the sky.*

Finally, ask students to look at the last discussion box on page 69.

The people in the photograph are enjoying the sunshine. Why will some people have to move in order to stay in the Sun for longer?

You could approach this discussion with the whole class and ask for volunteers to tell you what they think. Remind them that they have learned about how the Earth moves around the Sun in the solar system in Year 5.

> *Possible response: Students should suggest that as the Earth moves on its axis the amount of light from the Sun changes. The shadows cast by the trees will move position so some people who are currently in sunlight will be in shadow.*

Workbook activities

Key words (page 68)

This activity supports understanding of the key vocabulary through a crossword puzzle. Ask students to read out each of the key words and then all of the definitions before they start the activity. This will encourage them to process the information slowly and to allocate the correct word to its definition. Then ask them to complete the crossword puzzle.

> *Answer: Across: 4 beam; 5 transparent; 6 mirror; 7 ray; 9 reflect; 10 silhouette. Down: 1 intensity; 2 opaque; 3 source; 5 translucent; 8 shadow*

Rainbows (page 69)

This activity is a light art activity to elicit prior learning about white light and how raindrops can cause a rainbow from sunlight, which is covered later in this unit.

The white light from the Sun contains all the colours of a rainbow. If students have not covered the light spectrum before, remind them of the rainbow colours: red, orange, yellow, green, blue, indigo and violet. These can be remembered with the acronym ROYGBIV.

Ask them to colour in the rainbow and label each colour, and then to complete the labels on the rainbow illustration below.

Answer: Students should colour the lines from the outside edge to the centre in the following colours: red, orange, yellow, green, blue, indigo and violet. They label the diagram as follows: the source in the centre above the rainbow where the white light is emitted is the Sun. The drops are labelled rain drops.

Extra activities

1 Students can keep a diary of every time they use a source of light over a full day.

2 **Maths link:** Students can count how many shadows they see in one hour when they are outside.

3 Ask students to draw a silhouette of a partner. Allow them to set up a light source so they can cast a shadow and remind them to use the drawing on page 68 of the Student Book as a clue. Display the silhouettes and see if people can guess who they represent.

Sight

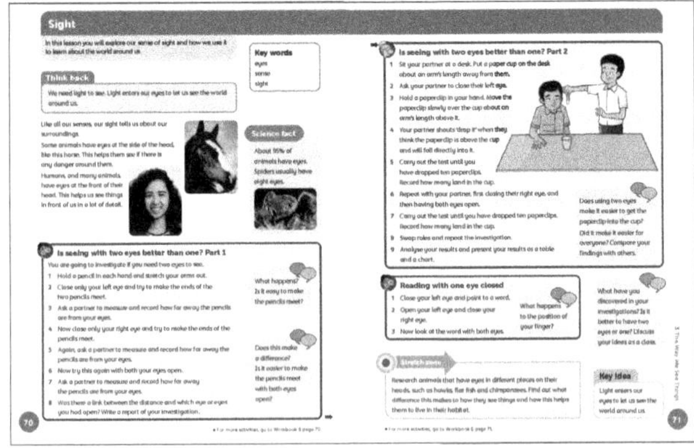

Supporting activities are in Workbook 6 pages 70–71.

Getting started

In this lesson students consider the sense of sight and how we use this to make sense of the world around us. They explore the eyes of humans and other animals, observing the position of the eyes and considering why they are situated where they are. They are encouraged to use scientific procedures to record their observations. Students also investigate if two eyes make our sense of sight better. Remind students that some people cannot see well and some cannot see at all.

Language support

To help develop language skills discuss what the word 'sight' means. Tell students that this explains how we see things. They will also experience the word 'sense' or 'senses'. This has been covered before in earlier years, but remind them that sight is a sense and that there are other senses, such as touch or hearing. Explain that we use our senses to help us to understand the world around us. You could ask students how they are using their sense of sight now. Ask them how seeing can keep them safe.

Resources

Student Book: pencils; rulers or tape measures; paper cups; paperclips.

Workbook: range of food samples; blindfolds; walking sticks.

Key words

eyes   sense   sight

Other words in the lesson

allergy   blindfold   hypothesis

Scientific enquiry key words

Plan and/or carry out enquiries to answer questions

Make observations

Take measurements, using equipment accurately

Record data and results

Analyse data, notice patterns and group or classify things

Report and present findings

Draw conclusions and give explanations

Identify causal relationships

Lesson at a glance

The key teaching points for students in this lesson are:

- sight is a sense that tells us about our surroundings
- we have two eyes that work together.

In the next lesson, students will learn about optical illusions.

Think back: We need light to see. Light enters our eyes to let us see the world around us.

Ask students to discuss with a partner the fact that we need light to see things. This topic has been covered in previous years. Ask them if they can recall how light results in us seeing things.

> **Possible response:** *Students should suggest that light travels from a light source into our eyes. Our brain makes sense of the light and informs us what is around us. For example, light is reflected from objects such as this book and enters the eye. The brain works out that it is a book in front of us.*

Read through the text at the top of page 70 and point out the photographs of the horse and the person. Like all of our senses, our sight tells us about our surroundings. Some animals have eyes at the side of their head, like the horse. This helps them see if there is any danger around them. Humans, and many other animals, have eyes at the front of their head. This helps us see things in front of us in a lot of detail.

Science fact About 95% of animals have eyes. Spiders usually have eight eyes.

Read out the Science fact or ask a volunteer to read it out. Ask students to discuss why so many animals have eyes. Ask them to think about what kinds of animals in the remaining 5% wouldn't have eyes. Explain that some animals that live in darkness do not need eyes because there isn't any light.

Investigation: Is seeing with two eyes better than one? Part 1

Students work with a partner for the investigations Parts 1 and 2. For Part 1, they will need a ruler or tape measure to measure the distance from the pencil to the eyes. Ask them to follow the instructions in the Student Book carefully and to make accurate measurements. They should carry out this investigation independently – provide prompts and support only if needed. After the first two steps encourage reflection using the discussion question.

What happens? Is it easy to make the pencils meet?

Ask students to discuss this with a partner then encourage them to share their ideas with the class.

> **Possible response:** *Students should suggest that when only one eye is open it is harder to make the pencils meet.*

Explain that when only one eye is open their range of vision is not very good. This makes it harder to make the pencils meet.

After the sixth step, pause again and ask pairs to discuss the questions.

Does this make a difference? Is it easier to make the pencils meet with both eyes open?

> **Possible response:** *Students should suggest that using two eyes makes the task easier.*

Point out that when both eyes are open their range of vision is improved. This makes it easier to make the pencils meet. They should conclude whether there was a link between the distance and which eye or eyes were open and then write a report of their investigation.

Investigation: Is seeing with two eyes better than one? Part 2

Students can work with the same partner for Part 2. They will carry out another investigation to explore whether two eyes are more efficient than one. This time they will use a paperclip and a paper cup. Again, allow students to follow the instructions independently, with your prompts only when necessary.

Students will analyse their results and present them as a table and a chart. You can offer support by suggesting that they need to have a column for the number of paperclips that land in the cup and a column for how many paperclips miss the cup. They will need three rows – one for the titles at the top and two for recording the results for each person in the pair.

Use the discussion box at the end of the investigation to prompt a discussion about their findings.

Does using two eyes make it easier to get the paperclip into the cup? Did it make it easier for everyone? Compare your findings with others.

Students can continue to work with their partner. Point out that when we use one eye the vision changes. This is why when we measure using a scale, we should never close one eye and should always look directly at the scale and not from one side or the other. You can demonstrate this with a measuring jug.

Possible response: Students should suggest that most people would have collected more paperclips when both eyes are open. Some people might have a vision impairment where their sight is better in one eye. This could affect the results and some sensitivity might be required here.

The final investigation in this lesson is intended to corroborate the findings in the previous activities.

Investigation: Reading with one eye closed

Students continue to work with a partner. They will point at a word with one eye closed and then two eyes open. They will observe where the finger is pointing each time.

Their partner should see them point slightly away from the word with only one eye open. They might need to focus on the first letter of the word only to make the comparison clearer. Use the discussion to encourage students to evaluate and reflect on the investigation.

What happens to the position of your finger?

Possible response: Students should suggest that the position of the finger is slightly to the side of the word with only one eye open. When they look with their right eye, their finger will be slightly to the right of the word, and the opposite with their left eye. With both eyes open they can point more accurately to the word.

The final discussion on page 71 encourages students to evaluate and review their investigation findings in the lesson.

What have you discovered in your investigations? Is it better to have two eyes or one? Discuss your ideas as a class.

Students should reflect on each of the investigations and draw their conclusion using all of their results. Ask students to work with their partner to write down two things they have learned from their investigations. They can pin these around the room and then all move around to compare ideas. Remind them that the investigations they carried out were to answer the question 'Is seeing with two eyes better than one?' – scientific enquiry usually starts with a question. Refer back to the flow diagram on Student Book page 6 as a reminder.

Answer: Students should suggest that all of the investigations support the idea that two eyes are better than one. Seeing with one eye is acceptable but is not as good as seeing with two eyes.

Stretch zone: Research animals that have eyes in different places on their heads, such as hawks, flat fish and chimpanzees. Find out what difference this makes to how they see things and how this helps them to live in their habitat.

Read out the Stretch zone task to students and ask them to work in teams to research answers to the task. Access to the internet would be useful or you can download information sheets to ensure the correct level of detail. Ask students to find out about different animals and why they have their eyes positioned in ways different from humans. Remind them about their prior learning on adaptation in Unit 1. Encourage a class discussion about their findings.

Possible response: Students should suggest that different animals' eyes are positioned because of the environment they live in and the activities they are involved in. A hawk's eyes are positioned more centrally on their head which means they can focus on prey more efficiently. Flat fish have an eye on each side of their head which allows them to see if they lay on their side. Chimpanzees and other primates have large eyes in relation to their size. They are positioned very close together at the front of the head which gives them 3D or stereoscopic vision.

Key idea

Light enters our eyes to let us see the world around us.

Read through the key idea or ask a volunteer to read it out to the class. Ask students to close their books and then list from memory what they have learned about sight and having two eyes. You can also ask them to tell a partner how they investigated if seeing is better with two eyes.

Workbook activities

Is there a link between sight and taste? (page 70)

Warning! Tell your teacher if you have any food allergies. Why is this important?

Read out the warning at the start of the activity. Ask volunteers to explain what an allergy is and how this may be dangerous. Talk to students about the risk involved and let them know that it is their responsibility to make sure everyone is safe in any investigation they are doing.

This activity is based on the idea that scientists believe that the appearance of food can influence how we taste it.

Students are asked to test this hypothesis using a set of instructions and then to write a short summary to give their ideas about any possible links between sight and taste.

> *Possible response: Students should suggest that seeing food helps them to identify the flavours. Being able to see foods helps our senses to assess whether or not they are going to be safe to eat and whether or not they are going to taste good.*

How important is sight? (page 71)

Students can work with a partner or in a small group for this activity. They will take it in turns to wear a blindfold or close their eyes and move around the room with and without a walking stick. Their partner or another member of their group will make sure they are safe. They are asked to record how it feels, what they hear and whether the stick helps them.

> *Possible response: Students should suggest that moving around blindfolded is a bit worrying. Moving when we can't see makes us worry about bumping into things or falling and tripping. The sound from the stick gives us confidence that we are not going to bump into anything because the stick will warn us of obstacles in our path. Ask students to imagine how difficult it can be for people who are sight impaired to move around.*

Review and reflect

Encourage students to identify aspects they may not have completed fully and help them to identify improvements. For example, they can look at the results of their investigations and reflect on the discussions they had. This will help students to develop a positive approach to learning by understanding that learning is a process that will improve with practice and reflection.

Extra activities

1 **Maths link:** Students can investigate how many paperclips the people at home can drop into a cup using each eye and compare this to using both eyes. They could work out the percentage of paperclips they drop into the cup for each test.

2 Play a game where a partner guides the other blindfolded person to get a ball or piece of paper into a basket two metres away. This will encourage students to develop their communication skills and also introduces the idea that being able to hear is very helpful when you can't see properly.

Differentiation

Supporting: Allow students time to practise using each eye individually and both eyes together in the three Student Book investigations.

Consolidating: Students could test other people to find if everyone can drop more paperclips into a cup using both eyes to give extra data to evidence their conclusions.

Extending: Students could research animals that have the best vision. Do they have two eyes, where are they positioned, and why?

Differentiated outcomes	
All students	should be able to understand the benefit of being able to see and recall how light is needed for us to see things
Most students	will be able to explain how two eyes help us to see better
Some students	may be able to explain how the position of human and animal eyes has adapted to their habitats

Brain tricks

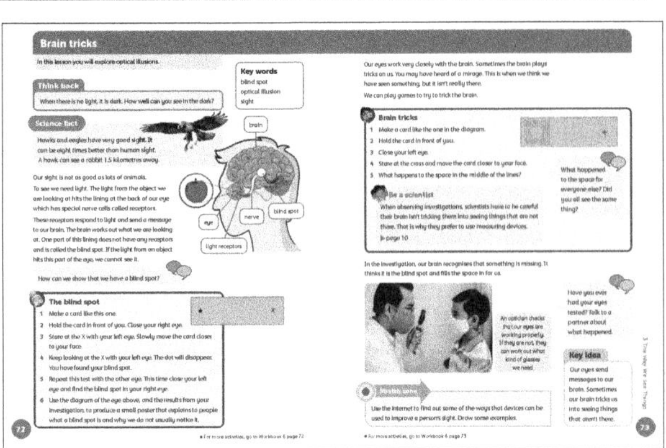

Supporting activities are in Workbook 6 pages 72–73.

Getting started

This lesson introduces students to the idea that our brain can play tricks on us and optical illusions demonstrate this. Students explore the steps in how we see things. They also explore the blind spot which demonstrates how our eyes have some weaknesses. Students investigate optical illusions and experience how we can see things differently to how they are in reality.

Language support

Write the key words on the board or on word cards; you can also use the eBook so students can hear them pronounced. Ask students what 'optical' means. Discuss how it relates to light and seeing. They could think of other words that are similar, such as optician, and the optical illusions they will investigate. Students could make a word plate with optical at the centre and pictures drawn or cut out of magazines to help them define and recall the meaning of the word.

Resources

Student Book: pieces of cardboard; writing materials, materials to make posters.

Workbook: sets of sticky notes; writing materials; smartphones or digital cameras.

> **Key words**
>
> blind spot   optical illusion   sight
>
> **Other words in the lesson**
>
> animation   horizontal   nerve   optician
> receptors

> **Scientific enquiry key words**
>
> Plan and/or carry out enquiries to answer questions
>
> Make observations
>
> Report and present findings
>
> Draw conclusions and give explanations

Lesson at a glance

The key teaching points for students in this lesson are:

- we have a blind spot, which means our vision is not as good as some animals
- our brain can play tricks on us and we don't always see what things really look like.

In the next lesson, students will study more about how we need light to see things.

Think back: When there is no light, it is dark. How well can you see in the dark?

Ask students to discuss the Think back question. They could discuss what things look like when there isn't any light. Ask them when they might have experienced this, but be sensitive as some children are uncomfortable in the dark.

> *Possible response: Students should suggest that they cannot see at night unless they switch a light on. This demonstrates that we need light to see things.*

Science fact Hawks and eagles have very good sight. It can be eight times better than human sight. A hawk can see a rabbit 1.5 kilometres away.

Read the Science fact to the class. Ask, 'Why do hawks and eagles need such good sight? How does this help them to survive?' Relate the discussion to adaptation, which students have covered in Year 4 and in Unit 1 of Year 6.

Ask students to read the text and study the diagram at the top of page 72. Check understanding by asking, 'What must be available for us to see? How does light enter our eyes? What are the cells called in the lining at the back of our eyes? What do the receptors do? How is our brain involved?'

Explain that one part of this lining does not have any receptors and is called the blind spot. If the light from an object hits this part of the eye, we cannot see it.

How can we show that we have a blind spot?

Ask students to discuss what the blind spot is. Ask them to discuss with a partner how they could investigate this.

Possible response: Students should suggest that they could test each other's eyes using an object and moving it closer and further away from the eye until it disappears.

Explain that they will now investigate this.

Investigation: The blind spot

Students work with a partner for this investigation. Give them a piece of card or provide them with cards already prepared. Some students could have visual impairments so be sensitive to this and find out before the investigation if this affects any students.

Allow them to follow the instructions in the Student Book independently. Only offer support if you need to. They should use the diagram of the eye and the results from their investigation to produce a small poster that explains to people what a blind spot is and why we do not usually notice it. Display the posters in the room and give students a chance to circulate and evaluate each other's work.

Possible response: Students should suggest that they could find their blind spot using this investigation. This is evidence to prove that what they have read is correct.

Ask students to read the text at the top of page 73. This describes how the brain can play tricks on us. We can think we see things, or something might look different to how it is in reality. Sometimes this is done purposefully using optical illusions.

Investigation: Brain tricks

Organise students into pairs for this investigation and provide them with card to copy the diagram onto. They will make a card like the one in the picture and hold it in front of them. They will then close their left eye, stare at the cross, move the card closer and assess what happens to the space in the middle of the lines. Students might need to repeat this several times as our brain tells us we should be seeing something different.

Possible response: Students should suggest that the brain completes the diagram and the space fills with lines.

What happened to the space for everyone else? Did you all see the same thing?

Ask students who worked together on the investigation to discuss what they saw. Then encourage them to share their observations with the class to see if they all saw the same thing.

Possible response: Students should suggest that they all saw the empty space become filled with the lines as they concentrated on the cross.

 **Be a scientist:** When observing investigations, scientists have to be careful their brain isn't tricking them into seeing things that are not there. That is why they prefer to use measuring devices. (Student Book, page 10)

Read out the Be a scientist text. Explain that this is evident when scientists are measuring using a scale. Use a thermometer scale as an example of this. You could allow students to test this if you have some thermometers.

Have you ever had your eyes tested? Talk to a partner about what happened.

Organise students into pairs or small groups, where one of them has had an eye test, to recall what happened.

Ask students to look at the photograph at the bottom of page 73 to show them what an eye test is like. The person is called an optician and they can check to make sure that the eyes are working as they should be. Sometimes they might prescribe glasses. Some students might wear these in the class. Be sensitive to them. Follow up this discussion by asking the same groups to attempt the Stretch zone activity.

Stretch zone: Use the internet to find out some of the ways that devices can be used to improve a person's sight. Draw some examples.

Possible response: Students should suggest that they found different glasses/spectacles, lenses or contact lenses. They might also have found out about corrective surgery.

Encourage students to display their drawings and diagrams to share their findings.

Key idea

Our eyes send messages to our brain. Sometimes our brain tricks us into seeing things that aren't there.

Ask students to summarise what they have learned with a partner. Let them share their ideas, and read out and discuss the key idea. This will remind students of how our brain can play tricks on us and we don't always see what is really there.

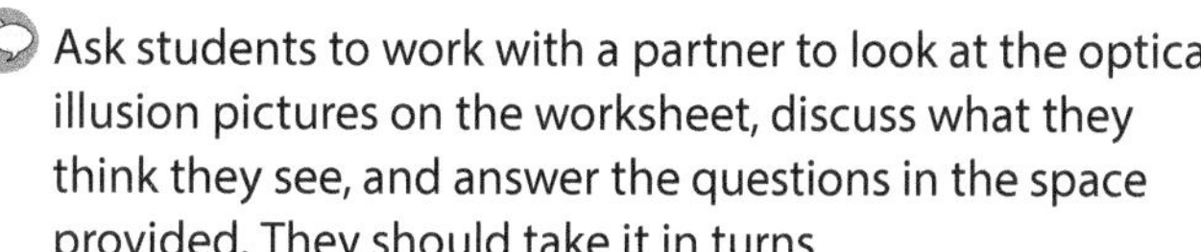

Tricking your eyes (page 72)

Ask students to work with a partner to look at the optical illusion pictures on the worksheet, discuss what they think they see, and answer the questions in the space provided. They should take it in turns.

Remind them that visual tricks like these are called optical illusions. If your brain does not get a full picture, it tries to fill in any gaps.

Finally, ask pairs to attempt the Stretch zone question to research and print out three more examples of optical illusions to challenge their friends.

Making a science animated film (page 73)

This activity gives students instructions on how to make an animated film using sticky notes or a small notepad. They draw two circles as instructed on 10 sticky notes and flick through them. They observe the outcome and are asked to write a hypothesis about the outcome.

The second part of the activity asks students to make an animation using this technique by choosing an idea in science that involves changes – for example, water and water vapour moving in the water cycle or a butterfly life cycle. They can film the animation with a smartphone or digital camera and create a film festival. This is a creative end to the lesson, which will remind students about their learning.

 **Review and reflect**

Encourage students to reflect on their own learning by pausing for a few moments and thinking about which parts of the work they found tricky. Talk about how they managed this and suggest that next time they stop for a short while and take a few deep breaths. Also remind them that when they are doing challenging work, they are training their brain and this will help their future learning.

Conclude the lesson by asking students to complete the first statement in the 'What I have learned about the way we see things' activity on page 102 of the Workbook. (See also the teaching notes in this Teacher's Guide, pages 134–136.)

Extra activities

1   Ask students to research how opticians and doctors can help people with problems with their eyesight. Encourage them to include partially sighted and unsighted people and how they can be helped.

2   Students can investigate what a mirage is and how it occurs.

Differentiation

Supporting: Display the diagram of how light enters the human eye on page 72 to help understanding of the blind spot.

Consolidating: Ask students to describe to you how the blind spot happens so you can correct any misconceptions.

Extending: Students can draw a simple diagram of the journey of light to their brain and mark the point where the blind spot happens.

Differentiated outcomes	
All students	should be able to identify that we have a blind spot
Most students	will be able to state that the blind spot can cause optical illusions
Some students	may be able to explain how optical illusions happen

How do our eyes see things?

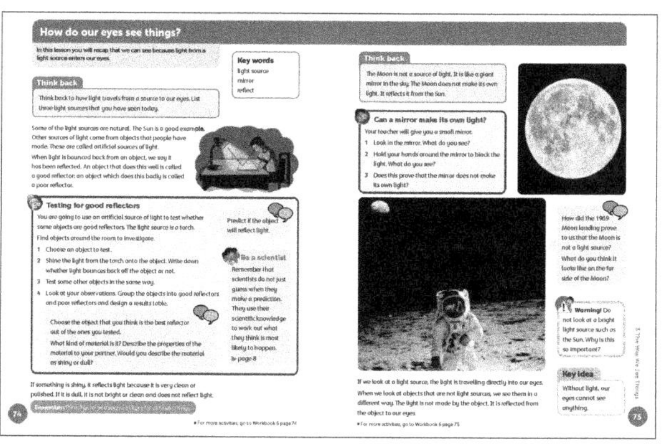

Supporting activities are in Workbook 6 pages 74–75.

Getting started

This lesson reminds students what a source of light is. They have covered this already in previous years. Students will observe natural sources of light – using the Sun as a good example – and human-made sources, for example a lamp. Students become aware that when light is reflected from an object, we see it. Students explore and observe sources of light and how some materials can reflect them better than others. They are reminded that mirrors are not a source of light but are just good at reflecting it. They compare this to the Moon, which reflects the light from the Sun.

Language support

Read through the key words with the class. Ask students to define some of the words based on their previous learning. They can add to any glossaries or Word walls that they might have started. Discuss examples of light sources that they are familiar with, for example the Sun, a candle or lamps. Discuss how some materials are not a source of light but reflect it. Ask them to give any examples that they might know. Explain that a mirror is a good reflector of light and that is how we can use them to see ourselves.

Resources

Student Book: torches; small mirrors.

Workbook: torches.

Key words

light source   mirror   reflect

Other words in the lesson

artificial   Moon   natural   reflector

Lesson at a glance

The key teaching points for students in this lesson are:

- some sources of light are natural and others are artificial
- a mirror is not a source of light, it just reflects it.

In the next lesson, students will learn about how beams or rays of light can be reflected by surfaces.

Think back: Think back to how light travels from a source to our eyes. List three light sources that you have seen today.

Ask students to discuss how light travels from a source to our eyes. They can refer back to the previous lesson if they need to. Ask them to look around the room to remind them of all of the sources of light that they might have seen.

> *Possible response: Students should suggest lamps, bulbs, the Sun, candles, streetlights, car headlights or the light from a smartphone.*

Ask students to read the text and study the picture at the top of page 74. This tells them about natural and artificial light sources. After they have read the text, ask them to list two examples of natural light sources and two examples of artificial light sources. Explain that when light is bounced back from an object, we say it has been reflected. Ask them to study the picture and tell a partner how reflected light helps them to read a book.

Investigation: Testing for good reflectors

Students work with a partner for this investigation. They will use an artificial source of light – a torch – to test whether some objects are good reflectors. They are asked to find objects around the room to investigate and follow the instructions given. The worksheet on page 74 of the Workbook supports this investigation.

 **Be a scientist:** Remember that scientists do not just guess when they make a prediction. They use their scientific knowledge to work out what they think is most likely to happen. (Student Book, page 8)

Remind students how to make a prediction. Use examples such as: I predict I am having an apple at lunch time. This is not a guess. Perhaps they have an apple everyday so the prediction they made is based on their experience. Remind them to make their predictions at the start of the investigation, and to record their observations and compare them with others in the class.

 **Predict if the object will reflect light.**

Remind students that it is important that they record their predictions because this is what they are testing in the investigation.

> *Possible response:* Students should use their ideas to make predictions. They should suggest that good reflectors have a shiny and smooth surface like a mirror. The ones that are poor reflectors are rough and dull, for example a rug or a brick wall.

 **Choose the object that you think is the best reflector out of the ones you tested. What kind of material is it? Describe the properties of the material to your partner. Would you describe the material as shiny or dull?**

Encourage students to have this discussion with their investigation partner. It is important that students learn to review and reflect upon their findings from an investigation.

> *Possible response:* Students should suggest that the properties of their best reflector choice were smooth and shiny.

Ask students to read the information about the Moon at the top of page 75 and to look at the photograph. They might recall learning about the Moon in previous years.

Think back: The Moon is not a source of light. It is like a giant mirror in the sky. The Moon does not make its own light. It reflects it from the Sun.

Make sure students realise that the Moon is acting like a huge mirror and does not produce its own light. Explain that they are going to collect some evidence to prove this in the next investigation.

Investigation: Can a mirror make its own light?

Give a group of students a small mirror. They will look in the mirror and explain what they see. Then they will hold their hands around the mirror to block the light and then explain what they see. Ask them to use their observations to explain how this proves the mirror does not make its own light.

> *Possible response:* Students should report that they cannot see anything when the mirror is covered by their hands: their hands are blocking the light reaching the mirror. This is evidence that the mirror is not a source of light.

Ask students to look at the photograph of the astronaut on the Moon on page 75. In pairs, they should discuss the questions in the discussion box. As prompts before they start, you can ask, 'What planet can you see in the distance? What do you think the 'far side of the Moon' means?' Elicit that this is the side that is further away from the Sun. Does the background of the photograph look light or dark?

How did the 1969 Moon landing prove to us that the Moon is not a light source? What do you think it looks like on the far side of the Moon?

> *Possible response:* Students should suggest that the fact there was a Moon landing at all is evidence, but photographs, film footage and reports from astronauts of the 1969 Moon landings are also evidence that the Moon must not be a source of light. It is dark on the far side of the Moon because the Sun would not be visible. This is further evidence that the Moon is not a source of light.

Warning! Do not look at a bright light source such as the Sun. Why is this so important?

Ask students to think back to their prior learning about how to stay safe when looking at the sky. Ensure they recall that the Sun emits huge amounts of heat and light energy. This could damage the eyes or even result in blindness.

Ask students to read the text at the bottom of page 75. Ask for volunteers to share their ideas with the class about the difference between seeing a star, a streetlight or a lit candle and seeing a book or the Moon. Elicit that the former are light sources and the book and Moon are only seen as they reflect light from another source. The book and Moon do not make their own light.

Key idea

Without light, our eyes cannot see anything.

Read through the key idea or ask a volunteer to read it out to the class. Ask students to summarise what they have learned in the lesson with a partner. Let them share their ideas, and read out and discuss the key idea.

Workbook activities

Shining back (page 74)

This activity supports the investigation on page 74 of the Student Book where students assess whether materials are good or poor reflectors.

You will need to make the room darker to allow students to test the six different objects.

Warning! Do not shine a torch directly at someone else. Discuss why this is important.

Before starting the investigation, read out the safety warning. Ask a student to explain why they need to act safely and not shine a torch directly at someone. Elicit that bright lights can damage eyes.

The table provides a structure for students to record their predictions and observations for each of the objects. They are then asked to assess how accurate their predictions were. Encourage them to discuss why they made the predictions and if their results supported or refuted them.

> *Possible response: Students should record that objects made from shiny and smooth materials reflected the light well. They should find that their predictions were quite accurate.*

Artificial light (page 75)

Remind students that sources of light can be natural or artificial. Ask them to write down one example of a natural light source and some examples of artificial light sources. They can use their own experiences and also ideas in the photograph. Next, ask them to look at the artificial sources of light they listed and ask them to explain why a natural source of light cannot be used. Elicit that at night the main natural light source (the Sun) is not visible in the sky and others, such as volcanoes and lightning, are neither safe nor reliable sources.

Computing link: Finally, allow students access to the internet so they can research the development of lamps and torches over the past 200 years. They can download pictures and present these in a timeline.

Review and reflect

Use the discussion tasks, the research investigation activities and the Workbook tasks to encourage students to think about what they understand and what they are finding less straightforward. Discuss the outcomes of each task with students to correct misconceptions and identify improvements.

Extra activities

1 Students investigate different coloured mirrors to find out if they reflect the light in the same way. They can colour plastic mirror strips with felt pens or use tissue paper as a filter.

2 Students record how many mirrors there are at home. They should comment on how the mirrors are used. Sometimes they are used to reflect light into a dark space.

Differentiation

Supporting: Provide photographs or cut-outs of pictures from magazines to help students identify natural and artificial light sources.

Consolidating: Give students a wide range of materials to find out whether or not they reflect light.

Extending: Ask students to find out how we use reflection apart from seeing ourselves in a mirror.

Differentiated outcomes	
All students	should be able to identify natural and artificial sources of light
Most students	will be able to predict and identify materials that reflect light
Some students	may be able to list ways that reflection of light is used

The journey of light

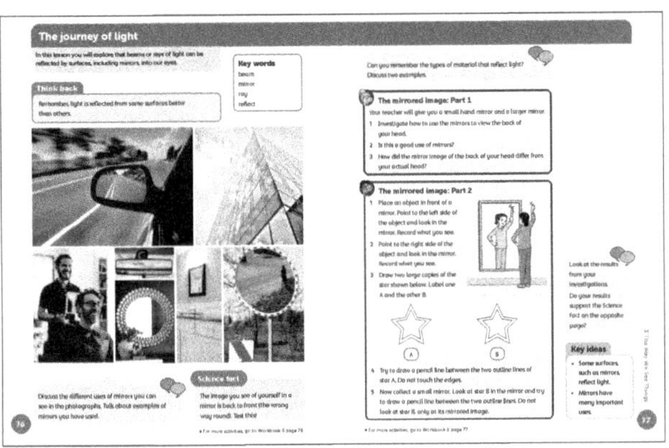

Supporting activities are in Workbook 6 pages 76–77.

Getting started

In this lesson students will explore how some materials reflect light better than others. Students learn that beams or rays of light are reflected by shiny surfaces into our eyes or onto other objects. They investigate the image that they see in a mirror. They explore how they can see the back of their head using two mirrors. They also investigate the mirrored image and how this is opposite, and how to draw using a mirrored image.

Language support

A strategy you can use is to write the key words on the board but leave some gaps so letters are missing. Students can copy the words and then discuss which letters are missing. They can self-check by referring to the key words box on page 76 of the Student Book. They can also share examples of when they have heard of the words before so you can elicit prior understanding. Encourage students to make sentences up using some of the key words. Ask for volunteers to read out the sentences they have constructed.

Resources

Student Book: small mirrors; large mirrors.

Workbook: small mirrors; large mirrors.

Key words

beam mirror ray reflect

Other words in the lesson

mirrored image reflected

Scientific enquiry key words

Plan and/or carry out enquiries to answer questions

Make observations

Record data and results

Analyse data, notice patterns and group or classify things

Report and present findings

Draw conclusions and give explanations

Lesson at a glance

The key teaching points for students in this lesson are:

- beams/rays of light can be reflected by surfaces including mirrors
- reflected light enters our eyes and we see the object.

In the next lesson, students will learn about the uses of reflection from surfaces.

Think back: Remember, light is reflected from some surfaces better than others.

Ask students to individually read the Think back text and to recall the surfaces that reflected light better. They have carried out these investigations in previous lessons. They might suggest smooth or shiny surfaces.

Discuss the different uses of mirrors you can see in the photographs. Talk about examples of mirrors you have used.

Students can work in pairs. Ask them to look at and discuss all seven of the photographs. They show light being reflected by mirrors and shiny glass surfaces. Ask for volunteer pairs to tell you about the use of the mirror in each photograph.

> *Possible response: Students should suggest that mirrors have many uses including seeing vehicles that are behind and seeing the back of the head. They are also useful in seeing things that are out of view, or for reflecting light from a source to a darker place.*

Science fact The image you see of yourself in a mirror is back to front (the wrong way round). Test this!

Ask students if they have experienced this. Tell them that they will investigate this later in the lesson.

Can you remember the types of material that reflect light? Discuss two examples.

Students have already identified that the properties needed to reflect light are being smooth and being shiny. Now ask students to discuss with their partner which materials they found to be more reflective. They should compare their ideas with other pairs.

Possible response: Students should suggest glass and metals are most reflective. Some plastics and fabrics can also be reflective.

Investigation: The mirrored image: Part 1

Students work with a partner for this investigation. They investigate how to hold the smaller mirror while they stand in front of a larger mirror in order to see the back of their head. This is a well-used practice in hairdressing. They then discuss the questions together.

Possible response: Students should suggest that this is a good use of mirrors. The back of the head looks the opposite way around and this can be difficult when arranging a piece of hair, for example, but it is a good use of mirrors.

Investigation: The mirrored image: Part 2

Encourage students to work with a partner. They will discuss what happens when they try to point to either side of their head or an object. Ask them to record what they see in writing or as a drawing. They then explore this further using two large copies of the stars shown. The worksheet on page 76 of the Workbook supports this investigation.

Students are asked to draw two large copies of the stars shown and label one 'A' and the other 'B'. They are then instructed to compare how easy it is to draw a pencil line between the two outline lines of star A and a mirrored image of star B.

Possible response: Students should suggest that when they point to the left side it appears on the right side in the mirror, and vice versa. Drawing around star A was relatively easy but drawing around star B caused more problems as everything was in reverse.

Look at the results from your investigations. Do your results support the Science fact on the opposite page?

Students should discuss the question with their investigation partner. Encourage them to use the terms 'evidence' and 'proof' in their discussions, so they are learning how they are using scientific evidence to support or refute an idea or hypothesis.

Answer: Students should suggest that their investigation findings support the Science fact.

Key ideas

* *Some surfaces, such as mirrors, reflect light.*
* *Mirrors have many important uses.*

Read through the key ideas or ask a volunteer to read them out to the class. Ask students to turn to a partner and describe one use of mirrors. They can then discuss why the left side of a person looks like the right side in a mirror image. Ask volunteers to share their ideas with the class.

The mirrored image (page 76)

This activity supports Part 2 of the investigation on page 77 of the Student Book.

Students will use evidence from this investigation to prove that a mirrored image is back to front. They are asked to point to each side of an object in the mirror and record their observations. They also compare the difficulty of drawing a star from a normal and a mirrored image, and explain their findings. Ask students to follow the instructions on the worksheet.

Possible response: Students should suggest that when they point to one side the other side is reflected in the mirror. Students should find it quite difficult to draw around star B using the mirror.

Mirror writing (page 77)

Point out the mirror writing at the top of the page and ask students if they can read it. Hand out small mirrors or ask students to take the writing to a larger mirror. Ask them to read the reflected text and to check if they were correct. Next, challenge students to use a mirror to help them produce their own secret writing. Point out that this should be a sentence to describe some science they have done at school. Encourage them to look in the mirror as they write. Explain that a famous scientist, Leonardo da Vinci, wrote his notes using a mirror.

Allow students to check with classmates if they can read their mirror writing secret message.

Finally, ask students to draw an object in the classroom while looking at it in a mirror and then think about how the real object is different from the image in the mirror.

Possible response: Students should suggest that they found it difficult to make sense of the sentence without the mirror but could read it using a mirror. The object in the mirror is the reverse of the real object.

Encourage students to find a quiet place to think about their learning in this lesson. Ask them if they found any part of it challenging. Remind them that when they overcome a challenge, they are making their brain work harder. This is like going for a run to make your lungs work harder. Ask students how they felt when the learning was challenging and how they overcame the challenge. Ask them to record the strategies they used as this will help them in future situations. If all students found something more difficult than others, you might want to review this as a class before moving onto the next lesson.

Extra activities

1 Ask students to use a mirror to walk backwards. They will need another student to make sure they do not trip over anything. Students could make a maze by placing cones or bags on the ground for students to navigate around.

2 Students sit in a chair and hold a mirror in front of them. They can only look at the mirror. The other person holds an object behind them and they have to describe what they see in the mirror. This models how rear-view mirrors are used.

Differentiation

Supporting: Display photographs of some uses of mirrors and fix a mirror above the door of your room so students can use it regularly.

Consolidating: Demonstrate to students that the image in a mirror is reversed by asking them to hold a word card with 'right' written on it in their right hand and look at themselves in a mirror. Ask them what should be written on the card in the image.

Extending: Ask students to investigate why right becomes left in a mirror but top does not become bottom.

Differentiated outcomes	
All students	should be able to give examples of how mirrors are used
Most students	will be able to describe the image seen in a mirror
Some students	may be able to investigate how a mirror image differs from the real object

Uses of reflection

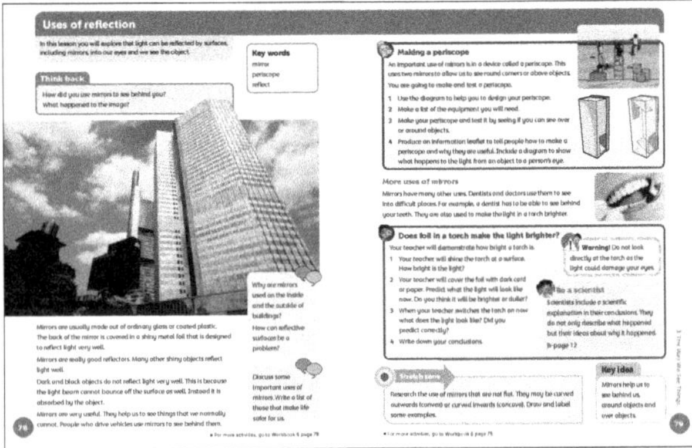

Supporting activities are in Workbook 6 pages 78–79.

Getting started

This lesson extends students' learning that some of the properties of materials can be used to reflect light. Students learn that many mirrors and reflective surfaces are made out of glass. Students explore how mirrors can be used to see over objects and to intensify the light produced by a source of light.

Language support

Read out the key words to students and ask volunteers to explain each one. They should recall the words 'mirror' and 'reflect' from prior learning. Show them pictures of a periscope or a model periscope and spell the word out for them on the board.

Resources

Student Book: small cardboard boxes; small mirrors; pieces of thin card; scissors; sticky tape; torch; foil; dark card; materials to make leaflets.

Workbook: small cardboard boxes; small mirrors; pieces of thin card; scissors; sticky tape; torch; shoe boxes or other narrow boxes made of thin card.

> **Key words**
>
> mirror   periscope   reflect
>
> **Other words in the lesson**
>
> concave   convex   dentist   doctor   obstacle course

> **Scientific enquiry key words**
>
> Plan and/or carry out enquiries to answer questions
>
> Make predictions
>
> Make observations

Record data and results

Report and present findings

Draw conclusions and give explanations

Lesson at a glance

The key teaching points for students in this lesson are:

- mirrors are made from shiny materials, usually glass
- mirrors have many uses.

In the next lesson, students will study rays of light and how they change direction when reflected from a surface.

Think back: How did you use mirrors to see behind you? What happened to the image?

Ask students to read the Think back questions and say what they found out in the investigations in the last lessons. This can be done as a whole-class activity to make sure all students understand the concepts covered so far in the unit.

Possible response: Students should suggest that the mirror was held in a position to reflect an image from another mirror of the back of the head. They might recall that the image was reversed.

Ask students to read the information at the bottom of page 78. This explains how mirrors are constructed, usually from coated glass. Students might not have realised that the back of a mirror is covered in another material that will reflect the light more. Ask them to read about why darker materials do not reflect light well. This is because the light is absorbed by the object so not as much is reflected.

Why are mirrors used on the inside and the outside of buildings? How can reflective surfaces be a problem?

Ask students to observe the photograph of the glass building. They should discuss the questions in pairs. These should stimulate a detailed discussion within the classroom.

Possible response: Students should suggest that mirrors can be used to reflect light. They can be used inside and on the outside of buildings. Reflective surfaces can cause some problems as bright lights could startle people or temporarily blind drivers. Some students may also suggest that when sunlight is reflected, a place may get hot.

Discuss some important uses of mirrors. Write a list of those that make life safer for us.

This task consolidates learning about mirrors. Ask students to discuss with their partner and to record their points and ideas as a list. Students should compare their list with other pairs.

Possible response: Students should suggest that we use mirrors to see behind us when driving and when brushing our hair, for example; dentists and doctors use mirrors to see in difficult places. Mirrors are also used to reflect light.

Investigation: Making a periscope

An important use of mirrors is in a device called a periscope. This investigation allows students to construct and explore how a periscope works. Clean drinks cartons are ideal for this construction. Page 78 of the Workbook supports this investigation.

Students might need some support. If the periscope doesn't work, it is usually because the angles of the mirrors are not correct. These might need some adjusting.

Students should investigate their periscope and evaluate the effectiveness of it. They are then asked to produce an information leaflet to tell people how to make a periscope and why they are useful. Remind them to include a picture to show what happens to the light going from an object to a person's eye.

More uses of mirrors

Ask students to read about more uses of mirrors and how doctors and dentists use them to see into difficult places. Point out the photograph of the mouth and how the dentist is using a mirror to see the back of the person's teeth.

Investigation: Does foil in a torch make the light brighter?

Warning! Do not look directly at the torch as the light could damage your eyes.

Before the investigation, read out the warning and ask students to make sure they follow this instruction so that everybody is safe.

This is a demonstration. You will take the torch apart to cover the foil behind the bulb in the torch with paper or dark card and students will observe the effect. Students should be encouraged to make a prediction, record their observation, assess whether they predicted correctly, and write their conclusions.

Be a scientist: Scientists include a scientific explanation in their conclusions. They do not only describe what happened but their ideas about why it happened. (Student Book, page 12)

Read out the Be a scientist feature. Remind students that they should always back up their claims with scientific information. Encourage students to share their predictions and to justify them using their knowledge about mirrors and reflection.

You can challenge students who have grasped the concepts to learn more about different types of mirrors or lenses. Ask them to work in pairs or small groups to attempt the Stretch zone activity.

 **Stretch zone:** Research the use of mirrors that are not flat. They may be curved outwards (convex) or curved inwards (concave). Draw and label some examples.

Key idea

Mirrors help us to see behind us, around objects and over objects.

Summarise the lesson by asking students what they have learned. Let them share their ideas. Ask one student to read out the key idea. This will remind students of the ways that mirrors help us. They should recall that mirrors help drivers to see the cars behind them, and that mirrors can be used to see into difficult places and over objects.

Workbook activities

Making a periscope (page 78)

This activity supports the investigation on page 79 of the Student Book. Students can use the diagrams and instructions to help them make their periscope. If you use CDs, read out the warning feature and highlight to students the risks involved.

Warning! You can use pieces from an old CD if you do not have any small mirrors. Ask an adult to cut the CDs. Be careful – mirrors and cut CDs have sharp edges.

They can then answer questions 3 and 4 to record their findings and understanding.

Guiding light (page 79)

This activity extends learning about the journey of light. Students make an obstacle course to test how light travels. They can follow instructions in steps 1 to 4

and then record a prediction in step 5. Encourage them to include an explanation of their prediction before they add mirrors to guide the light through their obstacle course in step 6.

 ## Review and reflect

Print out numerous pieces of paper with the title 'Evaluation report' and on them should be two statements: 'Two things I think you have done well are …' and 'One suggestion I have is …'. Make the gaps large enough for students to write their comments in. Ask students to walk around to observe the periscope models made by other students. They should leave an evaluation report at each periscope.

Students should then return to their own periscope and read the evaluation reports. They can then enjoy the feedback and think of a target based on the suggestions.

Conclude the lesson by asking students to complete the second statement in the 'What I have learned about the way we see things' activity on page 102 of the Workbook. (See also the teaching notes in this Teacher's Guide, pages 134–136.)

Extra activities

1. Students can demonstrate to the people at home how the mirrors work in a periscope.

2. Students stand at the edge of a doorway. They use their periscope to describe what they can see around the other side.

Differentiation

Supporting: Give students time to practise directing the light from a source to other places using a mirror.

Consolidating: Students can use the worksheet on page 78 of the Workbook to help structure and record the periscope investigation.

Extending: Give students some specific secondary sources to use to research how curved mirrors reflect light differently.

Differentiated outcomes	
All students	should be able to predict which materials will reflect light
Most students	will be able to redirect light using more than one mirror
Some students	may be able to explain how different shaped mirrors are used

Ray diagrams

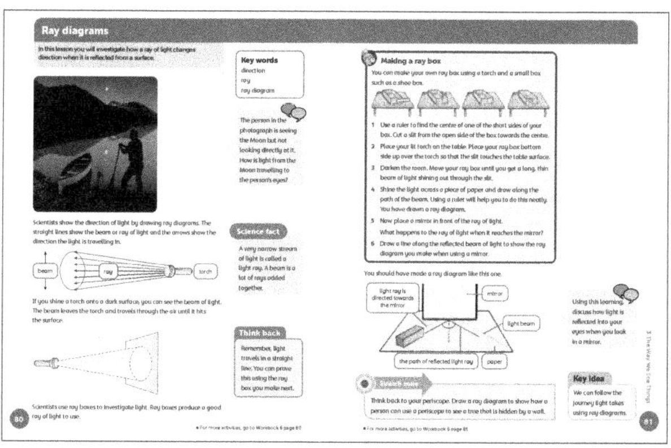

Supporting activities are in Workbook 6 pages 80–81.

Getting started

The focus of this lesson is to observe how light travels in beams or rays of light. Students will construct a ray box and observe how the light behaves when it leaves it. They will explore how ray diagrams can be drawn to represent the way that light is directed. They will also explore how the rays can be changed by using mirrors. Finally, they represent reflected light from a mirror using ray diagrams.

Language support

Ask students to read out the key words and discuss them with a partner. There are some unusual words in this lesson (ray and ray diagram), which students might not have experienced before. Use the diagrams on page 80 to explain 'ray' and 'beam'. Explain that a beam is made up of many rays but we can split a beam into smaller (narrower) rays. Encourage students to use the correct key words throughout the lesson.

Resources

Student Book: torches, small boxes such as shoe boxes; rulers; scissors; mirrors; large sheets of paper.

Workbook: magazines showing interiors with mirrors (or access to the internet); shoe boxes or other narrow boxes made of thin card; tissue paper; scissors; glue or sticky tape; lamps or candles.

Key words

direction   ray   ray diagram

Other words in the lesson

design   interior   pinhole camera

Scientific enquiry key words

Plan and/or carry out enquiries to answer questions

Make observations

Analyse data, notice patterns and group or classify things

Report and present findings

Lesson at a glance

The key teaching points for students in this lesson are:

- observe how rays can change direction when they are reflected from a mirror
- use ray diagrams to show the journey of light.

In the next lesson, students will investigate the angle of incidence and the angle of reflection.

The person in the photograph is seeing the Moon but not looking directly at it. How is light from the Moon travelling to the person's eyes?

Ask students to study the photograph showing the person at night and discuss the question in small groups.

> **Possible response:** *Students should suggest that the light from the Moon is being reflected by the smooth surface of the water.*

Ask students to read the information on page 80. They should use the diagrams to help them understand the information about how scientists show the direction of light using ray diagrams. The text also introduces the ray box. This is a piece of equipment that directs a ray of light.

Science fact A very narrow stream of light is called a light ray. A beam is a lot of rays added together.

Ask students to observe the diagram that clearly shows the beam and rays of light. To check understanding, you could draw a simple version on the board and ask students to guide you on labelling a ray and a beam.

Think back: Remember, light travels in a straight line. You can prove this using the ray box you make next.

Ask students to tell you how light travels. They should recall that it travels in straight lines. Ask them to tell you how they know this. Ask, 'What scientific evidence do you have to support your ideas?'

Investigation: Making a ray box

This investigation guides students on making their own ray box using a torch and a small box. You might want to show a scientific ray box or a picture of one from the internet to show what they are constructing.

"

Ask small groups to work together. They can follow the instructions independently, and you can support if needed. When all groups have prepared their ray box, darken the room. They need to move each ray box until they get a long, thin beam of light shining out through the slit.

Next, they shine the light across a piece of paper and draw along the path of the beam – this is a ray diagram. Encourage them to use a ruler.

Now they will place a mirror in front of the ray of light and observe what happens to the ray before drawing the reflected beam of light.

> *Possible response: Students should see the ray of light leaving the ray box in a straight line. They should observe the ray of light being reflected back in a straight line but at an angle from the mirror.*

Ask students to study the diagram at the bottom of page 81 showing the light leaving the ray box and how it is reflected from the mirror. Ask them if this represents their observations. They can label their own diagrams using the labels in the book.

Using this learning, discuss how light is reflected into your eyes when you look in a mirror.

Ask students to continue to work in their small groups. Encourage them to use the key words from the lesson in their discussion.

> *Possible response: Students should suggest that the light is directed towards the mirror, where the light rays are reflected and the path of the reflected light ray enters the person's eyes.*

Stretch zone: Think back to your periscope. Draw a ray diagram to show how a person can use a periscope to see a tree that is hidden by a wall.

The final activity in this lesson is the Stretch zone. Ask small groups to read the Stretch zone text and draw the scenario using a ray diagram. It would be useful to give students the periscopes they used in the last lesson to help with this.

> *Answer: Students should draw the light ray from the tree being reflected from the first mirror to the second one and then reflected to the eye by the second mirror.*

Key idea

We can follow the journey light takes using ray diagrams.

Read out the key idea or ask a volunteer to read it out. Ask another volunteer to describe how they used a ray box in the lesson. They should describe how the light behaves when it leaves the ray box and meets the surface of a mirror.

 **Mirrors and design (page 80)**

Mirrors are used for making the insides and outsides of houses and buildings look bigger and brighter. Inside buildings this is a part of the interior design. This activity directly covers the learning about light rays and reflection from the Student Book lesson.

Students are asked to find examples of mirrors used in interior design. They can cut out pictures from magazines (or print them from the internet) showing mirrors that help to make a room bigger or brighter.

Next, students are asked to find the main source of light in the picture they have chosen and to draw a ray diagram onto the picture to show how the light travels from the light source and where it travels. Then they explain how the mirror makes the room bigger or brighter.

> *Answer: Students should suggest that the light is reflected back into the room from the mirror. They should draw a straight line from the source of light to the surface of the mirror and then from the mirror to the eye. The mirror can make the room look bigger because it reflects the other side of the room which makes it appear to be extended.*

Finally, students can undertake research for the Stretch zone activity which is a revision of Student Book page 78.

> *Possible response: Students should suggest that we use mirrors to see behind us when driving and when brushing our hair, for example; dentists and doctors use mirrors to see in difficult places. Mirrors are also used to reflect light.*

 **Make a pinhole camera (page 81)**

Warning! Do not point your camera at the Sun.

Start by reading out the warning information and ask students to discuss why looking at the Sun would be so harmful.

Explain that the pinhole camera is one of the earliest types of camera. Tell students they are going to work in small groups to make their own camera.

Hand out a cardboard box, such as a shoe box, to each group and ask them to follow the instructions. Point out that the small hole (the pinhole) must be very small. They can always make it bigger if they need to but they must start with a very small hole. Explain that the tissue paper is acting as a screen. You can tell students that inside real pinhole cameras was light sensitive paper that captured the image and this could be developed into a photograph. Tell students they can observe the image on their screen but it will not be permanent or fixed. They could try to take a digital photograph of the screen with a smartphone though.

Once the pinhole cameras are constructed, allow students to take them outside and point them at bright objects. Ask them to draw what they notice about the images of the objects.

Possible response: Students should suggest that they can see the object in the camera but that it is upside down. Show students the diagram at the bottom of the page and explain that the light from the bottom of an object travels in a straight line through the pinhole and keeps going until it hits the screen. Light from the top of the object does the same. That is why top becomes bottom and bottom becomes top in the image.

Review and reflect

Ask students to write a note to themselves to remind them how they overcame some of the challenges in this lesson so that they can refer to this when they are challenged in future lessons.

Extra activities

1 Students draw two ray diagrams to show the direction of a ray of light from objects in the room.

2 Students take their ray box home and show people how it works. They record the rays of light leaving the ray box and how they react to a mirror and one other object.

Differentiation

Supporting: You can provide students with pre-made ray boxes that shine a very thin beam of light to make observations easier.

Consolidating: Students can practise drawing ray diagrams and label the rays that enter the mirror and the rays that leave the mirror.

Extending: Students could explore what happens to light that hits mirrors that are not flat.

Differentiated outcomes	
All students	should be able to state that light changes direction when reflected from a surface
Most students	will be able to show reflection of light as a ray diagram
Some students	may be able to label ray diagrams including the path of reflected light

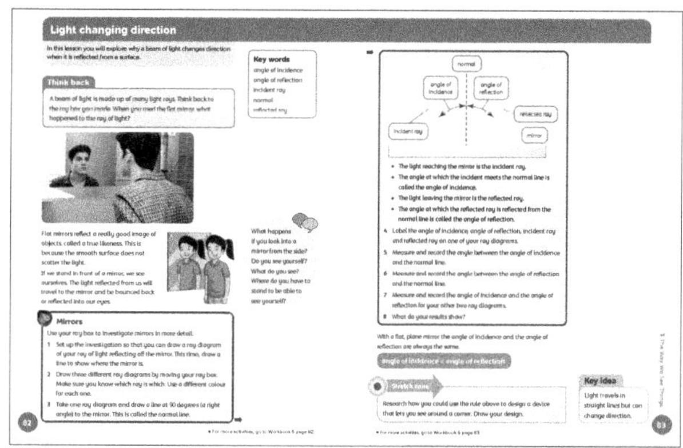

Supporting activities are in Workbook 6 pages 82–83.

Getting started

In this lesson students look at how we can use mirrors to reflect light. The image reflected is a true likeness to the objects using a normal or flat mirror. Students explore the normal line when using a mirror and how this is shown on a diagram. They draw and label ray diagrams showing the normal line, angle of incidence and angle of reflection. They also explore how the image is not a true reflection when the mirror is not flat.

Language support

You could create a wall display with the help of students to display the key words. This could be visually stimulating. For example, download and display pictures of ray diagrams. Help students to tell the difference between incident ray and reflected ray by pointing out that the incident ray goes into the mirror.

Resources

Student Book: ray boxes made previously; writing materials; large sheets of paper; protractors or other angle measurers; rulers.

Workbook: flexible mirrors or mirrors on a roll; pieces of thick card; a hole punch or pencil; modelling clay or reusable adhesive; torches or lit bulbs; a small area of blank wall or large pieces of card covered with paper.

> **Key words**
>
> angle of incidence   angle of reflection
>
> incident ray   normal   reflected ray
>
> **Other words in the lesson**
>
> likeness   scatter

Scientific enquiry key words

Plan and/or carry out enquiries to answer questions

Recognise and control variables

Make observations

Take measurements, using equipment accurately

Record data and results

Analyse data, notice patterns and group or classify things

Report and present findings

Draw conclusions and give explanations

Identify causal relationships

Lesson at a glance

The key teaching points for students in this lesson are:

- flat mirrors reflect light to give a true reflection
- the angle of reflection and angle of incidence can be identified on a ray diagram.

In the next lesson, students will study refraction and colour.

Think back: A beam of light is made up of many light rays. Think back to the ray box you made. When you used the flat mirror, what happened to the ray of light?

Ask students to read the Think back text and ask volunteers to answer the question. This will help to make sure all students are clear about their prior learning.

> **Possible response:** *Students should suggest that the ray of light was reflected back from the mirror.*

Ask students to read the text above the investigation box on page 82 and study the pictures. Point out that when a mirror is flat with a smooth surface the image reflected is known as a true image. Ask students where the light comes from when we look in a mirror. Elicit that the light from a light source is reflected from us to the mirror.

What happens if you look into a mirror from the side? Do you see yourself? What do you see? Where do you have to stand to be able to see yourself?

Students can observe the picture of the student sideways onto the mirror. This should act as a hint to address the discussion questions.

> **Possible response:** *Students should suggest that when they look from the side they see a side view of themselves only if they change their position so that they can see the mirror.*

Investigation: Mirrors

Students will use the ray box they made in the previous lesson to investigate mirrors in more detail. Ask them to follow the instructions and only offer support when needed. This will encourage independent learning. Remind students that they have set up a ray box before and encourage them to draw any lines with a ruler and to measure any angles accurately.

Explain that to make sure the ray reaching the mirror (incident ray) hits the mirror at the same place every time, a line is drawn called the normal line. This is at 90° to the mirror and can be used as the target point and also the reference point for measuring the angles for the rays. You may wish to demonstrate this for the first incident and reflected ray.

> **Possible response:** *Students should suggest that they draw the mirror as a straight horizontal line. This can be drawn with slanted lines to show that it is a mirror. The normal line is at 90° to the mirror. Students should observe that the angle of the ray reaching the mirror (incident ray) is always the same as the angle of the ray leaving the mirror (reflected ray).*

Point out the text at the bottom of page 83 and ask students whether or not the results of their investigation support this idea. They should agree that the angle of incidence and the angle of reflection are the same. Show students how this is often written like a mathematics equation: angle of incidence = angle of reflection. Ask students why this only occurs with a very flat (plane) mirror. Elicit that with a rough or curved mirror the rays will bounce off at other angles.

Stretch zone: Research how you could use the rule above to design a device that lets you see around a corner. Draw your design.

Ask students to read the Stretch zone task. Allow them to work with a partner to discuss their ideas and then ask them to produce some diagrams to share with the class. Remind them of the periscope they made and investigated earlier in the unit.

> **Possible response:** *Students should suggest that the mirrors should be positioned so that they are at 45° to each other.*

Key idea

Light travels in straight lines but can change direction.

Read through the key idea or ask a volunteer to read it out to the class. Check understanding by asking, 'Describe how the mirror reflects light. How would you show how the ray of light is reflected from a normal mirror? What angle should the normal line be at to the mirror and what fact do you know about the angles of incidence and reflection?'

Having fun with mirrors (page 82)

Ask students to look at the photograph of the person in front of the mirror. Arrange students in pairs so they can discuss what has happened to the mirror image. Encourage them to look at the shape of the mirror and think back to their work with plane mirrors and reflection in the Student Book investigation. Explain that students are going to use some flexible mirrors or mirrors on a roll to investigate the images produced when mirrors are not flat.

Allow students to bend the mirrors in different ways and observe the reflections. Ask them to record their findings in the table.

Once the table is completed students can complete the Stretch zone task. Ask them to look for some examples of shaped mirrors and then research why some are called concave and others are called convex.

> *Possible response: Students should suggest that any mirror that does not have a flat surface reflects an image that is not a true likeness. The reflected image can be changed by bending the mirror. A concave mirror bends inward like a cave and a convex mirror bends outwards.*

Does light travel in straight lines? (page 83)

Show students the diagram at the top of the page and explain that they are going to use the pieces of card and a torch or light bulb to investigate how light travels. Ask students to work with a partner or in small groups and to follow the instructions. Once the cards are made and set up, ask students to shine a torch or a lit bulb onto the cards. They can then investigate by moving the cards around until light can shine through all of the cards. They should then explain their answer and draw their apparatus so it shows the light as a beam leaving the light source and hitting the wall or paper screen.

> *Possible response: Students should suggest the holes have to be exactly aligned to allow the light to pass through. If they are not aligned, the light will stop as it does not go around corners to find the hole.*

Review and reflect

Ask students to reflect on the investigations they carried out and think about how well they worked as a team member and how well they communicated with each other. Ask them to think of one way of working that they are proud of and one thing they could improve on. Tell them this is their target and they can see if they can meet this target in the next few lessons.

Conclude the lesson by asking students to complete question 5 in the 'What have I learned about the way we see things?' activity on page 103 of the Student Book and the third statement on page 102 of the Workbook. (See also the teaching notes in this Teacher's Guide, pages 134–136.)

Extra activities

1 Ask students to research how designers and architects use mirrors inside buildings to allow light through the building and to give a feeling of space. They could download photographs of some examples.

2 Students change the shape of bendable mirrors and observe how it changes the shape of their face and the faces of the people at home. They see who can make the funniest reflection.

Differentiation

Supporting: While students observe their reflections, talk to them about how the light is being reflected from each surface.

Consolidating: Allow students plenty of time to practise measuring the normal line and angles of incidence and reflection in the investigation.

Extending: Encourage students to observe reflected images using concave and convex mirrors to extend their understanding of how reflections change if it is not a flat mirror.

Differentiated outcomes	
All students	should be able to explain how an image is reflected using a flat mirror
Most students	will be able to draw a flat mirror, a normal line and angles of incidence and reflection on a ray diagram
Some students	may be able to explain how angles of reflection change when mirrors are not flat

More on light

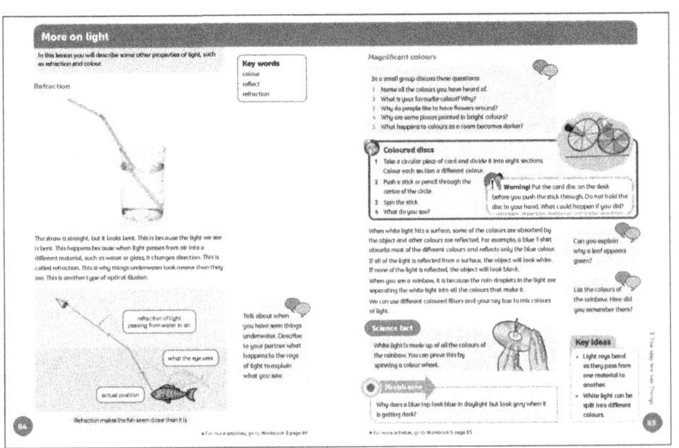

Supporting activities are in Workbook 6 pages 84–85.

Getting started

In this lesson students learn about other properties of light. They observe refraction and how water bends the light or refracts it. They learn that light travels differently in different materials. Students also investigate how white light is made up of all the colours of the spectrum. They make discs and spin them to show how they appear white. This can be clarified by spinning the discs faster. You could use an electric drill to demonstrate this.

Language support

You could produce a collection of key word cards for the lesson that students can stick next to images that you have provided. Alternatively, you could use flashcards with the key words and ask students to remind you what each word is and what it means.

Resources

Student Book: writing materials; circular pieces of card; pieces of dowelling or pencils.

Workbook: coins, cups; water; torch; coloured objects; black object; coloured acetate or filter; writing materials.

> ### Key words
>
> colour   reflect   refraction
>
> **Other words in the lesson**
>
> bent   filter   optical illusion   rainbow   separating   white light

Lesson at a glance

The key teaching points for students in this lesson are:

- refraction is where light is bent or changes direction at the boundary of a new material
- white light is made up of the different colours of the spectrum.

In the next lesson, students will learn about transparent, translucent and opaque materials.

Refraction

Ask students to read the information about refraction and observe the photograph and diagram on page 84. You could demonstrate the straw to show how the light is changed by the water. In air it looks different to when it goes into the water. This demonstrates refraction, which is a property of light.

Set up a pencil or a straw in a glass half filled with water and let students look at it from the side. Point out that when we look at the bottom part of the pencil the light has travelled from the pencil and through the water – but then it has to pass through the surface of the water and this makes it bend or refract. To avoid future misconceptions, stress that any refraction happens as light passes from one substance to another – such as glass to air or water to air. It might help if you ask students to imagine these surfaces are like weak mirrors – they cannot bounce the light back (reflect it) but they can bend it as it passes through.

 **Talk about when you have seen things underwater. Describe to your partner what happens to the rays of light to explain what you saw.**

Remind students to use the ray diagram of refraction at the bottom of page 84. You can ask students to draw their own ray diagram but with the object they are thinking of in place of the fish.

Magnificent colours

The first part of this topic is a discussion task. This is intended to check on students' understanding about colour and light before they carry out the investigation. Split the class into small groups to discuss the questions one by one, and you can circulate to check on their baseline knowledge about colour. You can write the colours students mention on the board as a reminder for the investigation. It is important that you clarify the answer to question 5 to the class.

In a small group discuss these questions:

1 **Name all the colours you have heard of.**

2 **What is your favourite colour? Why?**

3 **Why do people like to have flowers around?**

4 **Why are some places painted in bright colours?**

5 **What happens to colours as a room becomes darker?**

Investigation: Coloured discs

Warning! Put the card disc on the desk before you push the stick through. Do not hold the disc in your hand. What could happen if you did?

This investigation encourages students to think about the relationship between light and colour. This can be an independent task. Before they start, ask a volunteer to read out the warning information. Ask students to tell you what could happen if they hold the disc in their hand.

Students can then follow the instructions to divide the circular card into eight sections and colour each section a different colour. They then push a stick or pencil through the centre of the circle and spin the stick. Encourage them to talk about their observations.

You might want to attach a coloured disk to a drill bit on an electric drill. Make sure the card is securely in place. The drill will be able to spin the disc so fast that students can see the card looks white.

Ask students to read the information below the investigation box on page 85. Explain that some colours are absorbed by an object and we see the colours that are reflected. White objects reflect all of the colours. There are many things – including raindrops – that can disperse or separate white light and this is how we see the colours of the rainbow.

Can you explain why a leaf appears green?

Ask students to work with a partner. You can allow them to look at leaves outside or on a potted plant in the room. Ask them to think about which colours in light are hitting the leaf and which colour or colours are reflecting from the leaf into their eyes. Remind students of the rainbow activity they completed on page 69 of the Workbook and let them look back at this. They could cover all of the colours other than green to give them a clue.

List the colours of the rainbow. How did you remember them?

If students have referred back to page 69 of the Workbook to help with the discussion above, then they should find it easier to recall the colours. Suggest that they use a mnemonic to help. Point out that this is a saying that is easy to remember where the first letter of each word is the first letter of the words they are trying to remember. For example, Really Orangey Yoghurt Gives Best In Value. Another way is to think of an imaginary person called Roy G Biv.

Science fact White light is made up of all the colours of the rainbow. You can prove this by spinning a colour wheel.

Ask a volunteer to read out the Science fact. This acts as a summary of the learning about colours in the lesson.

Stretch zone: Why does a blue top look blue in daylight but look grey when it is getting dark?

Ask students to discuss the Stretch zone question with a partner. You can allow them to test their ideas by giving them a coloured object and then darkening the room by closing blinds or curtains. Remind them that colours are seen when only some of the colours in light reflect from a surface. Ask some volunteers to stand up and share their ideas with the class.

> *Possible response: Students should suggest that in bright daylight all of the colours of the rainbow are hitting the blue top and only the blue part of light is reflecting to their eyes, but when it is getting dark the lighter parts of the spectrum are lost first so only the darker colours (indigo and violet) reflect into the eyes. Students may be interested to know that the human eye cannot see purples as well as lighter colours so objects appear grey in poor light.*

Key ideas

- *Light rays bend as they pass from one material to another.*
- *White light can be split into different colours.*

Read out the key ideas or ask volunteers to read out a key idea each. Ask students to think back over the lesson and then on a blank piece of paper write a description of what white light is. Ask them how they would show that the light we see is actually made up of the rainbow or spectrum of colour. Ask students to list the seven colours of the rainbow, which we accept as being the parts of white light, from memory.

Workbook activities

Investigating refraction (page 84)

Students will use refraction to see a hidden object: a coin in a cup. They should follow the instructions on the worksheet and observe what happens. They are asked to use their knowledge of refraction to explain why adding water makes a difference to the appearance of the coin. They are also asked to draw ray lines on the diagram to show their ideas. They can use the drawing on page 84 of the Student Book to help with this.

> *Possible response: Students should explain that adding water causes the light to refract because light travels differently in different materials. They should draw a diagram that shows the light travelling from the coin in a straight line to the edge of the water. It then changes direction slightly when it hits the air and travels into the air in a different direction, closer to the eye than without water.*

Students extend their learning by researching how refraction can make objects underwater appear much closer than they are. They should draw some examples and you can display their drawings.

> *Possible response: Refraction can make objects look closer than they actually are because we view them in a different medium to air which is how we would normally see them.*

Colour investigation (page 85)

This activity consolidates learning about colour. Students investigate colours that are reflected from objects using a torch, coloured objects and colour filters. They can follow the instructions to complete the table and compare what they observe when a torch is shone on  coloured objects and a black object. They then observe the same objects when a coloured acetate or filter is used to cover the torch.

Encourage students to think about how light is absorbed and reflected when they are carrying out the steps. Point out to students that black is the absence of light and not a colour.

> *Possible response: Students should suggest that the colours will reflect their true colour when using white light. When using coloured filters, they can look different. A red filter on a red object will reflect red. A red filter on any other object will look black. This is because all of the colours are absorbed into the object so none is reflected making it look black.*

Review and reflect

Ask students to sit opposite a partner. Ask them to think about what they have learned and then take it in turns to tell their partner about two examples. They should start by saying, 'One thing I have learned this lesson is …' and the second time they say, 'Another thing I have learned is …'.

Conclude the lesson by asking students to complete question 3 in the 'What have I learned about the way we see things?' activity on page 102 of the Student Book and the fourth statement on page 102 of the Workbook. (See also the teaching notes in this Teacher's Guide, pages 134–136.)

Extra activities

1 Students carry out a magic show for the people at home. They select some coloured objects and a range of coloured tissue paper or fabrics. They say, 'I can magic away the colour of this object.' They ask for comments and then show what happens.

2 Students make a poster about white light on the top half of the paper and then the absence of light (black) on the bottom half. They include how objects look and facts and information about the properties of light.

Differentiation

Supporting: Students should practise spinning their disc until they see it turn white. Give them lots of opportunities to observe how objects look in air and water.

Consolidating: The Workbook activities can be used to help students understand refraction in water and air, and to support learning about colour and what white light is.

Extending: Students can practise making objects appear black using a range of filters. They observe refraction using different liquids, for example baby oil.

Differentiated outcomes

All students	should be able to observe refraction and white light
Most students	will be able to describe what refraction is and what white light is
Some students	may be able to explain how refraction changes the image of an object and how to make a coloured object appear to have no colour

Can we see through it?

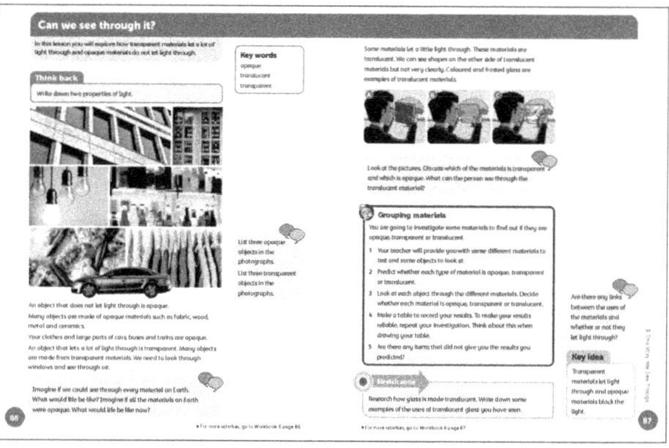

Supporting activities are in Workbook 6 pages 86–87.

Getting started

This lesson introduces students to the idea that some materials allow light to pass through them and others block light. Materials are described as transparent if they allow light to pass through them and opaque if they block it completely. Students also investigate materials that allow some light to pass through – these are described as translucent. Students group or classify materials into these categories and begin to consider how these properties are used.

Language support

The key word 'translucent' might be unknown to most students. They might have heard of 'transparent' and they studied opaque materials in Year 3 but check they fully understand what they mean. These words are not easy to spell. A strategy you can use is to write the words on the board but leave some gaps so letters are missing. Students can copy the words and then discuss which letters are missing. They can self-check by referring to the key words box on page 86 of the Student Book.

Resources

Student Book: samples of materials that are opaque, transparent and translucent; selection of small objects.

Workbook: selection of objects that are opaque, transparent or translucent.

> **Key words**
>
> opaque   translucent   transparent
>
> **Other words in the lesson**
>
> frosted

Scientific enquiry key words

Plan and/or carry out enquiries to answer questions

Make predictions

Make observations

Record data and results

Analyse data, notice patterns and group or classify things

Draw conclusions and give explanations

Lesson at a glance

The key teaching points for students in this lesson are:

- materials let different amounts of light pass through them
- materials can be classified or grouped as transparent, translucent or opaque.

In the next lesson, students will investigate shadows.

Think back: Write down two properties of light.

Ask students to write down their thoughts. This activity can be used to support them to recall prior learning about light.

> *Possible response: Students should suggest that light travels in a straight line, it can be reflected or refracted, and white light is made up of a spectrum of colour.*

List three opaque objects in the photographs. List three transparent objects in the photographs.

Arrange students into pairs or groups of three or four for discussion work. A useful strategy is to start with students in pairs and then move pairs together to make small groups so students can share their ideas and discussions with others. Ask students to study the photographs and then discuss the questions.

> *Possible response: Students may suggest: opaque = clothing, fruit, metal and rubber on the car, concrete in the building; transparent = windows in the buildings, bottles, light bulbs, plastic food bags, the car windscreen.*

Ask students to read through the text beneath the photographs on page 86. Volunteers could read out a sentence each. Ask them to use the information in the text to answer the discussion questions that follow.

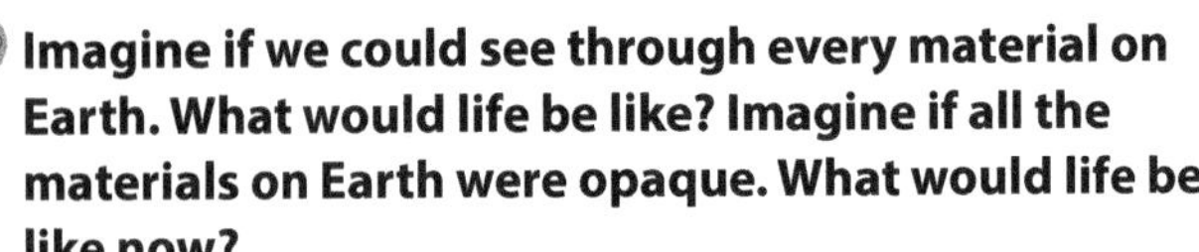

Imagine if we could see through every material on Earth. What would life be like? Imagine if all the materials on Earth were opaque. What would life be like now?

Allow students to work with a partner to talk about the questions.

> *Possible response: Students should suggest we wouldn't have any privacy if all of the materials were transparent. If they were all opaque, we wouldn't see outside or be able to drive in a car.*

Students can continue to work with their partner or in their group to read the information at the top of page 87 about translucent materials. If possible, show students some real examples of coloured and frosted glass. Ask them where they have seen examples before. Elicit that translucent materials are useful because they allow light through but still provide some privacy.

Look at the pictures. Discuss which of the materials is transparent and which is opaque. What can the person see through the translucent material?

Point out the drawings on page 87 and ask students to discuss the questions to review their understanding of these properties of materials.

> *Answer: A opaque; B translucent; C transparent. The person can see the outline of the tree but not all its detail through the translucent material.*

Investigation: Grouping materials

Ask students to work in pairs or groups of three of four. Explain that they are going to investigate some materials to find out if they are opaque, transparent or translucent. You should provide them with some different materials to test and some objects to look at. They should predict whether each type of material is opaque, transparent or translucent. Then they look at each object through the different materials to decide whether each material is opaque, transparent or translucent. Encourage them to record their results in a table. Ask them how they can make their results more reliable. Elicit that they should repeat the investigation. Finally, they analyse their results to assess whether any items did not give the results they predicted. Check their answers and discuss any misunderstandings.

Are there any links between the uses of the materials and whether or not they let light through?

Ask students to read through the discussion question and work with their group to consider their investigation results. Ask them to link the property of the material to its use.

Stretch zone: Research how glass is made translucent. Write down some examples of the uses of translucent glass you have seen.

Allow pairs access to the internet and any other suitable resources, such as catalogues, to research how glass is made translucent. Remind them that they don't need to write down the process, but they should write down examples of the uses of translucent glass. Encourage them to share their findings with other pairs.

Key idea

Transparent materials let light through and opaque materials block the light.

Read through the key idea or ask a volunteer to read it out to the class. Ask students to describe how we can group materials according to the amount of light they allow through. Ask them to give one example of each use of the three properties of material they have learned about in the lesson.

Workbook activities

Does it let light through or block light? (page 86)

This activity supports language development as well as helping to define the terms used to describe the properties of materials from the lesson. In the first task, students complete the sentences using the words in the box. Remind them that they will need to use each word more than once.

In the second task, they look at the pictures and label each picture with the correct word from the box to describe the material.

Translucent, transparent and opaque objects (page 87)

This worksheet encourages students to identify and discuss the three properties of materials and to survey their local area.

The first task is a discussion activity where students use the words 'translucent', 'transparent' and 'opaque' to describe different materials shown in a picture.

The second task asks students to apply their learning to survey and collect objects made from each type of material. Remind them to explain to the person they are working with why they know the object is made from opaque/transparent/translucent material. They are also asked to prove how that object has that property. Finally, they are asked to draw the two transparent objects they have found and label the drawings to explain what the objects are used for.

Review and reflect

Discuss the outcomes of the grouping materials task with students during and at the end of the process. Encourage them to identify improvements that they can make to their testing and point out that this process of testing and redesigning is a normal part of the way that scientists and engineers work. This will help students to develop a positive approach to learning by understanding that learning is a process that will improve with practice and reflection.

Conclude the lesson by asking students to complete question 1 in the 'What have I learned about the way we see things?' activity on page 102 of the Student Book and the fifth statement on page 102 of the Workbook. (See also the teaching notes in this Teacher's Guide, pages 134–136.)

Extra activities

1 Students survey their homes for the different categories of materials. They find the most unusual example for each category.

2 **Computing link:** Students find out if there are any buildings that are completely made out of transparent materials. They make a computer presentation including pictures and facts.

Differentiation

Supporting: Give students lots of different objects to practise grouping into transparent, translucent and opaque materials.

Consolidating: Encourage students to look for examples of the different categories of materials when they are in the playground.

Extending: Give students photographs or cut-outs of magazines and ask them to identify why materials with one of these properties were selected for use.

Differentiated outcomes	
All students	should be able to group materials into transparent, translucent and opaque
Most students	will be able to identify materials with transparent, translucent and opaque properties
Some students	may be able to explain why the property of the material is useful in specific examples

Making shadows

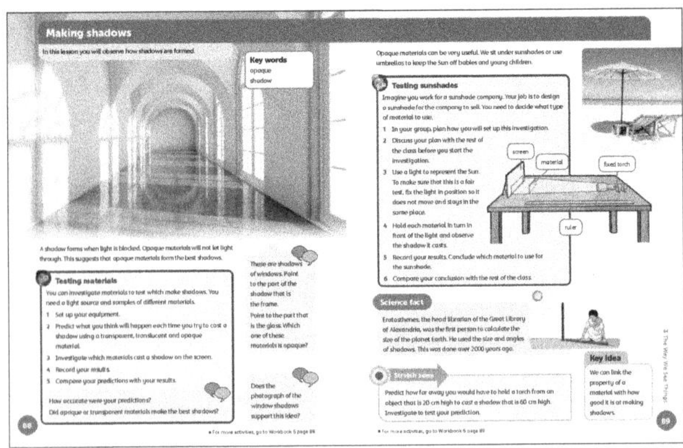

Supporting activities are in Workbook 6 pages 88–89.

Getting started

This lesson builds upon the previous lesson and considers what happens when objects block light. Students review work from Year 3 and investigate the shadows cast by transparent, translucent and opaque objects. Students are given an opportunity to apply their knowledge of shadows in designing and testing sunshades.

Language support

Create a set of matching cards with words and definitions to ensure that students are confident in the use of the main ideas covered in this lesson. Words should include opaque, translucent, transparent and shadow.

Resources

Student Book: torches; cardboard or other material to use as a screen; samples of materials that are opaque, transparent or translucent; rulers.

Workbook: torches or lamps; selection of objects that are opaque, transparent or translucent; large sheets of white paper or a white wall.

Key words

opaque   shadow

Other words in the lesson

translucent   transparent

Scientific enquiry key words

Plan and/or carry out enquiries to answer questions

Make predictions

Make observations

Record data and results

Analyse data, notice patterns and group or classify things

Report and present findings

Draw conclusions and give explanations

Identify causal relationships

Lesson at a glance

The key teaching points for students in this lesson are:

- when light is blocked a shadow is formed
- opaque materials make shadows.

In the next lesson, students will apply their knowledge of shadows to develop shadow games.

These are shadows of windows. Point to the part of the shadow that is the frame. Point to the part that is the glass. Which one of these materials is opaque?

Arrange students into pairs or groups of three or four. Ask them to study the photograph at the top of page 88 and read the information about how opaque materials block light and form shadows.

Answer: Students should suggest the shadow is the opaque window frame and the white part is the transparent glass.

Investigation: Testing materials

Ask students to work with their group to investigate materials to test which make shadows. They will need a light source and samples of different materials.

They will need to plan and set up their equipment. Allow independent working here and prompt only if needed. Students should predict what they think will happen each time they try to cast a shadow using a transparent, translucent and opaque material, and then carry out the investigation. Remind students to record their predictions after they have observed the equipment set up and the samples they will investigate so they can compare their predictions with their results.

Answer: Students should predict and find that the opaque materials will make the best shadows. Their results should support their predictions.

How accurate were your predictions? Did opaque or transparent materials make the best shadows?

Ask students to discuss these questions with their investigation group to assess their skills at making predictions based on their scientific knowledge.

Possible response: Students should suggest their predictions were quite accurate. However, some translucent materials might have made shadows and students may have predicted that they wouldn't.

Does the photograph of the window shadows support this idea?

Students can work with their investigation group to talk about the shadows formed in the photograph of the windows.

Possible response: Students should suggest that this supports their predictions and ideas. Some parts of the window frame have not made shadows as well as others.

Read to students the top two lines of text on page 89. Tell them they are now going to apply their learning and become product designers: they are going to test the best material to make a sunshade.

Investigation: Testing sunshades

Students should imagine their job is to design a sunshade for a company to sell. They need to decide what type of material to use. In their group, they will plan how to set up this investigation, and will communicate, collaborate and problem solve as a group.

They should follow the instructions given. Make sure they discuss their plan with the rest of the class before they start the investigation. They could use the diagram to help them set up the investigation. Give them prompts to make sure that they plan a fair test. Ask them to make sure they record their results clearly so they can conclude which is the best material to use for the sunshade. Compare the conclusions as a class discussion.

Possible response: Students should suggest that an opaque material that is thick and dark in colour would make the best shadow on the screen and will therefore make the best sunshade.

Science fact Eratosthenes, the head librarian of the Great Library of Alexandria, was the first person to calculate the size of the planet Earth. He used the size and angles of shadows. This was done over 2000 years ago.

Read out the Science fact or ask a volunteer to read it out. Ask students if they think it would be possible to find the shadow of the Earth. Then ask them to look at the diagram which shows how Eratosthenes used the shadows cast by a stick on the ground.

Stretch zone: Predict how far away you would have to hold a torch from an object that is 20 cm high to cast a shadow that is 60 cm high. Investigate to test your prediction.

This Stretch zone activity is pre-learning for the next few lessons on shadows in this unit. Students can work with their investigation group to plan and carry out this activity. Ask them to measure the distance from the source of light and the size of the shadow to help investigate their predictions.

> *Possible response: Students should suggest the object would need to be quite close to the torch. The shadow would be larger than the object but fuzzy around the edges and not very clear.*

Key idea

We can link the property of a material with how good it is at making shadows.

Read out the key idea and ask students to turn to a partner and take it in turns to describe how we can make shadows using opaque materials. Elicit that the shadow is cast when the object blocks the light. Light cannot go around corners, so a shadow is cast as the light stops at the boundary of the opaque materials.

Workbook activities

Does it make a good shadow? (page 88)

The worksheet is made up of two activities, firstly a discussion task about how shadows are made, what kind of material makes the best shadow and why transparent materials do not make good shadows.

> *Possible response: Students should recall that opaque materials cast good shadows and transparent materials do not. This is because shadows are only made when light is blocked – so transparent materials, which allow light to pass through them, do not cast shadows. You can remind students that there are materials between opaque and transparent that let some light through and therefore cast some weak shadows. These are translucent materials.*

Second, the investigation allows students to test which materials make a good shadow. They are asked to find five objects to predict and test whether each object will make a good shadow or a bad shadow. They use the table provided to complete their predictions and their test results. They then decide if any of the objects made good shadows and if any of the objects surprised them.

> *Possible response: Students should suggest that opaque materials make the best shadows.*

Finally, students are asked to think about where they see shadows around the school and link this to their investigation.

> *Possible response: Students will observe shadows anywhere in the school where there is a source of light and opaque objects stopping the light. Outside they might see shadows made by buildings blocking sunlight. This should support the idea that opaque objects make good shadows.*

Investigating translucent and opaque materials (page 89)

This activity supports the investigation on page 88 of the Student Book. Students choose three translucent materials and three opaque materials. The worksheet takes them through the process of planning the investigation and drawing a diagram of the equipment. Ask, 'Are the shadows made by the opaque and translucent materials the same?' Students record their thoughts and are asked to then draw pictures to show the shadow a translucent object would make and the shadow an opaque object would make.

> *Possible response: Students should suggest that opaque materials make better shadows than translucent materials.*

The Stretch zone activity asks students to suggest one situation where a translucent material would be better than an opaque one. They can record their answer on the page.

> *Possible response: Translucent materials are useful in places where light is useful but you don't want anyone to be able to have a clear view inside. Bathroom windows are a good use of translucent glass.*

Review and reflect

You can ask students to produce a scientific report of their investigation. This can include a prediction, method, results and their conclusions. You can read through these and look for evidence of sensible predictions and an understanding of variables.

Encourage students to celebrate success during the lesson as this raises confidence and engagement. It also reminds them that learning is a process and they can learn from the things that did not go as well as they hoped.

Conclude the lesson by asking students to complete question 4 in the 'What have I learned about the way we see things?' activity on page 102 of the Student Book. (See also the teaching notes in this Teacher's Guide, pages 134–136.)

Extra activities

1 Students stand outside in sunlight. They investigate how to make their shadows smaller and bigger. They make a poster to show how they did this. Point out to students that this will help them to recall work in Year 5 when they studied how shadows change during the day. Also tell students they will get a chance to study this in more detail in a forthcoming lesson.

2 Students make a collage of pictures cut out from magazines to show what materials make the best shadows.

Differentiation

Supporting: Display photographs of sunshades, blinds and curtains being used to create shade to show students some examples in context.

Consolidating: Students can make sketches of areas of shadow around the school and label the materials that are casting the shadows.

Extending: Students can research the uses of translucent materials based on their property of letting some light through.

Differentiated outcomes	
All students	should be able to predict which objects will cast shadows
Most students	will be able to state that opaque materials cast the best shadows as they block light
Some students	may be able to relate the properties of transparent, translucent and opaque materials to the type of shadow they will cast

Shadow games

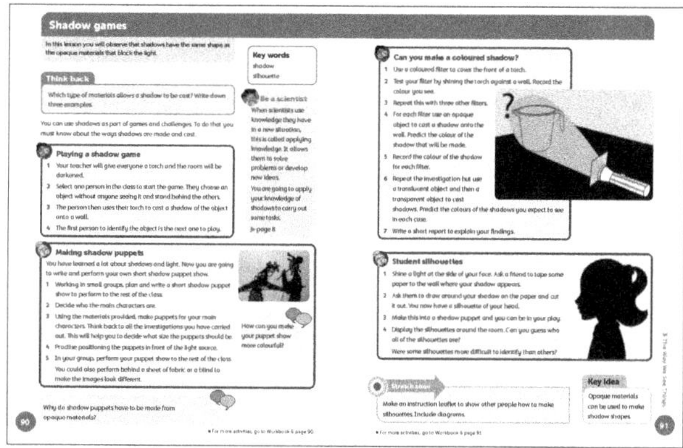

Supporting activities are in Workbook 6 pages 90–91.

Getting started

The aim of this lesson is to show students how scientists plan an investigation from what they know and observe. Students will learn that scientists can look at more than one factor or variable at a time but they still need to control their investigations. Students will explore how shadows can be used to make games, puppet shows and silhouettes.

Language support

Students are familiar with the word 'shadow', but write the word 'silhouette' on the board and point out that they saw an example of this in the unit's introductory lesson. Remind them it is named after the artist who invented it and let them practise saying the word out loud.

Resources

Student Book: torches; materials to make puppets for a shadow puppet show; sheets or blinds (optional); coloured filters; selection of opaque, transparent and translucent objects; large sheets of paper.

Workbook: sharp pencils; large pieces of paper; torches or lamps; pieces of black card, scissors; sticky tape.

> ### Key words
>
> shadow   silhouette
>
> **Other words in the lesson**
>
> filter   opaque   puppet

Scientific enquiry key words

Plan and/or carry out enquiries to answer questions

Make predictions

Make observations

Record data and results

Report and present findings

Lesson at a glance

The key teaching points for students in this lesson are:

- shadows have many different uses
- coloured shadows can be made using filters.

In the next lesson, students will investigate how the size of a shadow changes when it is nearer or further away from the light source.

Think back: Which type of materials allows a shadow to be cast? Write down three examples.

Ask students to work with a partner to recall the properties of materials that allow shadows to form. Prompt them to recall their investigations in previous lessons.

Answer: Students should suggest that opaque materials cast shadows. They can give any three examples that are made from opaque materials, for example a plastic toy, wooden block or stone.

This lesson is made up of four investigations and a final challenge activity.

Be a scientist: When scientists use knowledge they have in a new situation, this is called applying knowledge. It allows them to solve problems or develop new ideas. You are going to apply your knowledge of shadows to carry out some tasks. (Student Book, page 8)

Read out the Be a scientist feature to let students understand that they are going to use what they have learned in the previous lessons. They are going to use or apply their learning to play some games with shadows.

Investigation: Playing a shadow game

Students should work in large groups, or as a whole class. Provide torches and objects for students to make shadows with. Make sure you have cleared the floor of any objects that could be a trip risk. Students take turns to cast a shadow while the others guess the object.

Select one person in the class to start the game. The first person to identify the object is the next one to play.

Investigation: Making shadow puppets

Students work with a group to make up a shadow puppet show. This is an opportunity for students to work as a team and respect the decisions of others as they plan, write and perform together.

Encourage them to use the materials provided to make puppets for the main characters. They should think back to all the investigations they have carried out to help them to decide what size the puppets should be. They will need to practise positioning the puppets in front of the light source and will then perform the puppet show to the rest of the class. If you have a sheet of fabric or a blind available, they can perform the show behind it to make the images look different.

How can you make your puppet show more colourful?

Ask students to work with their group to reflect on their puppet show and discuss if they observed any other colours. Ask how they could introduce colour into their show.

Possible response: Students should suggest the only colours they observed were shades of black because all of the light was blocked by the puppet. They may suggest using a colour filter or a coloured sheet or blind to make the show more colourful.

Why do shadow puppets have to be made from opaque materials?

This discussion can be led by you with the whole class as a quick recap before moving onto the next investigation. Encourage students to think back to their earlier work on shadows and the materials they tested.

Possible response: Students should suggest that the material has to be opaque to cast a good shadow. Light travels in straight lines so it cannot curve around the puppet. The opaque material stops the light rays and so a shadow is formed.

Investigation: Can you make a coloured shadow?

This investigation challenges students to try to make a coloured shadow. The only way to make a coloured shadow would be to block some of the spectrum of colour. This would be a very difficult task. The principle of making a shadow is that the light is blocked completely. Students will find that they cannot successfully make a coloured shadow.

They will use four coloured filters to cover the front of a torch and test their filters by shining the torch against a wall. They need to record the colour.

For each filter, they will use an opaque object to cast a shadow onto the wall. They should predict the colour of the shadow that will be made then test it and record the actual colour of the shadow for each filter. Remind students that an opaque object forms a shadow by blocking all of the light and this should act as a powerful clue.

They will repeat the investigations using a translucent object and then a transparent object to cast shadows. Again, they should predict the colours of the shadows they expect to see in each case. Remind students that transparent materials allow all of the light to pass through and ask them if they expect to see a shadow. You can also ask, 'How much light does a translucent material allow through? If this casts a weak shadow, what colour might it be?'

Students are asked to write a short report to explain their findings. This is a tricky concept, so you may need to discuss the challenges of this activity as a class before they write a report.

Investigation: Student silhouettes

Students will work with a partner to produce their own silhouettes. Some students find this very tricky and resilience plays a huge part in producing a good silhouette. You may need to support pairs in following the instructions.

Students might find it difficult to create a true likeness using a silhouette. Encourage them to use other clues to support them in identifying their classmates.

Display some silhouettes around the room and discuss the technique. Concentrate on the process and principle of forming shadows rather than the quality of the silhouette.

Stretch zone: Make an instruction leaflet to show other people how to make silhouettes. Include diagrams.

Ask students to work with their group to carry out this task. Remind them of the steps they took in the investigation. This activity should help them to consolidate their understanding of this technique.

Key idea

Opaque materials can be used to make shadow shapes.

Ask students to close their books and count to 10. Then you can ask them to turn to a partner and tell them what they predict the Key idea for the lesson should be. After this, they can open the book and read the Key idea to check if their prediction was correct.

Workbook activities

Making shadows (page 90)

This activity supports the shadow puppets investigation on page 90 of the Student Book. It shows students different ways to make animal shadow puppets with their hands. They are then asked to write a story using their hand puppets and act out the story for the rest of the class.

Making silhouettes (page 91)

This activity supports the student silhouettes investigation on page 91 of the Student Book. The technique is tricky and time consuming, but with patience students can create a good silhouette. The worksheet will support students in following the method

at their own pace. Students can also use this to model their instructions for others to use in the Stretch zone exercise on page 91 of the Student Book.

The first part of the activity asks students to recall what they know about silhouettes. Refer to pages 68 and 91 of the Student Book. Students then follow the more detailed instructions to help them to produce a silhouette. Remind them to ask permission before taping the paper to the wall or screen. Finally, they should tell the person they have drawn how Étienne de Silhouette made pictures of people in the 1700s in this way.

Review and reflect

Remember to praise the process of learning as much as, if not more than, the outcomes. Encourage students to ask questions and to look back through prior work on light and shadows to review and revise. Use every opportunity to allow students to work together to check understanding and share ideas. This can take place after each of the investigations on shadows and silhouettes and also after the Stretch zone task.

Conclude the lesson by asking students to complete the sixth statement in the 'What I have learned about the way we see things' activity on page 102 of the Workbook. (See also the teaching notes in this Teacher's Guide, pages 134–136.)

Extra activities

1 Students can make a silhouette of a person at home. They explain to the person each part of the method and why they are doing this.

2 Students research Étienne de Silhouette and make a poster about his work. They should include some of the silhouettes that he made.

Differentiation

Supporting: Use the Workbook activities to help students to make their shadow puppets and silhouettes.

Consolidating: Students could make their own hand-shadow shapes and explain to their classmates how to make them.

Extending: Encourage students to explore how to change the size and shape of their shadows.

Differentiated outcomes	
All students	should be able to make a shadow of a character using one technique
Most students	will be able to use different techniques to make a shadow character
Some students	may be able to explain how shadows are used to make puppets and silhouettes

Growing and shrinking shadows

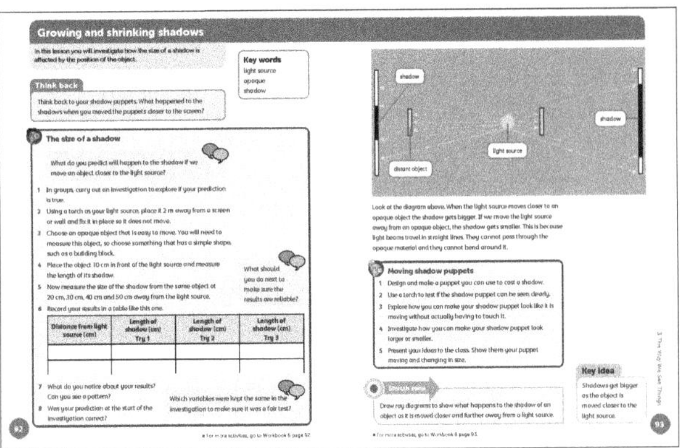

Supporting activities are in Workbook 6 pages 92–93.

Scientific enquiry key words

Plan and/or carry out enquiries to answer questions

Recognise and control variables

Make observations

Take measurements, using equipment accurately

Record data and results

Analyse data, notice patterns and group or classify things

Report and present findings

Draw conclusions and give explanations

Identify causal relationships

Getting started

This lesson explores how the size of a shadow can be changed by moving the object nearer to or further away from the source of light. Students investigate this by measuring and recording the distance the object is from the source of light and the size of the shadow. They carry out a fair test and collect reliable results. They design a puppet and use this to explore how to make its shadow smaller and bigger.

Language support

Students have been introduced to the key words 'shadow' and 'opaque'. Ask them to define these words using the glossary. They should also be familiar with the key term 'light source'. Ask them to describe a light source in the room.

Resources

Student Book: torches; large screens or a blank wall; small opaque objects such as building blocks; materials to make shadow puppets.

Workbook: shadow puppets made previously; torches; large sheets of white paper or a white wall.

Key words

light source   opaque   shadow

Other words in the lesson

ray diagram

Lesson at a glance

The key teaching points for students in this lesson are:

- the distance an object is from a light source changes the size of the shadow
- shadows get bigger if the object moves closer to the light source.

In the next lesson, students will investigate how shadows can be used to track the position of the Sun.

Think back: Think back to your shadow puppets. What happened to the shadows when you moved the puppets closer to the screen?

Ask students to read the Think back question and discuss it as a class to make sure students are reminded about their shadow puppets. You could demonstrate what students suggest to show them what really happens.

Answer: When the puppet was closer to the light source, the shadow was bigger.

Investigation: The size of a shadow

Students work in groups for this activity. Start the investigation by asking students to discuss the question in the discussion box. They will then carry out the investigation to test their prediction and use the data or results to refute or support their predictions.

What do you predict will happen to the shadow if we move an object closer to the light source?

Answer: Students should suggest that the shadow increases in length/size as it is closer to the light source.

Groups can carry out the investigation independently, asking for your help if needed. Remind them to think about the variables and how they are going to make sure they carry out a fair test. They should record their results in a table like the one provided. Encourage students to make sure they have reliable data by repeating the tests. They could practise their maths skills and calculate averages for shadow sizes.

What should you do next to make sure the results are reliable?

Students should discuss this as they begin to plan their investigation. Remind them that reliability is associated with an investigation that is repeatable. They should be able to repeat the method and so should other scientists.

> *Possible response: Students should calculate averages using the data they have collected. Some students might begin to ignore anomalous results in these calculations.*

Steps 7 and 8 of the investigation guide students to analyse their results and reach a conclusion. Encourage them to discuss the questions in their group and share their thoughts with the class. At the end of the investigation, ask them to agree in their group which variables were kept the same.

Which variables were kept the same in the investigation to make sure it was a fair test?

Ask students to agree on the independent, dependent and any control variables in this test. How did they use their knowledge of variables to make sure the test was fair?

> *Possible response: Students should suggest that the position of the light source and the size of the object casting the shadow were kept the same (the control variables) with the distance from the light source changing (the independent variable). The dependent variable is the length of the shadow.*

Ask students to study the diagram and text at the top of page 93. Read through the text or ask volunteers to read a sentence each. Ask students to study the diagram to help them understand the text. Relate it to their investigation set-up. The diagram looks complicated but that is because two sides are used so a large and small shadow can be shown. Ask students to cover up the right side of the diagram first to only look at the small shadow side. Ask, 'How far away from the light source is the object? How close is the object to the screen?' Remind them that light travels in a straight line and ask them to follow some of the light rays with their finger from the light source to the screen. They can then cover the left side of the diagram and look at the larger shadow side. Ask them the same questions. They should realise that the nearer the object is to the

light source, the bigger the shadow because the object is blocking more of the light rays. The next investigation consolidates this concept.

Investigation: Moving shadow puppets

Students can work on their own or with a partner if you want them to collaborate to design and make a puppet to cast a shadow. They will then explore how to make their shadow puppet look like it is moving, look larger and look smaller without actually having to touch it. They will practise their presentation skills by telling the class their ideas and showing their puppet moving and changing size. The worksheet on page 93 of the Workbook supports this investigation.

Tell students that moving puppets like this – so the shadow appears to move closer and further away from the audience – is a technique used in shadow theatres.

Stretch zone: Draw ray diagrams to show what happens to the shadow of an object as it is moved closer and further away from a light source.

Students can work on their own or with a partner in this Stretch zone activity. They can use the diagram at the top of page 93 to help them. This will link their work on shadows to drawings of ray diagrams.

> *Answer: Students should suggest that the light is blocked by the object. The rays of light stop at the boundary of the object. The closer it is to the source the more light is blocked and so the shadow is darker and larger.*

Key idea

Shadows get bigger as the object is moved closer to the light source.

Read through the key idea or ask a volunteer to read it out to the class. Ask students to describe how we can make shadows change size.

Workbook activities

Investigate the size of shadows (page 92)

This activity gives students practice at making shadows with different-sized objects and in measuring the shadows. The instructions and questions guide students through the worksheet. They reach a conclusion whether a bigger object casts a bigger shadow. There is a possible misconception to address here. Students may think that the size of the object is the only reason shadows are of different sizes – big objects cast big shadows and small objects cast small shadows. The hands-on nature of this task and the investigation in the lesson should provide evidence for students that addresses this. They will see that the position of the object relative to the light source affects shadow size.

 Moving shadow puppets (page 93)

This activity supports the investigation on page 93 of the Student Book. Students will explore how to make shadow puppets appear to move when the light source is moved various distances. They measure how much the shadows move and record their results in the table.

Review and reflect

Ask students to sit down for a few minutes and think about what they have learned from the lesson. They can then think of one idea from the lesson that they would like to investigate and find out more about. Ask them to make a plan about how they would do this. This planning time will allow students to think up some imaginative and original things to do and you could let them carry out these projects.

Conclude the lesson by asking students to complete question 6 in the 'What have I learned about the way we see things?' activity on page 103 of the Student Book. (See also the teaching notes in this Teacher's Guide, pages 134–136.)

Extra activities

1 Students make a shadow puppet story at home. They make a scary monster or a hero as big as possible by moving the source of light.

2 Students find objects of the same shape but different sizes and reinforce their understanding of how they form a shadow. They could use a shoe box and a large crate or box for example. Set students the challenge of trying to make the shadow from the shoe box larger than the shadow from the large crate or box. They will need a darkened room and a light source.

Differentiation

Supporting: Students can practise moving their puppets closer and further away from the source of light to reinforce their understanding.

Consolidating: Students can practise measuring the size of shadows and the distance from the light source.

Extending: Students can explore how to make a shadow appear to increase in size by moving the light source.

Differentiated outcomes	
All students	should be able to change the size of a shadow
Most students	will be able to measure the distance of an object from the light source and the size of the shadow cast
Some students	may be able to predict how far the light source should be from the object to make different-sized shadows

Tracking moving shadows

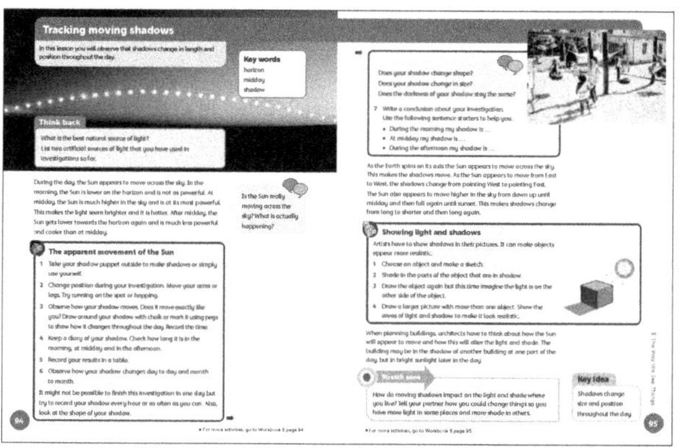

Supporting activities are in Workbook 6 pages 94–95.

Getting started

This lesson looks at how shadows cast by the Sun change in length and intensity throughout the day. The Sun appears to move across the sky but this is not the case; students learned this in Year 5. In the morning, the Sun is low on the horizon and is not as bright as at midday. At midday, it is at its highest and brightest. Students investigate and track the Sun and observe how shadows change throughout the day.

Language support

Ask students to discuss the key words. They can use the eBook to help with pronouncing them. They can say a sentence to their partner using each of the words. Remind them that midday is at 1200 hours. Ask them what they do at midday. This will help them to understand the word. Then ask them to look at the horizon and use this word in a sentence to their partner. They might say 'I can see a tree on the horizon.'

Resources

Student Book: access to outdoors at various sunny times (perhaps over more than one day); shadow puppets made previously; selection of small opaque objects; drawing materials.

Workbook: access to outdoors at sunny times; cameras.

Key words

horizon   midday   shadow

Other words in the lesson

architect   artificial   axis   east/west
natural   spin

Lesson at a glance

The key teaching points for students in this lesson are:

- shadows change in length and position during the day
- the Sun is highest in the sky at midday.

In the next lesson, students will investigate more about the changes in the length and position of shadows.

Think back: What is the best natural source of light? List two artificial sources of light that you have used in investigations so far.

Start by asking students to discuss what a source of light is and which ones are natural. Then ask them if the Sun is a natural source of light. They should discuss what artificial sources of light are and why we need them.

Possible response: Students should suggest that the Sun is the best natural source of light. We use many sources of artificial light including lamps and light bulbs.

Ask students to read the first paragraph of text on page 94. It explains about the different positions of the Sun during a day. Ask them the time of day when the Sun is most powerful. Students then discuss the questions alongside the paragraphs and share their answers as a whole class. This is a revision of prior learning.

Is the Sun really moving across the sky? What is actually happening?

Possible response: Students should suggest that the Sun does not move across the sky. The Earth is spinning on its axis which changes its position in front of the Sun. This makes the Sun appear to move across the sky.

Investigation: The apparent movement of the Sun

Ask students to work in a small group to carry out this investigation. They are asked to observe how their shadow changes day to day and month to month so it would be useful to plan this activity over a few weeks.

They will use the shadow puppet they made before, or their own shadow, to investigate how shadows move position, shape and size during a day: in the morning, at midday and in the afternoon.

Students can follow the instructions, and you can support when needed. Remind them to record their results in a table.

Ask them to discuss the questions in the discussion box to help them review their investigation results and to write a conclusion. They can use the sentence starters given to help them.

Does your shadow change shape? Does your shadow change in size? Does the darkness of your shadow stay the same?

> *Possible response: Students should suggest that our shadows change shape and size during the day when the source of light is the Sun. Some shadows can appear darker at midday. In the morning their shadows are longer. At midday the shadows are the shortest. In the afternoon their shadows are longer and thinner.*

Read out the information under the investigation on page 95. Explain how in the morning the Sun reaches the Earth at an angle which makes shadows look long and thin. At midday, it is directly above so shadows are shorter. Ask them to look at the photograph at the top of page 95 and to suggest whether the children are playing in the morning/afternoon or at midday. Ask them to explain why they reached their conclusion.

Explain that a shadow is created because of the absence of light. This means the darkness of it does not change. The amount of light surrounding an object, however, can make the shadow appear darker. So, when the light is brighter at midday the contrast will be greater so the shadow could appear darker.

Investigation: Showing light and shadows

This investigation gives students a chance to be creative and use their artistic skills. Artists often show shadows in their pictures. It can make objects appear more realistic. Students choose an object and make a sketch by shading in the parts of the object that are in shadow. Then they draw the object again but this time with the light on the other side of the object. Ask them if the shadow changed. Ask them why they think the shadow changed.

> *Possible response: Students should suggest that the shadow will be in line with the source of light but at the other side of the object.*

Point out the text below the investigation box and ask students to read it and then discuss how architects need to think about the Sun and shadows. Ask, 'Why might an architect want to let a lot of light into a building? Why is shade so important in buildings?' You could ask students if they would like to use their science knowledge to help them to design buildings in the future.

Stretch zone: How do moving shadows impact on the light and shade where you live? Tell your partner how you could change things so you have more light in some places and more shade in others.

Students can read the question in the Stretch zone and discuss with a partner how to change the light and shadows where they live.

> *Possible response: Students should suggest that they can move opaque objects that block the light from places where they want more light, and put them into places where shade is needed.*

Key idea

Shadows change size and position throughout the day.

Read out the key idea or ask a volunteer to read it out. Ask students to turn to a partner and tell them how we can track the movement of the Sun using shadows. Ask students to suggest why we think the Sun moves across the sky during the day and what scientific evidence we know about that refutes this idea.

Workbook activities

Shadow hide and seek (page 94)

This activity encourages outdoor learning. Students use shadows to play a game of 'Shadow hide and seek'. They can play the game in a small group or with the whole class. They take it in turns to be the seeker and the hiders. The seeker looks for people's shadows and guesses who the person is by the shape of the shadow.

> *Possible response: Students should suggest that they could guess some students more than others. Students who have distinguishing characteristics, such as being taller or who wear their hair in a ponytail, for example, will be more easily identified.*

Does the darkness of shadows change during the day? (page 95)

This activity supplements the investigation on pages 94–95 of the Student Book. Students investigate whether the darkness of shadows changes during the day. They will need to take measurements in the morning, midday and afternoon and record their findings in the table given. They are also encouraged to take photographs of their observations.

The questions on the page help them to write a conclusion about their observations and to consider a scenario about fair testing.

Possible response: Students should suggest that the shadows appear darker at midday compared to the morning or afternoon. This is because the light from the Sun is brighter so the comparison of the bright surroundings makes the shadows look darker.

Students are finally challenged on whether their results are reliable and how they could make them more reliable. They can write their explanation in the space provided.

Possible response: Students could suggest that shadow observations could be taken at exactly the same time every day, not just morning, midday and afternoon. They could also take more than three measurements in a day and repeat each measurement three times to help make the readings more accurate. They could use a light meter to measure how dark the shadows are. Some may suggest carrying out the investigation at different times of the year as day length may alter.

Review and reflect

The pictures and written responses created in the lesson can be used for you and your students to evaluate learning. These show evidence of understanding and creativity. You can give feedback about each one – verbally or in writing – and also encourage students to pass on positive comments and suggestions for improvements.

Let students sit quietly for a few minutes and think about what they believe they have done well and what single thing they could take as a learning target into the next lessons.

Extra activities

1 Students observe the position of the Sun at a specific time in the morning, midday and afternoon for two days. They compare the position of the shadows on both days. Are they always the same? Remember to tell students never to look directly at the Sun.

2 Students draw their shadows on the ground with chalk in the morning, midday and afternoon. They take a picture of the shadows and compare them.

Differentiation

Supporting: Allow students to model the apparent movement of the Sun and its effects on shadows by using a torch as the Sun and a football as the Earth.

Consolidating: Ask students to track the movement of a shadow across the room by adding a string outline every 30 minutes to show how it changes.

Extending: Students could research the difference in shadow length in summer, spring, autumn and winter in a northern hemisphere country such as Sweden or Canada.

Differentiated outcomes	
All students	should be able to observe the apparent movement of the Sun and how shadows change during the day
Most students	will be able to use shadows to show how the position of the Sun appears to move
Some students	may be able to predict the time of day shadows were made

Shadow investigations

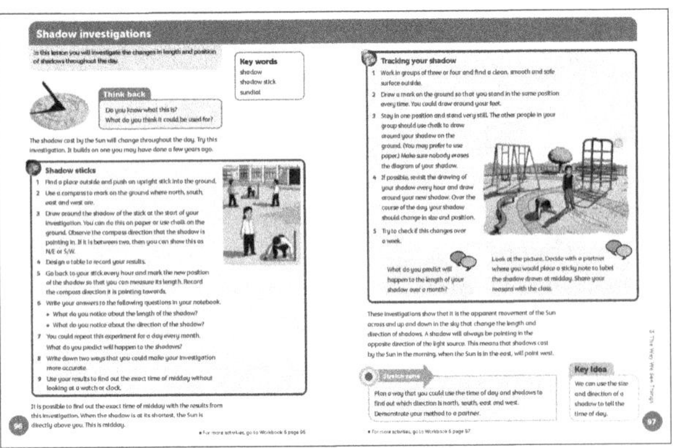

Supporting activities are in Workbook 6 pages 96–97.

Getting started

This lesson explores how we can investigate shadow sticks. These can be used to track the changes in the sunlight throughout the day. The shadows can be used to tell us the time of day and this is how sundials work. Shadows are investigated over the course of a day and students try to find midday using their investigations.

Language support

Students are introduced to two new key terms within this lesson – 'sundial' and 'shadow stick'. It is important that they are able to recognise these words and understand their meanings. As new words are introduced, allow students time to practise the spelling and pronunciation. Remind them that they have already learned the word 'shadow'.

Resources

Student Book: 1-metre sticks; compasses; chalk; writing materials; paper; access to outdoor area with clean, hard surface.

Workbook: paper plates; pencils; writing materials; access to outdoors at sunny times.

> ### Key words
>
> shadow   shadow stick   sundial
>
> **Other words in the lesson**
>
> midday

Scientific enquiry key words

Plan and/or carry out enquiries to answer questions

Make observations

Take measurements, using equipment accurately

Record data and results

Analyse data, notice patterns and group or classify things

Report and present findings

Draw conclusions and give explanations

Lesson at a glance

The key teaching points for students in this lesson are:

- shadow sticks can be used to track the Sun
- sundials use shadows to tell the time.

In the next lesson, students will find out that light intensity can be measured.

Think back: Do you know what this is? What do you think it could be used for?

Ask students to look carefully at the photograph at the top of page 96. Ask them if they have ever seen one of these. It is a sundial. Encourage a discussion about their experiences of them.

> **Possible response:** Students should suggest that the picture is a sundial that uses shadows cast by the Sun to tell the time of day.

Investigation: Shadow sticks

This investigation builds on learning in Year 5. It would be useful to plan to do this activity over the course of a month.

Students can work in pairs or small groups. They can follow the instructions in the Student Book independently. They will use a compass and measuring tape to record the direction and size of shadows every hour during the day. They will need to design a suitable table to record their results. Check the design of their table and make sure students are confident using a compass and a tape measure.

Students are asked to repeat this experiment for a day every month and predict what will happen to the shadows. They are also asked to evaluate their work and write down two ways to make their investigation more accurate.

Finally, they use their results to find out the exact time of midday without looking at a watch or clock.

Possible response: Students should be able to estimate the time of day using the position of the shadow cast. They could check the accuracy using a clock or watch.

Read out the text at the bottom of page 96 or ask a volunteer to read it. Ask students if they were certain about finding the time of midday using their results.

Investigation: Tracking your shadow

Students can work in groups of three or four. This investigation will need to be carried out over a week. Ask students to read through the steps and ask you to explain if they have any questions about the task.

Students can then follow the instructions to produce the marking point for their investigation. The picture on page 97 explains the process. If possible, they should revisit the drawing of the shadow every hour and draw around the new shadow. Over the course of the day, their shadow should change in size and position. Students are asked to try to check if this changes over a week.

Possible response: Students should suggest that the size and position of the shadow changes throughout the day. It should be short at midday and longer earlier in the morning and later in the afternoon.

After they have carried out the investigation over a week, ask each group to make a prediction about what will happen over a whole month.

What do you predict will happen to the length of your shadow over a month?

Possible response: Students should suggest that their shadow will change a little because of the changes in sunlight between seasons.

Look at the picture. Decide with a partner where you would place a sticky note to label the shadow drawn at midday. Share your reasons with the class.

Students should study the picture before deciding where the sticky note should go. Remind them what the shape of the shadow should be like at midday. Direct them to the information at the bottom of page 96 as a reminder.

Answer: Students should suggest that the sticky note should be added to the blue shadow as it is the shortest.

Ask volunteers to read out the text beneath the investigation box. They could read a sentence each. Explain that 'apparent' means 'seems to be' and stress that the Sun does not move across the sky. You could demonstrate using a ball or even a globe as the

Earth – ask a student to slowly turn it round and round. Ask another student to be the Sun – they stand still and shine a torch on the model Earth. You can stop the Earth once or twice to point out which parts are in sunlight and which are in shade, and ask students to imagine what shadows would look like in different places. You can fix small pieces of modelling clay to the surface to cast shadows and test students' predictions.

Stretch zone: Plan a way that you could use the time of day and shadows to find out which direction is north, south, east and west. Demonstrate your method to a partner.

Unless you are going to carry out the investigation as an extra activity (see below) then tell students this Stretch zone will challenge their planning skills. Ask them to discuss how they would find out the positions and show their partner how they will carry out the method.

Possible response: Students should suggest that a shadow always points in the opposite direction to where the Sun appears to be in the sky. They should write into their plan that in the morning the Sun appears to rise in the east so shadows will point west. In an evening, the Sun sets in the west so shadows point east. With these two compass points they can find the points between them to locate north and south. They can then use a compass to check their ideas.

Key idea

We can use the size and direction of a shadow to tell the time of day.

Read through the key idea or ask a volunteer to read it out to the class. Ask students to describe how a sundial works to tell us the time. They should comment on the accuracy of using a sundial.

Workbook activities

Make a sundial (page 96)

Explain to students that they are going to make and test a sundial. Arrange them into groups of three or four and ask them to follow the instructions to make their sundials. Suggest they use the diagrams as a clue. Once the sundials are constructed they can take them outside and place them in the sunshine, on the ground or a flat surface. Students should observe where the shadow of the pencil is cast onto the plate. Explain that they should then turn the plate so the shadow is lined up with the time of day they wrote on the plate. This was checked with a watch so it should be accurate. Allow them to predict, check and mark where the line will be every hour thereafter. Ask students to test their finished sundial throughout the next day and decide if their sundial is accurate in telling the time.

Ask students to evaluate their sundial. Allow them to discuss some ways to improve it with other students in the class.

> *Possible response: Students should suggest that the sundial is accurate at telling us when it is midday or early morning or evening. It is not so accurate at telling the time by the hour. This would become more accurate if more readings were taken and comparisons made. The time would be more reliable.*

Telling the time with shadows (page 97)

Ask students to work with a partner and describe to each other how they made a sundial. Point out that they should talk about how they used shadows to tell the time. Remind them to discuss the improvements they would like to make to the sundial.

Explain that in this activity they are going to take their evaluation ideas to make and test an improved sundial. Allow students to test their improved sundial for a few days and record their results in tables. Once they have completed their testing ask them to discuss if their improvements worked.

> *Possible response: Students may suggest moving the position of the sundial so that the light is better. They need to check there are no other sources of light to pollute the light and affect the shadow cast.*

Review and reflect

Use the activities in the Student Book and Workbook to encourage students to identify aspects they have not completed correctly and help them to identify improvements. For example, they can walk around and look at the sundials made by other students to identify what they have done well and what they can learn from others as targets to improve their work.

Conclude the lesson by asking students to complete question 2 in the 'What have I learned about the way we see things?' activity on page102 of the Student Book and the seventh statement on page 102 of the Workbook. (See also the teaching notes in this Teacher's Guide, pages 134–136.)

Extra activities

1 Students can take their sundial from the Workbook activity home and continue to measure the time over the weekend.

2 Students can write an advertising leaflet for sundials. They discuss the benefits and pitfalls of using them instead of a watch.

Differentiation

Supporting: Allow students time to practise using shadow sticks and recording the position of the shadow.

Consolidating: Allow students time to practise recording the position of the shadow and comparing this to the time on a clock.

Extending: Students can determine midday by observing shadows on a sundial.

Differentiated outcomes	
All students	should be able to use changes in the position of shadows through the day to estimate the time
Most students	will be able to investigate the shadows at different times of the day
Some students	may be able to predict the time using the position of a shadow

Light intensity

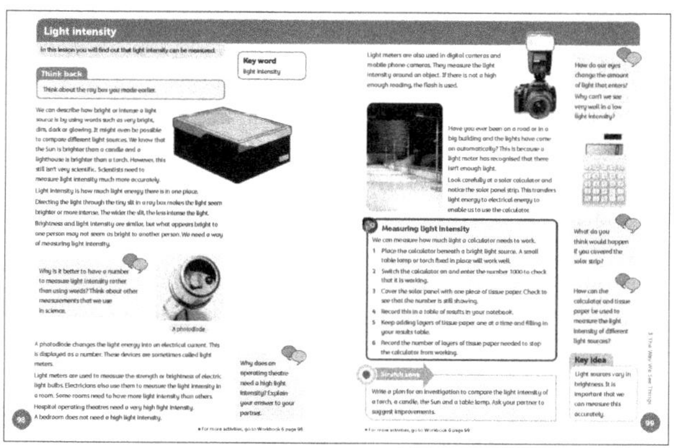

Supporting activities are in Workbook 6 pages 98–99.

Getting started

In this lesson students learn how we can measure how much light there is. This is called light intensity. Students explore how instruments can transfer light energy into an electrical current and this provides us with a number to measure the intensity of the light. Students investigate the amount of light by using a solar-powered calculator.

Language support

Students are familiar with the word 'light', but write the word 'intensity' on the board and point out that this means the amount or strength of something. Ask students to stand up and walk slowly. Tell them they are exercising with low intensity. Then ask them to move quickly and tell them that they are now exercising with high intensity.

Resources

Student Book: bright light sources; solar-powered calculators; tissue paper.

Workbook: bright light sources; solar-powered calculators; tissue paper.

> ### Key word
>
> light intensity
>
> **Other words in the lesson**
>
> bright   dark   dim   intense   lighthouse   meter
> photodiode   ray   solar

Lesson at a glance

The key teaching points for students in this lesson are:

- the brightness of light can be measured
- we can investigate the amount of light using a solar calculator.

In the next lesson, students will learn about how light intensity can be measured using scientific methods.

Think back: Think about the ray box you made earlier.

Ask students to work with a partner to reflect on their earlier work with a ray box. Point out the picture on page 98 to help them.

Read through the text at the top of page 98 with students or ask volunteers to read out a sentence each. Ask students to summarise the information to a partner without looking at the text. The other person checks using the text if they have recalled everything correctly.

Why is it better to have a number to measure light intensity rather than using words? Think about other measurements that we use in science.

Arrange students into pairs or groups of three or four and ask them to discuss the question. Encourage them to think about how they might describe how warm or cold it is. Some people will say it is hot and others just warm. This is why we use instruments to measure things like temperature and light intensity as we do not always agree on the words we use.

> *Possible response: Students should suggest that measurements are accurate and cannot be questioned in normal circumstances.*

Ask students to continue to work together to read the text at the bottom of page 98. Discuss the photodiode and how this transfers the light energy it detects into a reading. You might show students an example of one if you have one. Light meters are used by many professionals including art restorers so that the light doesn't damage expensive paintings. Students might have seen or used a light meter in a digital camera.

Why does an operating theatre need a high light intensity? Explain your answer to your partner.

Allow students to continue to work with their partner or group. Ask them to think about the importance of the accurate work that is carried out in an operating theatre.

> *Answer: Students should suggest that the people in an operating theatre need to be able to see well to carry out very detailed work.*

Ask students to read through the text and look at the photographs at the top of page 99. Ask them if they have ever used a digital camera. These react to the light and apply a bright flash of light if the levels are low. The lens of a camera works a bit like the eye: it checks the level of light intensity.

How do our eyes change the amount of light that enters? Why can't we see very well in a low light intensity?

> *Possible response: Students should suggest that the pupil opens when it is dark to allow lots of light in so that we can see. When it is bright, the lens closes to protect the eye. We need light to see so when it is dull it is hard to see.*

What do you think would happen if you covered the solar strip?

Ask students to look at the photograph of the solar calculator and think about how this works. Ask where the energy comes from to run the machine.

> *Possible response: Students should suggest that when light does not react with the strip there isn't any electricity so the calculator doesn't work.*

Investigation: Measuring light intensity

Students can work with a small group for this investigation. The worksheet on page 98 of the Workbook supports this investigation.

Students will place the calculator beneath a bright light source and then cover the solar panel with increasing numbers of pieces of tissue paper to see if the calculator still works. They record their results until the calculator stops working.

> *Possible response: Students should suggest that the more layers that were added the fainter the calculator display was, as not enough energy is entering the solar panel. Eventually, the display will not work when the light is completely blocked.*

How can the calculator and tissue paper be used to measure the light intensity of different light sources?

Students can work with their investigation group to think about how this investigation could measure light intensity.

> *Possible response: Students should suggest that they carry out the same method but using sunlight, a lamp and a candle as sources of different light intensity, for example.*

Stretch zone: Write a plan for an investigation to compare the light intensity of a torch, a candle, the Sun and a table lamp. Ask your partner to suggest improvements.

Allow students to work with a partner or their investigation group to plan the Stretch zone investigation. They could use the same method that they used in the light intensity investigation or design one of their own. Remind them that this needs to be a fair test. The independent variable is the light source, the dependent variable is how many sheets of tissue paper blocked the light to the solar panel. Ask students to use their own scientific knowledge to predict what will happen.

> *Possible response: Students can suggest the method. They should predict which sources will be brighter and so these will need more sheets to block the light. A candle is not very bright, so will use fewer pieces of paper.*

Key idea

Light sources vary in brightness. It is important that we can measure this accurately.

Summarise the lesson by asking students what they have learned. Let them share their ideas. Ask one student to read out the key idea. Ask students to describe how we can measure how much light there is using instruments to give us readings. Light meters are used in many devices, including cameras, that work like our eyes to control the light entering the lens. Students should describe how they have investigated the light intensity from different sources using a calculator.

Measuring light intensity (page 98)

This activity supports the investigation on page 99 of the Student Book. Students are asked to recall how to use a solar-powered calculator as a light meter. They then plan an investigation to compare a torch, a candle, the Sun and a table lamp, as well as two other sources of their own to compare. They can use the space provided to write the instructions and diagrams to help others carry out their investigation.

Possible response: Students should write a repeatable method to investigate the different light sources using a calculator.

Which light source will be brightest? (page 99)

This activity can be used as support for the Stretch zone activity on page 99 of the Student Book. Students predict which light source will be the brightest and then carry out an investigation to find out how many sheets of tissue paper are needed to stop the calculator display working with each source. Students can record their results in the table provided and are asked to rank the light sources in order of brightness. Finally, they are asked to evaluate their prediction.

Possible response: Students should suggest that the brightest source should be the Sun but that the light is polluted by clouds and dust by the time it reaches Earth. So, lamps and bulbs will probably be the brightest followed by sunlight, or torches and moonlight. This depends on the wattage of the bulbs used and the phase of the Moon when the investigation is carried out.

Review and reflect

Hold up word cards for the words used in the lesson: light, intensity, ray, solar, dim, dark, bright, photodiode, meters. Each time you hold up a word ask students to turn to a partner and use the word in a sentence. They should write down any of the words they are unsure of and look these up.

Extra activities

1 Students demonstrate how a solar calculator can be used to measure light intensity with the people at home.

2 Students ask the people at home to describe the light intensity in different places. They record the place and their description. They compare their responses to show how we are not very reliable at describing light intensity and to demonstrate the need to measure it.

Differentiation

Supporting: Demonstrate different light intensities by using different light sources so students can relate to high and low light intensity.

Consolidating: Allow students to measure light intensity around the school by using the improvised light meter they made in the investigation.

Extending: Ask students to research how light intensity measurements are used in places of work and to measure stars.

Differentiated outcomes	
All students	should be able to understand the need to measure light intensity
Most students	will be able to compare light intensity measurements from investigations
Some students	may be able to compare light intensity using different light sources

Using scientific methods to measure light intensity

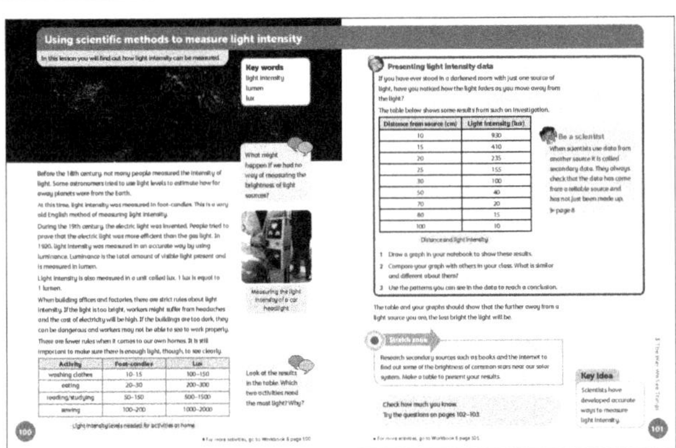

Supporting activities are in Workbook 6 pages 100–101.

Getting started

This lesson looks at how we can measure light intensity using scientific methods. Measuring light intensity is important to many people in current times, including architects and artists. This lesson is challenging because it synthesises the information that students have learned into the history of how these measurements have developed. Students investigate and present their ideas like scientists.

Language support

Write down the words dark, dim, shade and bright on the board. Explain that terms like these are not scientific – they are subjective. In other words, dim and bright relate to how something seems to someone and people may have different opinions. Point out that scientists prefer to measure the exact intensity of light, and lux and lumen are two units that have been used for measuring this.

Resources

Student Book: access to the internet or books with information on the brightness of stars.

Workbook: access to the internet or books with information on measuring light intensity; graph paper.

> ### Key words
>
> light intensity   lumen   lux
>
> **Other words in the lesson**
>
> astronomers   foot-candles   luminance
> solar system

Lesson at a glance

The key teaching points for students in this lesson are:

- light intensity can be measured scientifically
- light intensity used to be measured in the units of foot-candles but is now measured in units called lux or lumen.

In the next lesson, students will review their knowledge and understanding of the unit.

Ask students to read the information on page 100 with a partner. There is a lot of challenging text to work with. One method that could be used is that students cover the page with a blank piece of paper and take it in turns to reveal and read a paragraph at a time. They take turns to tell each other the context of what they have read. The other person comments on the accuracy of their description. They take turns to reveal, read and describe in this way to the bottom of page 100.

Point out the photograph of the person measuring the light intensity of a car headlight. Ask, 'Why is it important to know how bright a car headlight is? What would happen if car headlights were too bright?' Explain that the glass disk allows light to enter the sensor and this has a light metre inside. Point out that smaller light metres are found in cameras, automatic light switches and smartphones.

What might happen if we had no way of measuring the brightness of light sources?

Allow students to work with a partner to talk about the discussion question. There is a lot of information for students to process. Discussing the key points will help them to understand the content.

Possible response: Students should suggest that workers might suffer from headaches and even damage their eyes. Too much light would not be cost effective, but if there isn't enough light some people might not be able to work well, for example surgeons. We would not be able to measure how far the planets are from Earth or how big the universe is if we could not measure light accurately.

Look at the results in the table. Which two activities need the most light? Why?

Encourage students to study the table and to analyse the data by answering the discussion question.

> *Answer: Reading and sewing need the most light. This is because they involve lots of small detail in the text characters and stitches.*

Investigation: Presenting light intensity data

The investigation gives students practice in presenting and analysing data about how light intensity changes with distance from the source. They are asked to draw a graph and to identify patterns in the data to reach a conclusion. The worksheet on page 101 of the Workbook supports this. Encourage students to compare their graph with others in the class and to identify what is similar and different about them. Explain that a line graph is the best choice for the type of graph. Point out that both the *x* (horizontal) axis and the *y* (vertical) axis have numbers. If the *x*-axis was categories – such as type of plant or favourite colour – a bar chart would be more suitable.

> *Possible response: Students should choose to draw a line graph. The independent variable or distance from the source should be plotted on the x-axis and the light intensity on the y-axis. They should report that the trend or pattern in the data shows that as the distance from the source increases the light intensity decreases. Students should conclude that the greater the distance from a source, the lower the light intensity.*

Be a scientist: When scientists use data from another source it is called secondary data. They always check that the data has come from a reliable source and has not just been made up. (Student Book, page 8)

Discuss with students what a reliable source is. They should check who has posted the information and when, to make sure the data is correct and has been checked by others to make it reliable. Ask, 'What might happen if we use unreliable data in science?'

Stretch zone: Research secondary sources such as books and the internet to find out the brightness of some common stars near our solar system. Make a table to present your results.

Allow students to have access to the internet or print out some pages relevant to the Stretch zone task. Ask students to think about the Be a scientist information when they are carrying out this research. Encourage them to present their table in order of brightness.

> *Answer: Students should find Sirius, Canopus, Rigil Kentaurus, Arcturus, Vega, Capella and Rigel as examples of Earth's brightest stars. They may make a table with more stars.*

Key idea

Scientists have developed accurate ways to measure light intensity.

Ask students to read the key idea or ask a volunteer to read it out to the class. Ask students to describe how we can measure how much light there is using units of measurement. Ask them to tell their partner what the modern unit of measurement of light intensity is, and how we used to measure light intensity and why this wouldn't be good enough today. Ask students to explain the pattern in light intensity as the distance from the source of light increases.

Workbook activities

Light intensity timeline (page 100)

Students work in a small team to make a timeline of methods of measuring light intensity. They are given an example of a timeline to help. They can use the internet or books or find some information in the Student Book. They are given some tips to help. They should display the timeline in the school so students can compare and evaluate their research with others.

Students then describe the methods used in the space provided. Encourage them to use diagrams and pictures.

> *Possible response: Students should suggest various different methods and units of measuring light intensity, including foot-candles, luminance, lumen and lux.*

Distance and light intensity (page 101)

This worksheet supports the investigation on page 101 of the Student Book. The steps support students in using the data to draw a line graph and in identifying a pattern showing what happens to light intensity as the distance increases.

> *Answer: Students should suggest that the distance from the light source should be plotted on the x-axis and light intensity on the y-axis.*

As a challenge, students are asked to use their graph to determine the light intensity at a distance of 18 cm and 75 cm. They can write their answers in the space provided.

> *Answer: At 18 cm the light intensity would be approximately 340 lux and at 75 cm it would be approximately 23 lux.*

Review and reflect

This a good time to encourage students to reflect on their learning for the lesson and for the whole unit. Ask them to look back through their books and any notes and to write down two things they have done in the unit that have made them proud. This could be something they have learned, a skill they have developed, or a model or presentation they have completed. Then you can ask them to think about one thing they feel they could have done better. Ask them to think about one way of learning from this.

Conclude the lesson by asking students to complete question 7 in the 'What have I learned about the way we see things?' activity on page 103 of the Student Book and the final statement on page 102 of the Workbook. (See also the teaching notes in this Teacher's Guide, pages 134–136.)

Extra activities

1   Students try to find products using the internet that use lux in their product information.

2   Students make a summary of the information from these pages using a wiki page template focused on using scientific methods to measure light intensity.

Differentiation

Supporting: Use the Workbook activity on page 101 to support students in drawing the graph of distance from a light source and light intensity.

Consolidating: Support students in looking for patterns in the data in the table and then on the graph. Ask them which pattern was easier to see and why.

Extending: Students could extend their historical research by adding to the timeline the names of scientists involved in the development of units.

Differentiated outcomes	
All students	should be able to name the correct unit of measurement for light intensity
Most students	will be able to explain why we need to use units of measurement to measure light intensity
Some students	may be able to name units of measurement that have been used historically

What have I learned about the way we see things?

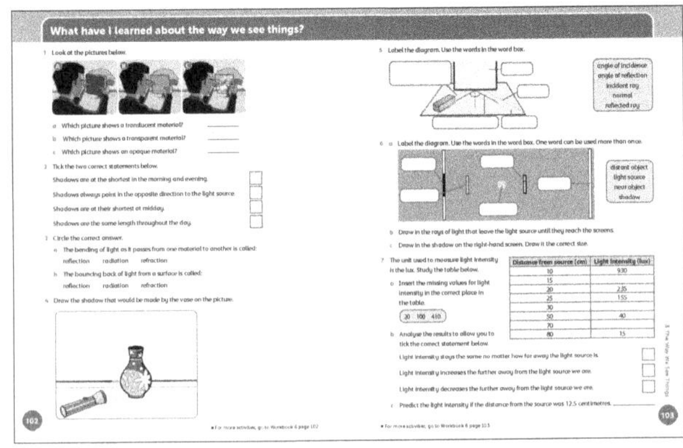

Supporting activities are in Workbook 6 pages 102–103.

Getting started

The aim of this section is to encourage students to review their learning after all of the lessons in the unit. All the lessons in the unit have questions and discussion tasks for students. You will have been using these formatively during lessons to help you assess students' knowledge and understanding of the topics.

On pages 102–103 of the Student Book there are seven questions related to the content of the unit. These will assess students' knowledge and understanding of the topics. The questions are arranged in increasing order of conceptual demand and not topic order. Students can tackle these one at a time after relevant lessons – specific advice on which questions are most appropriate is under the 'review and reflect' heading in the relevant lesson. The questions could also be answered as a single summative activity. This could be done by reading out the questions to the class and asking for volunteers to answer them, carrying out the activity as a group work task with students talking about each question, or as an individual written task. Whichever approach is adopted, the questions are designed to give you and the students feedback about progress and to help in identifying targets for development.

In addition to the questions in the Student Book there are some self-review statements on page 102 of the Workbook. Students can read through these statements and fill in the self-review boxes to show how confident they are about the unit's topics. You can ask students to reflect on each statement separately after the relevant lesson or carry out the review as an end-of-unit activity. Advice on which specific 'review and reflect' statement is most appropriate is given under the 'review and reflect' heading in the relevant lesson. It would be very useful to do both end-of-lesson and end-of-unit reviews, as you will be able to measure student levels of confidence throughout the unit to respond quickly and also gain a measure at the end of the unit to find out if this has changed for any topics.

It is important that students report areas that they are not confident with. This information is vital for you to provide support strategies in the end-of-unit summative assessment.

What have I learned about the way we see things? answers

1 Look at the pictures below.

   a Which picture shows a translucent material?

   b Which picture shows a transparent material?

   c Which picture shows an opaque material?

   Point out the three pictures of the person holding up the object. Explain that students need to decide which object matches each statement and then answer by writing in the most appropriate letter.

   Answer: *a B; b C; c A*

2 Tick the two correct statements below.

   Shadows are at their shortest in the morning and evening.

   Shadows always point in the opposite direction to the light source.

   Shadows are at their shortest at midday.

   Shadows are the same length throughout the day.

   Point out to students that an important word in the question is the word 'two'. Remind students to read questions carefully and this one makes it clear how many statements they have to tick.

   Answer: *Students should tick 'Shadows are at their shortest at midday' and 'Shadows always point in the opposite direction to the light source'.*

3 Circle the correct answer.

   a The bending of light as it passes from one material to another is called:

   reflection   radiation   refraction

   b The bouncing back of light from a surface is called:

   reflection   radiation   refraction

   Explain that these questions are multiple-choice questions so they are looking for only one of the three choices offered in each case. Suggest that if students are unsure, they could start by eliminating any obviously wrong choices.

   Answer: *a refraction; b reflection*

4 Draw the shadow that would be made by the vase on the picture.

   Remind students that light travels in straight lines and suggest they draw in the rays of light from the torch to help them to work out the answer.

   Answer: *Students should draw a representation of the shadow in line with the beam from the torch.*

5 Label the diagram. Use the words in the word box.

   Point out the word box and remind students to use words from here. Each word will only be used once so suggest students cross them out as they use them.

   Answer: *Labels from top left, clockwise: incident ray; reflected ray; angle of reflection; normal; angle of incidence.*

6 a Label the diagram. Use the words in the word box. One word can be used more than once.

   b Draw in the rays of light that leave the light source until they reach the screens.

   c Draw in the shadow on the right-hand screen. Draw it the correct size.

   Suggest that students use a ruler and a pencil to draw in the rays of light that would leave the light source and move towards each screen. Remind them that the object will block light.

   Answer: *a from left to right: shadow; distant object; light source; near object; shadow; b rays drawn as Student Book page 93; c shadow drawn as Student Book page 93 on right screen.*

7 The unit used to measure light intensity is the lux. Study the table below.

   a Insert the missing values for light intensity in the correct place in the table.

   b Analyse the results to allow you to tick the correct statements below.

   Light intensity stays the same no matter how far away the light source is.

   Light intensity increases the further away from the light source we are.

   Light intensity decreases the further away from the light source we are.

   c Predict the light intensity if the distance from the source was 12.5 centimetres.

   Ask students to locate the three gaps in the table first and suggest they put a small tick next to each one so they know which gaps they are looking for. Remind them to look for trends when working out the missing values.

Summative assessment

The review questions in the Student Book are a useful contribution when discussing the progress of each student individually. This information may also be used to create end-of-term reports for each student. You can score the questions as a summative test by awarding a mark for each answer. There are 23 marks in total (question 1 = 3; question 2 = 2; question 3 = 2; question 4 = 1; question 5 = 5; question 6 = 5; question 7 = 5).

It may also be useful to keep a record of individual and whole-class confidence levels for the self-review statements in the Workbook. This can help you and your students to identify areas that may need additional support later on. It will also help you to determine any concepts that have caused issues for many or all of the students. This will help you to reflect on how the topic was taught and so it helps to enhance your professional development.

Additionally, you can ask students to complete the Quiz Yourself questions for this unit to help check their understanding. These are on page 145 of the Workbook.

Investigate like a scientist

The 'Investigate like a scientist' task is designed to encourage students to apply their investigative and creative skills and review key aspects of the content of the unit. This task is found on page 103 of the Workbook.

 **Designing a model house of mirrors**

Resources: large cardboard boxes; flexible mirrors or mirrors on rolls; glue; sticky tape; scissors; small toys; measuring tapes long enough to measure an outside area; materials to create posters.

Provide students with the equipment for the investigation. Students will work as a team to design a fun house of mirrors. They use their understanding of how mirrors reflect light. They have observed how concave and convex mirrors reflect light differently. Remind students that concave mirrors are folded inwards (like caves) and convex mirrors bend outwards. They use this knowledge to design four different mirrors that will distort the image. They write a couple of sentences to explain how the mirrors are working. They can use small toys to model the house of mirrors. Students evaluate their model by answering the questions in steps 6a and 6b.

Being an architect

Provide students with the equipment for the investigation. Students will design an outside shaded area for the school. They observe the grounds and find a place where they would like to design the new area. They observe how the area is currently used and by whom and take measurements of the area and of the light in the area during the day. Students use their scientific skills to record their observations in a suitable method. They use their knowledge and understanding of light and shadows to draw designs of a shaded area. They can use the prompt questions to support them with their designs. Students produce an annotated poster to present their design and explain their ideas. Display these so that students can compare and evaluate their own and others' work to learn how to make improvements to their work in future.

4 Building Electrical Circuits

In this unit students will:

- revise that some materials are good electrical conductors, especially metals, and some are good electrical insulators

- associate the brightness of a lamp or the volume of a buzzer with the number and voltage of cells used in the circuit

- compare and give reasons for variations in how components function, including the brightness of bulbs, the loudness of buzzers and the on/off position of switches

- use recognised symbols when representing a simple circuit in a diagram.

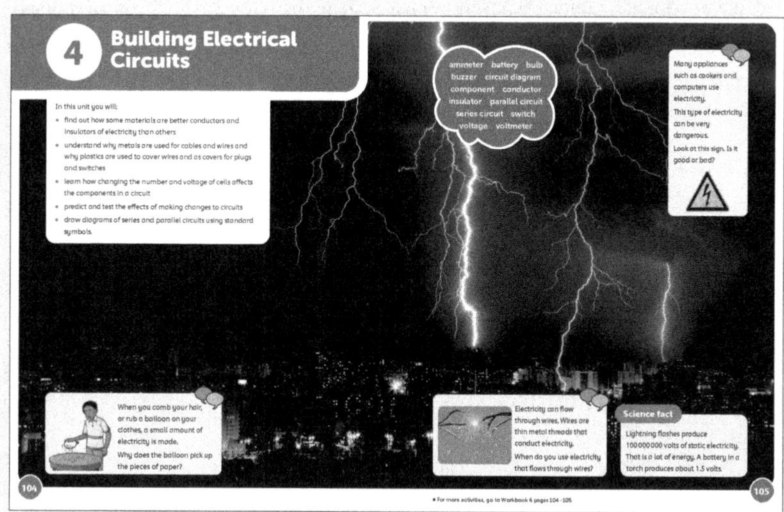

Supporting activities are in Workbook 6 pages 104–105.

Getting started

In this unit students review prior learning of electricity and then look in more detail at conductors and insulators. The content builds on work covered in Year 4.

The introductory lesson is designed to engage and inspire students. It reminds them of any prior learning and also introduces the ideas and concepts of electricity. In subsequent lessons, students investigate electrical conductors and insulators and move onto consider electrical circuits and building a test circuit. Students learn how to increase the number of components in a circuit and observe the effects. They learn the conventional symbols for drawing circuit diagrams. Finally, they measure the flow of electricity and link this to the brightness of bulbs and the number of batteries. They also understand the structure of wires. Circuit diagrams are drawn and interpreted.

Science in context

Use the lessons in this unit to encourage students to learn more about the importance of electricity in everyday life. Allow students to survey electrical appliances and other uses of electricity in school and their local area. Take them out to see important uses of electricity such as streetlights, traffic lights, shop and hospital lights and appliances such as refrigerators, computers and cookers. Students have studied electricity in earlier years, so remember to find out where they have visited in previous years to add to this experience rather than duplicate it. Encourage students to find out about how electricity is used to enhance life. Discuss the dangers of electricity and use examples of safety signs

and notices from the area. Invite people into school who use electricity as part of their work – such as electricians, electrical engineers, farmers and people who use portable lights such as climbers and cavers. Encourage students to keep a diary of electrical appliances and devices they use and to imagine what life would be like without electricity. Allow students to consider the development of our understanding of electricity and how appliances such as torches have changed over time. Also ensure that students are aware that electricity is very important but has negative impacts on the environment. Take them to see a power station, if possible, and discuss the fuels it uses; and visit a recycling plant to observe how appliances, devices and batteries are recycled.

Scientific enquiry skills

Students plan and carry out a range of investigations to help them test their ideas about electricity. They make predictions and write conclusions. They are expected to evaluate their enquiry work and ensure that they set up and carry out fair tests.

Prompt questions and discussions are included to support students as they plan and evaluate their investigations. Students also have opportunities to record findings in different ways and to interpret results.

You can use the Investigation master sheet on pages 4–5 to support investigative work. This provides prompts and structure to support students in planning and carrying out fair tests and in recording and drawing conclusions about their findings.

Student Book: batteries; wires; connectors; bulbs; range of everyday objects of varying conductivity to test (e.g. plastic and wooden rulers, paper, card, fabric, spoons, scissors); ammeters; nichrome wire in pieces of different thickness and length; switches; materials to make information sheets; voltmeters; writing materials; buzzers.

Workbook: materials for making leaflets, posters, information sheets and booklets; batteries; wires; connectors; bulbs; range of objects to test, some of which are good conductors and some are good insulators; access to information for research on Ampère; paperclips; six different metals to test for conductivity; nichrome wire in pieces of different thickness and length; switches; voltmeters; copper coins; aluminium foil; thin card; lemon juice or vinegar; salt; paper towels; sticky tape; LEDs with wires; string; paper plates or pieces of card; scissors; pencils; access to the internet and science books.

Key words for unit

Bold words are in the Word cloud and are included in the glossary.

ammeter appliance **battery** **bulb** **buzzer** cable cell circuit circuit breaker **circuit diagram** circuit symbol **component** **conductor** current fuse gauge **insulator** mains electricity metal **parallel circuit** plastic plug **series circuit** **switch** symbol **voltage** **voltmeter** wire

Scientific enquiry key words

Plan and/or carry out enquiries to answer questions

Make predictions

Recognise and control variables

Make observations

Take measurements, using equipment accurately

Record data and results

Analyse data, notice patterns and group or classify things

Report and present findings

Draw conclusions and give explanations

Identify causal relationships

The introductory lesson contains a Word cloud of the key words to be covered in the unit. These words could be placed on a Word wall in the classroom and students should add definitions of these words to their glossary when they have used the words and are familiar with how to use them. The eBook contains examples of the key words in context and students will hear how to pronounce them. The activity on page 104 of the Workbook is designed to support students in learning the key word 'circuit'.

Students should recall that a circuit in electricity is how all of the components join together. It has to be a complete circuit like a running track or the components will not work. A component is anything that is included in the circuit. Even a wire is a component.

'Voltage' is a more recent addition to the electricity vocabulary. Remind students that this is the push in a circuit. Some people think of it as the pressure pushing the current around the circuit.

You could collect science books and encyclopaedias to form a small class library as a resource for students to research definitions and also act as support for 'finding out' tasks. Consider downloading information sheets about the concepts to be covered in the unit, as well as allowing students access to the internet to help develop their computer skills.

Unit at a glance

The key teaching points for students in this unit are:

- to introduce the unit objectives
- to introduce the learning outcomes
- to engage students with the content of the unit
- to review and build on prior learning and understanding of the topics.

In the next lesson, students review prior learning of circuits from Year 4 and investigate insulators.

The introductory pages are designed to encourage students to think about examples of electricity they have seen. They are asked to think about lightning and static electricity, and about electricity in circuits. The lesson concludes with an emphasis on electricity and safety.

Read out the key words in the Word cloud on page 105 and remind students to add definitions of these words to their glossary throughout the unit. Encourage students to study the main photograph as this is meant to generate interest. The discussion tasks can be covered in any sequence, but a suggested order is shown below.

 **When you comb your hair, or rub a balloon on your clothes, a small amount of electricity is made. Why does the balloon pick up the pieces of paper?**

Ask students to work with a partner. Allow them to study the picture showing how the person is using a balloon to pick up small pieces of paper. They might recall investigations like this from their prior learning. Ask them to share examples of using this technique and then pairs can share their ideas with the class.

Possible response: Students might suggest that static electricity is causing the balloon to pick up the pieces of paper. This is different from the electricity that is used when we plug in an appliance. The charged particles attract oppositely charged particles.

Electricity can flow through wires. Wires are thin metal threads that conduct electricity. When do you use electricity that flows through wires?

Ask students to share their ideas about when they have used electrical devices with wires. Allow students to work in pairs or small groups. Elicit that many of the machines that help us at home, such as vacuum cleaners and refrigerators, use electricity that flows through wires. Ask if they can name any other appliances. Ask if they have seen wires connecting appliances to the mains supply. Use this opportunity to tell students to never touch wires, especially if the covering is broken.

Possible response: Students might suggest any appliances, for example computers, smartphone chargers or a games console.

Many appliances such as cookers and computers use electricity. This type of electricity can be very dangerous. Look at this sign. Is it good or bad?

Ask students to continue to work with their partner or small group. Ask students to look at the sign. Find out if any of them have seen this sign. You could ask if they have seen similar danger signs. Ask, 'Where will you see a sign like this? What does the sign tell you? What happens if we ignore the sign?'

Possible response: Students should suggest that the sign is a warning and they should stay away from the area.

Ask students to look again at the main photograph of lightning. Find out when they have seen storms like this. Ask, 'Did you know that lightning is an electrical spark?' Refer students to the science fact.

Science fact Lightning flashes produce 100 000 000 volts of static electricity. That is a lot of energy. A battery in a torch produces about 1.5 volts.

Read out the Science fact or ask a volunteer to read it out. Ask students to consider the electricity that is produced during a storm. Say the number out loud – one hundred million – and ask students why lightning can be so dangerous. Elicit that if a person is struck by lightning, the shock can kill that person, and if a building is struck by lightning, it can start a fire.

Key words (page 104)

Explain to students that this activity will let them explore the key word 'circuit'. This can be an individual or paired activity. Ask students to read the word at the top of the page out loud. They can then write some words beginning with 'circ' to help them make the link with a circuit being in a circle or at least a loop. Students should then write their own definition of the word 'circuit'. Students set up circuits in Year 4 so ask them to think back to this. Write down a definition of circuit on the board and ask students to compare their definition with yours. Allow them to change their definition if it doesn't match, but explain that this reviewing of answers and making changes is a normal part of learning. Ask students to write down their final definition and draw a picture of a circuit to help them remember the meaning.

Possible response: Students might list words like circle, circumference, circulate. They might define 'circuit' as a path that electricity can flow around. They could draw any picture including a source of electricity and wires.

Static electricity (page 105)

Explain that when people comb their hair, or rub a piece of plastic on their clothes, a small amount of electricity is made. Stress that this does not move anywhere. That is why it is called static electricity. Write 'static' on the board and explain it means 'not moving'.

Students can follow the instructions, record their observations and answer the questions. Explain that they need to make a list of when they have used devices and appliances with wires. This will help them to see the difference between static electricity and circuit electricity. Finally, ask them to describe how to keep safe when using electrical devices and appliances.

Possible response: The pieces of paper will be attracted to the comb because of the opposite charges on the comb and the paper. Students will have used many electrical devices including kettles, computers, smartphone chargers, hair dryers and a TV. In order to keep safe, we must not play with plug points, we must make sure electrical appliances are nowhere near water, and we must never use an appliance with a broken or damaged wire.

1 **Maths link:** Students could compare different materials by rubbing the comb with each material and counting which one can pick up the most pieces of paper. They should tear the pieces of paper into the same size so that it is a fair test. Ask them what else they should keep the same to make their investigation a fair test. Elicit that the material is the independent variable so will be changing, but the type of paper should be the same and they must try to keep the same distance between the paper and the comb.

2 Students make a small summary poster about everything they know about the uses of electricity. They can survey the room or the school and also download photographs of appliances and add them to their poster.

Revising electricity

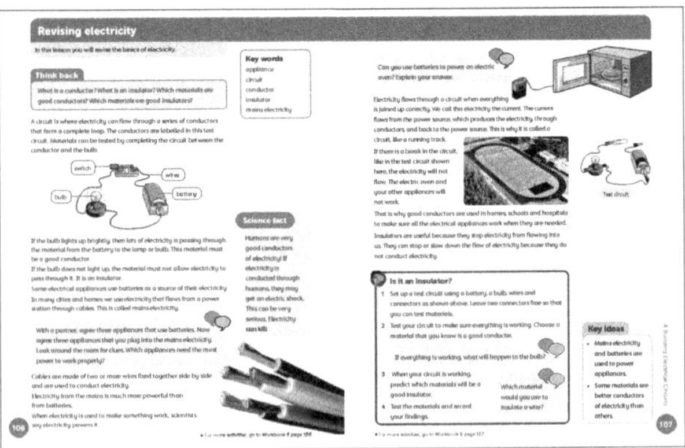

Supporting activities are in Workbook 6 pages 106–107.

Getting started

In this lesson students revise circuits, how electricity flows and how some materials conduct electricity and others do not. Students are reminded about the dangers of electricity. They make comparisons and predictions of electrical things that use batteries and those that use mains electricity. Finally, they set up a test circuit to investigate a selection of materials.

Language support

To explain the word 'conductor' ask students to look at the picture of the circuit on page 106. Ask, 'What makes the bulb light up?' Elicit that electricity flows from the battery and is carried in the wires around the circuit to the bulb. The wires conduct or carry electricity. The word 'insulator' means the opposite. You could also point out that these words are used for both electricity and heat.

Resources

Student Book: batteries; wires; connectors; bulbs; range of everyday objects of varying conductivity to test (e.g. plastic and wooden rulers, paper, card, fabric, spoons, scissors).

Workbook: materials for making leaflets; batteries; wires; connectors; bulbs; range of objects to test, some of which are good conductors and some are good insulators.

Key words

appliance   circuit   conductor   insulator   mains electricity

Other words in the lesson

cable   power   shock   wire

Scientific enquiry key words

Plan and/or carry out enquiries to answer questions

Make predictions

Recognise and control variables

Make observations

Take measurements, using equipment accurately

Record data and results

Analyse data, notice patterns and group or classify things

Report and present findings

Draw conclusions and give explanations

Identify causal relationships

Lesson at a glance

The key teaching points for students in this lesson are:

- mains electricity is much more powerful than batteries
- if electricity flows through a material, it is a conductor
- a circuit is where electricity can flow through a series of conductors that form a complete loop.

In the next lesson, students will learn more about how some materials are better conductors than others.

Think back: What is a conductor? What is an insulator? Which materials are good conductors? Which materials are good insulators?

Allow students to discuss their prior work on circuits and electricity with a partner. Ask them to share their ideas with another pair and then volunteers can tell the class.

Possible response: Students should recall that a conductor is a material that allows electricity to flow through it. An insulator stops or slows down the flow of electricity. There are many examples of good conductors and these are usually metals but graphite is also good. Insulators include examples such as rubber, plastic and wood.

Ask students if they remember any parts of a circuit from their earlier learning. Point out the diagram of a circuit on page 106 and ask students to study it and read through the text. Ask them why the bulb is lit in the circuit. Ask them what would happen if the connector to the battery was loose or the switch was open.

 **With a partner, agree three appliances that use batteries. Now agree three appliances that you plug into the mains electricity. Look around the room for clues. Which appliances need the most power to work properly?**

Students can work with their partner or form small groups to discuss the questions. Suggest they use the text they have just read to help them.

Possible response: Appliances that use batteries are smartphones, laptops, toy cars, remote controls, radios, watches and calculators, for example. Appliances that use mains electricity are washing machines, cookers, toasters, hair dryers and vacuum cleaners, for example. The appliances that use mains electricity need more electricity to work well.

Ask students to continue to read the text at the bottom of page 106. It is important that students understand that the metal wires inside cables are quite brittle and can be easily broken. They should take care of wires and not fold them or they could snap. Ask them what would happen if a wire did snap or break.

Science fact Humans are very good conductors of electricity! If electricity is conducted through humans, they may get an electric shock. This can be very serious. Electricity can kill!

Read out the science fact or ask a volunteer to read it out. Electrocution results in death and an electric shock can cause burns and pain. Ask students to draw from memory a sign they have seen that warns people of the danger of electricity.

 **Can you use batteries to power an electric oven? Explain your answer.**

Ask students to discuss the diagram of the oven connected to a battery. Use this to encourage a discussion about some things needing more electricity than a battery can provide. Ask students to discuss their ideas and responses to this question in small groups.

Answer: We would need lots of batteries to power an oven but it would be possible.

Ask students to read the text on page 107 and study the photograph and the circuit drawing. Ask, 'Why is a circuit called a circuit? What does each part in the test circuit do? Why wouldn't the bulb in the drawing light up? How could you change the circuit so that electricity can flow from the battery and light up the bulb?'

 **Investigation: Is it an insulator?**

Allow students to work in groups of three or four. Explain that they are going to set up a test circuit as shown on page 107 and then test materials to find out if they conduct electricity. The worksheet on page 107 of the Workbook supports this investigation.

Make sure students leave two connectors free so that there is a gap in the circuit where they can place the materials to be tested. After testing their circuit with a known conductor, ask them to test the other materials and record their findings.

> **Answer:** *Students should find that metals are good conductors and plastic, wood or rubber are good insulators.*

During the investigation, ask students to talk about the discussion tasks below.

If everything is working, what will happen to the bulb?

Allow students to discuss this after step 2 of the investigation. Ask them to use the diagrams to encourage their thinking.

> **Answer:** *The electricity will flow to the bulb and it will light up.*

Which material would you use to insulate a wire?

Use this discussion at the end of the investigation to help students think about their results and conclusions. Remind them that scientists analyse their results to help them reach conclusions.

> **Answer:** *Any of the insulators they tested could be used to insulate the wire. Plastic is usually used but rubber was used in the past.*

Key ideas

- *Mains electricity and batteries are used to power appliances.*
- *Some materials are better conductors of electricity than others.*

Read out the key ideas or ask a volunteer to read them out. These will remind students that some materials allow electricity to flow through them. We call these conductors. Others stop the flow and we call these insulators. Ask students to tell their partner some examples of conductors and insulators.

Workbook activities

Explain conductors and insulators (page 106)

Explain to students that they are going to design a leaflet to help younger children understand what conductors are and what insulators are. Point out that the information they put in the leaflet will have to be correct but easy to understand.

Ask students to use the checklist provided and to fill in the table of conductors and insulators to remind them what to include in their leaflet. Then let them design their leaflet and present it to the class. Finally, ask students to complete the sentences to help them remember the important safety rules.

> **Possible response:** *Students include that insulators slow down or stop the electricity flowing around a circuit; conductors allow the electricity to flow. They list the names of insulators, for example plastic and rubber; and conductors, for example metals and graphite. They complete the sentences with these words: broken; wet; water; electric socket.*

Is it an insulator? (page 107)

This activity supports the investigation on page 107 of the Student Book.

Explain that students are going to use a test circuit to investigate which materials are good insulators. Ask students to use this activity as additional support for carrying out the test. Point out that they should answer the questions as they work through the test and use the table to record their findings. Check their answers and discuss these. Ask them to decide which material they would use to insulate a wire.

> **Possible response:** *3 Students should name any conductor, for example metals. 4 The bulb will light. 5 Students may predict that good insulators are plastics and rubber. 7 Students will probably choose plastic to insulate a wire because it is cheap to use and very effective.*

Review and reflect

Encourage students to identify aspects they have not completed correctly and help them to identify improvements. This will help students to develop a positive approach to learning by understanding that learning is a process that will improve with practice and reflection. For example, they can walk around and look at any leaflets to identify what others have done well and what they can learn from others as targets to improve their work.

Conclude the lesson by asking students to complete question 4 in the 'What have I learned about building electrical circuits?' activity on page 124 of the Student Book. (See also the teaching notes in this Teacher's Guide, pages 167–169.)

Extra activities

1 Ask students to research different electrical appliances and tools to see which parts are made from conducting materials and which are made from insulating materials.

2 Ask students to carry out research to find out how some people use test circuits as part of their work. You could show them photographs of screwdrivers with LEDs used to test circuits and other test meters such as neon voltage testers and multimeters.

Differentiation

Supporting: Provide a test circuit already set up for students to use for testing insulators.

Consolidating: Allow students to set up a circuit and test materials around the room to become familiar with the workings of the test circuit.

Extending: Ask students to predict which materials will be good insulators based on their knowledge of materials before they test their predictions by setting up their own circuits.

Differentiated outcomes

All students	should be able to test materials using a test circuit
Most students	will be able to set up their own test circuit and test materials to find if they are insulators
Some students	may be able to predict the best insulators from a range of materials based on their knowledge of materials

Choose your conductor

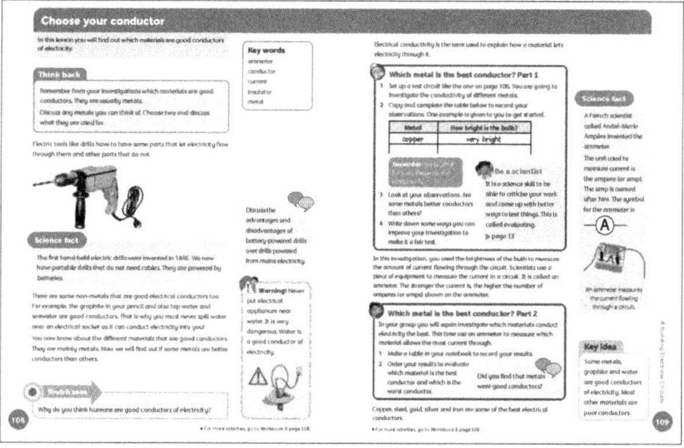

Supporting activities are in Workbook 6 pages 108–109.

Getting started

In this lesson students investigate conductors of electricity. The focus of this lesson is on the conductivity of a variety of metals. Students are introduced to the ammeter as a piece of equipment that measures current.

Language support

The Word cloud at the beginning of the unit could be used throughout to check students' understanding of concepts and introduce any unfamiliar words. Most students should be familiar with the key words in this topic, although 'ammeter' may be unfamiliar. Remind students that the word 'meter' is used for a device that measures something. They used forcemeters in Year 5, for example. Tell students that an ammeter measures the electrical current flowing around a circuit and the unit used is amps.

Resources

Student Book: batteries; wires; connectors; bulbs; range of metals to test for conductivity; ammeters.

Workbook: materials to make posters or leaflets; access to information for research on Ampère; batteries; wires; connectors; bulbs; paperclips; six different metals to test for conductivity.

> ### Key words
>
> ammeter   conductor   current   insulator   metal
>
> **Other words in the lesson**
>
> ampere (amp)   conductivity   copper   drill
>
> graphite   portable   powered   socket   tools

Scientific enquiry key words

Plan and/or carry out enquiries to answer questions

Make observations

Take measurements, using equipment accurately

Record data and results

Analyse data, notice patterns and group or classify things

Report and present findings

Draw conclusions and give explanations

Identify causal relationships

Lesson at a glance

The key teaching points for students in this lesson are:

- some materials are good conductors and others are not
- metals and some non-metals are good conductors
- a test circuit can be used to test if a material is a conductor.

In the next lesson, students will learn about the uses of metals and plastics in electrical circuits and appliances.

Think back: Remember from your investigations which materials are good conductors. They are usually metals. Discuss any metals you can think of. Choose two and discuss what they are used for.

Ask students to list as many different metals that they know of. They can do this individually and then compare their list with a partner and make a joint list. Let them choose two of the metals on their list. Ask, 'What do you think they are used for?' Elicit from students the materials they remember to be good conductors from their investigations. Write their answers on the board.

Discuss the advantages and disadvantages of battery-powered drills over drills powered from mains electricity.

Ask students to read the text and look at the photograph of the drill on page 108. Encourage them to look at the materials used in the construction of the drill. Ask them which ones would be insulators, using their knowledge from the investigation on page 107.

> *Answer: Battery-powered drills are not connected to mains electricity so they are more portable. The power stored in the battery will be used up so it will need to be recharged. Those powered by mains electricity carry on without needing to be recharged.*

Science fact The first hand-held electric drills were invented in 1895. We now have portable drills that do not need cables. They are powered by batteries.

Read out the Science fact or ask a volunteer to read it out. Ask students to think about the first drill that was invented. Ask them to think about what it might have been like. They might suggest that is was bigger and heavier as it was probably constructed out of metals and possibly wood.

Warning! Never put electrical appliances near water. It is very dangerous. Water is a good conductor of electricity.

Read out the safety warning and ask students why this is an important safety rule. Ask students to discuss why they should not have an electric heater above a bath or a socket very near to a shower.

Stretch zone: Why do you think humans are good conductors of electricity?

Ask students to talk about the Stretch zone question with a partner. Ask them to decide on a reason and then you can ask volunteers to share their ideas with the class.

> *Answer: Humans are made up of about 60% water, which is a good conductor of electricity. Salt water is an even better conductor of water and we also contain salts.*

Point out the sentence at the top of page 109 and ask a volunteer to read it out to the class. Ask, 'What do we call a material that does not let electricity through it? Why are these materials important?'

Investigation: Which metal is the best conductor? Part 1

Explain that in this investigation students are going to compare different metals to see if some are better conductors than others. They can work in groups of three or four so they can share roles and discuss ideas. The worksheet on page 109 of the Workbook supports this investigation.

Leave students to set up a test circuit as on page 107 and then test the different metals. Ask them to copy and complete the table given to record their observations. Remind students that the brighter the bulb, the better the conductor is. After testing all of the metals students should draw their conclusions.

Ask each group to discuss whether all the metals conducted electricity. Are some metals better conductors than others? Which metal did the group think was the best conductor?

Allow students to reflect on their investigation and then write down some ways they could improve it to make it a fair test. Did they make any mistakes? The observations will always result in some errors. The connections may be loose or the battery flat. Ask students if brightness is an accurate way of measuring the amount of electricity that the metal allows through to the bulb.

Be a scientist: It is a science skill to be able to criticise your work and come up with better ways to test things. This is called evaluating. (Student Book, page 13)

Ask students to read the Science fact as they are carrying out their investigation so it can encourage them to evaluate their work and suggest improvements.

Science fact A French scientist called André-Marie Ampère invented the ammeter. The unit used to measure current is the ampere (or amp). The amp is named after him. The symbol for the ammeter is —Ⓐ—.

Read out the Science fact or ask a volunteer to read it out. Ask students to discuss how using an ammeter would improve the investigation they have just carried out. Elicit that measuring the current flowing through each metal would give much more reliable results.

Ask students to read the information between the investigations on page 109. This describes how the ammeter could be used to make improvements to investigations.

Introduce students to an ammeter and how it is set up in a circuit. Show them the symbol as a circle with a capital 'A' in the centre. They can then carry out the second investigation.

Investigation: Which metal is the best conductor? Part 2

Ask students to repeat the earlier investigation. Explain that this time they will include the ammeter in the circuit as a more accurate measure of conductivity than the brightness of the bulb. Show students that the ammeter gives a number. A bigger number means more electricity is flowing through the metal. The units are amps. Allow students to test each metal and record the current in amps. They should make a table in their notebook and order the results to rank the conductors.

Did you find that metals were good conductors?

Ask students to work with their investigation group to discuss whether they found evidence to prove that metals were good conductors. Encourage them to look back at their results table to help develop the skill of analysing results and using data to help them draw conclusions. Point out that their science ideas should be supported by data.

Ask students to read the final sentence of text at the bottom of page 109. Ask them to think about which metals from the list they have tested.

Key idea

Some metals, graphite and water are good conductors of electricity. Most other materials are poor conductors.

Summarise the lesson by asking a volunteer to read out the key idea. Then ask students to close their books and on a piece of blank paper list two good conductors of electricity that they found from their investigations. They should also explain why using an ammeter improves their investigations.

Workbook activities

Research project: Ampère (page 108)

Point out to students that this activity will help them to learn about an important scientist who helped in discoveries about electricity over 200 years ago. Ask students to work with a partner to find out about Ampère. They should use the prompt questions and then present their findings on a poster or in an information leaflet. Arrange for them to have access to books, magazines, the internet or any other reliable source of information in their research. You can display the posters and information leaflets and allow students to walk around the exhibition to share ideas.

Investigate the conductivity of metals (page 109)

This activity supports the investigations on page 109 of the Student Book. The aim is to allow students to discover that metals are very good conductors but some are better than others.

Ask students to set up a test circuit and then follow the instructions to test their circuit. Then they use it to find out which metals are good conductors and which metals are not as good. Remind students to make sure the metals are properly connected to make the results reliable. Students should record their results in the first and second columns in the table and then number the metals in order of conductivity, from the best conductor (1) to the worst conductor (6).

Possible response: Students record their findings in the table. The results will vary according to the metals investigated but you could share the official order, from best to worst: silver, copper, gold, aluminium, zinc, nickel, brass, bronze.

Review and reflect

Encourage students to reflect on their own learning by pausing for a few moments and thinking about which parts of the work they found tricky. Talk about how they managed this and suggest that next time they stop for a short while and take a few deep breaths. Also remind them that when they are doing challenging work, they are training their brain and this will help their future learning. Allow them to walk around the displays of posters and information leaflets about Ampère to compare their work with others and pick up ideas for improvements.

Conclude the lesson by asking students to complete the first statement in the 'What I have learned about building electrical circuits' activity on page 124 of the Workbook. (See also the teaching notes in this Teacher's Guide, pages 167–169.)

Extra activities

1 **Computing link:** Ask students to research other scientists who have helped our understanding of electricity – for example the Ancient Greeks, Alessandro Volta, Edith Clarke, Maria Telkes. They can use books, the internet or printed information sheets you provide. They can find out who the scientists were, where they lived, and when and what they studied and found out. Ask, 'Did they invent or research anything else?'

2 **Computing link:** Students study their chart showing the order of conductivity and then select one metal from the list. They find out about the metal using books or by carrying out an internet search and they create an information card about it. They draw or download pictures of the metal being used and include any uses that depend on it being a conductor of electricity.

Differentiation

Supporting: Set up the ammeter in a circuit to demonstrate to students how it is used and how to take readings.

Consolidating: Ask students to make and re-assemble the test circuit and also to repeat tests to increase their confidence in building circuits.

Extending: Students can test the metals using an ammeter and then compare their list with secondary data to consider accuracy.

Differentiated outcomes	
All students	should be able to state that different metals do not conduct equally well
Most students	will be able to use an ammeter to measure the current flowing through different metals in a circuit
Some students	may be able to compare the ammeter readings of different metals in a circuit to secondary data

Using metals and plastics in electrical circuits

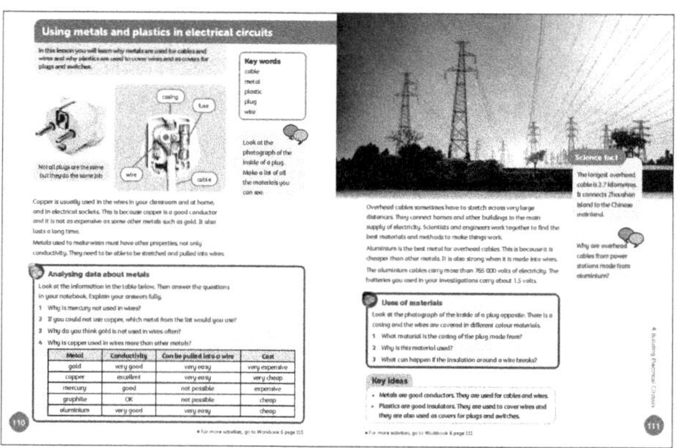

Supporting activities are in Workbook 6 pages 110–111.

Getting started

In this lesson students learn that metals are used to make cables and wires because they are good conductors of electricity and can also be pulled into long, thin shapes. They also understand that cables are covered in plastic because plastic is a good insulator.

Language support

'Cables' and 'overhead cables' may be unfamiliar words. Show students the image of them in the Student Book. Ask them if they have seen structures like these and if they know what they are used for. Most of the key words in this lesson will already be familiar to students.

Key words

cable   metal   plastic   plug   wire

Other words in the lesson

aluminium   conductivity   copper   gold
graphite   mercury   volts

Scientific enquiry key words

Make observations

Analyse data, notice patterns and group or classify things

Lesson at a glance

The key teaching points for students in this lesson are:

- as metals are good conductors they are used for cables and wires
- insulators are used to cover wires and to make covers for plugs and switches.

In the next lesson, students will predict and investigate the effects of making changes to circuits.

Look at the photograph of the inside of a plug. Make a list of all the materials you can see.

Show students a plug. Ask if they know what it is made of and what is inside. Direct them to look at the photograph of the plug and study the labels. Ask students to identify the materials in the plug and make a list. Ask, 'Why is copper used in the wire? Why is plastic used?' In most countries it is illegal to wire a plug unless you are an electrician. The idea of this discussion is to show the workings of a plug, not to learn how to wire one.

> **Answer:** *Students should observe plastic around the cable, plastic casing, plastic around metal wires, metal screws and metal fuse parts.*

Ask students to read the text on page 110 before starting the investigation. They should find that copper is commonly used in wires but it is not the best conductor of electricity.

Investigation: Analysing data about metals

Explain that the purpose of this investigation is to encourage students to analyse secondary data about metals. This can be an individual or a paired activity. Ask students to look at the information in the table and answer the questions. Elicit that the wires need to be cost effective, as well as good conductors. Volunteers can share their ideas with the class.

> **Answer:** *Mercury is a liquid at room temperature (and is classed as a poison). Students could use any of the other metals from the table. It would be very expensive to use gold in wires and cables. Copper is a good conductor and is easy to work with and it is much less expensive than gold.*

Ask students to continue to work with their partner to read the text on page 111 and look at the photograph. Read out the Science fact or ask a volunteer to read it out.

Science fact The longest overhead cable is 2.7 kilometres. It connects Zhoushan island to the Chinese mainland.

Why are overhead cables from power stations made from aluminium?

Discuss the difference between wires and cables (a cable is much thicker and may comprise many wires). Ask students to think about the distances cables need to stretch across. Ask them to look at the table on page 110 in pairs and decide which metal they would choose for an overhead cable. Use the Science fact to demonstrate that some cables need to cross very long distances.

Investigation: Uses of materials

The purpose of this investigation is to encourage students to consider how insulators are used in real-life objects to shield us from electricity. Ask students to look carefully at the plug on page 110 and decide which material the casing of the plug is made from. They can then discuss why this type of material is used and what would happen if the insulation around a wire breaks. They can write this down or ask volunteers to stand up and tell the class their ideas. Encourage them to think about their investigations on insulators and to consider the other properties of materials used to cover plugs and wires. For example, is it easy to shape? Is it expensive? Does it catch fire easily?

You could explain that before plastic was invented, plugs and wires were covered in other insulators. Rubber and sometimes wood were used. Take the opportunity to reinforce that students must never touch a broken wire, even if the appliance is not plugged in. Explain that electricity can still flow. Show them a computer laptop or a phone charger. Some have a light near the plug which stays illuminated for some time after it has been unplugged.

Key ideas

- *Metals are good conductors. They are used for cables and wires.*
- *Plastics are good insulators. They are used to cover wires and they are also used as covers for plugs and switches.*

Summarise the lesson by asking students to close their books and count to 20. After that they should turn to a partner and between them write down what they think the key ideas for the lesson should be. They can share their ideas with the class and then open up their books to check if they were correct. Once they have completed this you can ask students to tell their partner the types of materials that are used to cover wires and sockets, and to explain why. Ask them to say why we should never use cables where the insulation is damaged.

Workbook activities

Wiring a plug (page 110)

In most countries it is illegal to wire a plug unless you are an electrician. The idea of this activity is to show the workings of a plug, not to learn how to wire one.

Students can complete this activity as an individual or a paired task. Ask them to look at the diagram of the plug and label the parts. Point out that it is vital that the wires are connected properly – so they are colour-coded. Ask students to colour the wires in the correct colours and then recall what the different parts of the plug are made from. Explain that they need to label each part with 'C' for conductor or 'I' for insulator. Finally, ask students to choose two parts of the plug and state what each part does. They record their ideas by completing the table.

What have you learned so far? (page 111)

The purpose of this activity is to allow students to reflect on what they have learned about electricity so far. Ask them to read through each question and then fill in their answers in the boxes provided. You can then let students work with a partner to compare and discuss answers before reading out the correct answers and letting students self-assess their work.

Review and reflect

Use the activity on page 111 of the Workbook to encourage students to reflect on their learning. You can discuss this with individuals or the class. Ask students to note any questions they find difficult and encourage them to look back in their books for answers.

Conclude the lesson by asking students to complete the second statement in the 'What I have learned about building electrical circuits' activity on page 124 of the Workbook. (See also the teaching notes in this Teacher's Guide, pages 167–169.)

Extra activities

1 Ask students to research the parts of a plug. Ask them to explain what the different parts of the plug are used for. They should include the fuse and the Earth wire.

2 Students find out what kinds of plugs are used in their country. They make a poster explaining what the inside looks like and if the wires are coloured the same. They draw and label a diagram of their findings.

Differentiation

Supporting: Students can practise labelling the parts of the inside of a plug and you can allow them to look back at the picture on page 110 of the Student Book for support.

Consolidating: Students can draw a diagram of a plug and label the parts from memory.

Extending: Students can draw a diagram of a plug and label the parts and the materials used from memory.

Differentiated outcomes	
All students	should be able to label the parts of a plug
Most students	will be able to name good insulator and conductor materials used in plug
Some students	may be able to explain why some metals are used in wires and cables rather than others

Changing circuits

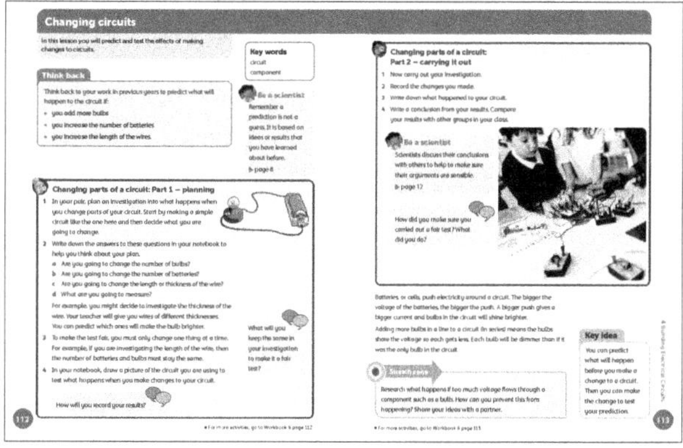

Supporting activities are in Workbook 6 pages 112–113.

Getting started

This lesson introduces students to the idea that when the components in a circuit are changed the circuit behaves differently. Students make predictions about the changes if more bulbs or batteries are added to a circuit. They also investigate if the thickness of a wire has any impact on a circuit. Students test their predictions. They practise using their results to draw conclusions. They also explore how in a series circuit the voltage from a battery or mains electricity is shared across all of the components.

Language support

Read through the key words with the class. Ask students to define 'circuit' and 'component' based on their prior learning. 'Component' is a difficult word – it has been covered in previous lessons but make sure you use the word interchangeably with 'electrical part'. Discuss the circuits they have used. Ask students to give examples of the components they have used, for example bulbs, batteries and switches. They can complete any glossaries or Word walls that they might have started as they experience the words and use them.

Resources

Student Book: batteries; wires; connectors; bulbs; nichrome wire in pieces of different thickness and length.

Workbook: batteries; wires; connectors; bulbs; nichrome wire in pieces of different thickness and length.

> **Key words**
>
> circuit   component
>
> **Other words in the lesson**
>
> batteries   bulbs   gauge   series   voltage   wire

Lesson at a glance

The key teaching points for students in this lesson are:

- changing the components in a circuit has an effect on the other components
- the changes in a circuit can be predicted.

In the next lesson, students will learn about the uses of circuit breakers such as switches.

Think back: Think back to your work in previous years to predict what will happen to the circuit if:

- you add more bulbs
- you increase the number of batteries
- you increase the length of the wires.

Ask students to work with a partner to discuss each of the Think back points. Ask them to recall the investigations they have carried out in prior learning. They could draw any diagrams to help them recall or list the components they used.

Possible response: Students should predict that when more bulbs are added, they will be dimmer. When the number of batteries is increased, the bulbs will be brighter. When the length of a wire is increased, the bulbs will be dimmer.

 **Be a scientist:** Remember a prediction is not a guess. It is based on ideas or results that you have learned about before. (Student Book, page 8)

Read out the Be a scientist information and ask students to think about what a prediction is before they carry out the investigations. Remind them that this is not a guess. It is based on knowledge and experience.

Ask students to work in pairs for the duration of the investigations in this lesson.

 **Investigation: Changing parts of a circuit: Part 1 – planning**

The purpose of this investigation is to help students explore what happens when changes are made to a circuit. Stress that in this first part they are only going to carry out the planning phase. Arrange students into pairs and ask them to follow the instructions to plan an investigation into what will happen if they change a simple circuit by altering the: number of bulbs; number of batteries; length of wire or thickness of wire. Ask the pairs to decide what they will change and what they will measure. If ammeters are available, the results will be easier to record.

Allow students to work independently and only step in if they need support. This will help them to develop perseverance and will help to increase their confidence and problem-solving skills. Also explain that to make the test fair, they should only change one thing at a time. Each time they should draw a picture of the circuit they are using to test what happens.

Tell students that when the components (bulb, ammeter) are connected one after another in a row it is called a series circuit.

If students require more support, you can let them use page 112 of the Workbook or structure the task by asking the following questions: (a) Are you going to change the number of bulbs? (b) Are you going to change the number of batteries? (c) Are you going to change the length/thickness of the wire? (d) What are you going to measure? Ask students to write down the answers as you ask the questions.

Possible response: Students will follow steps 1–4 to plan their investigation using all the prompts provided.

Ask students to talk about the discussion questions in their investigation groups before moving onto Part 2 of the investigation.

 **What will you keep the same in your investigation to make it a fair test?**

Ask students to discuss what makes a fair test and then to evaluate their investigation. Remind them that a fair test will have an independent variable (what they are changing), a dependent variable (what they are measuring) and control variables (what they are keeping the same).

Possible response: Students should recognise that to run a fair test they can only change one variable at a time. This is called the independent variable. They measure the outcome of the dependent variable.

 **How will you record your results?**

Ask students to think about how they will record their results. They should set up a table in their notebook or use the one on page 112 of the Workbook. Ensure that they prepare this prior to the investigation.

Investigation: Changing parts of a circuit: Part 2 – carrying it out

Students should now carry out their planned investigation. Remind them to record the changes they make and write down what happens in their circuit. Remind them to test their circuit and that all the components are working before beginning the investigation. Ask the pairs to carry out their investigation, changing either the number of bulbs or batteries or the length/thickness of the wire.

Ask the groups to discuss if there was any pattern in their results, for example as the length of wire increased what happened to the bulb? After analysing the results, they should write a conclusion and compare their results with other groups in the class.

Possible response: Students should use their results to write a conclusion. They should find that as more bulbs were added or the length of wire was increased the brightness of the bulb decreased. If they added more batteries, the brightness increased. They should be able to identify these patterns using their data.

Be a scientist: Scientists discuss their conclusions with others to help to make sure their arguments are sensible. (Student Book, page 12)

Ask students why it is so important to talk about their conclusions with other people. Elicit that by doing this they can check if their ideas make sense and also consider other people's ideas. Point out that this is why a good investigation plan has to be detailed enough so that other people can repeat it. This means they can share their data and see if they come up with the same conclusions.

How did you make sure you carried out a fair test? What did you do?

Ask students to talk about the questions to encourage them to think about their findings and how they carried out the investigation. Ask them to compare their methods and suggest any improvements to it.

Possible response: Students should have only changed one variable and controlled any others that could affect the dependent variable.

Ask students to work independently to read through the summary text under the investigation on page 113.

Stretch zone: Research what happens if too much voltage flows through a component such as a bulb. How can you prevent this from happening? Share your ideas with a partner.

Allow students to use the internet to find out about voltage and bulbs. Ask them to find different bulbs and record the voltage stated on them.

Possible response: If too much voltage flows through some filament bulbs, the wire heats up and breaks. Students should use the correct voltage recommended for all components to avoid breakages or even electrical fires.

Key idea

You can predict what will happen before you make a change to a circuit. Then you can make the change to test your prediction.

Ask a volunteer to read out the key idea. This will remind students of what happens to a circuit when components are changed. Ask them to explain to a partner what happens when more bulbs or batteries are added to a circuit. Ask them what they observed. You could draw a circuit on the board with three bulbs and two batteries and ask students to suggest one way of making the bulbs brighter and one way of making the bulbs less bright.

Workbook activities

Changing the components in a circuit (page 112)

This activity supports the investigations on pages 112–113 of the Student Book. Ask students to use it to structure their thinking as they plan their investigation. They should start by drawing a circuit diagram of the circuit they will use for their investigation. Ask students to follow the instructions and use the table to record their findings. If they use an ammeter, they can write in the number of amps. Finally, they can evaluate their results and decide if their predictions were correct and if any results surprised them.

Possible response: Students should find that when more bulbs are added they become less bright. When more batteries are added the bulbs become brighter. If the wire length is increased, the bulbs become less bright. Students should analyse their own results.

Investigating the thickness of a wire (page 113)

Explain that this investigation takes one of the ideas in the previous investigation and allows students to study it in more detail. Point out that they are going to use some wires of different thicknesses to test the effect on the brightness of the bulb in a circuit. Make sure they label each wire with a description of its thickness. It is sufficient to describe wires as very thin, thin, medium, wide or thick, for example. When you obtain the wires, they may have measurements in millimetres, so include this if so. Tell students that the thickness of a wire is called the gauge.

Ask students to order the wires according to how thick they look and to predict which wire will make the bulb light the brightest. They should then design an investigation to test and record their results. Allow them to use a digital camera or smartphone to take a photograph of the set-up and fix it in the space provided or draw a diagram. They can then write down their conclusions.

> **Possible response:** *Students should set up a simple series circuit and add the different thicknesses of wires using connectors. They should record their observations for each wire. They should find that the thickness of the wire affects the brightness of the bulb. The thicker the wire, the more electricity can flow around the circuit.*

 ## Review and reflect

Use the discussion tasks, the Stretch zone activity and the Workbook tasks to encourage students to think about what they understand and what they are finding less straightforward. Discuss the outcomes of each task with students. Encourage them to identify aspects they have not completed correctly and help them to identify improvements. This will help students to develop a positive approach to learning by understanding that learning is a process that will improve with practice and reflection.

Conclude the lesson by asking students to complete questions 2 and 9 in the 'What have I learned about building electrical circuits?' activity on **pages 124–125** of the Student Book and the third statement on **page 124** of the Workbook. (See also the teaching notes in this Teacher's Guide, **pages 167–169.**)

Extra activities

1 Students take pictures of their investigation and describe how they made this a fair test.

2 Students research the different thicknesses of wires and explain in a leaflet how they are used.

Differentiation

Supporting: Demonstrate a circuit so students can observe an example of what happens when one of the components in a circuit is changed.

Consolidating: Students can continue to add bulbs to a circuit until they no longer light to illustrate how the bulbs have to share the electricity and so each has less and less.

Extending: Students predict and test the effect of the thickness of a wire on a bulb. They repeat their observations.

Differentiated outcomes	
All students	should be able to observe what happens to a circuit when more bulbs are added
Most students	will be able to predict what will happen when more batteries are added to a circuit
Some students	may be able to make a number of predictions based on changes to a circuit

Circuit breakers

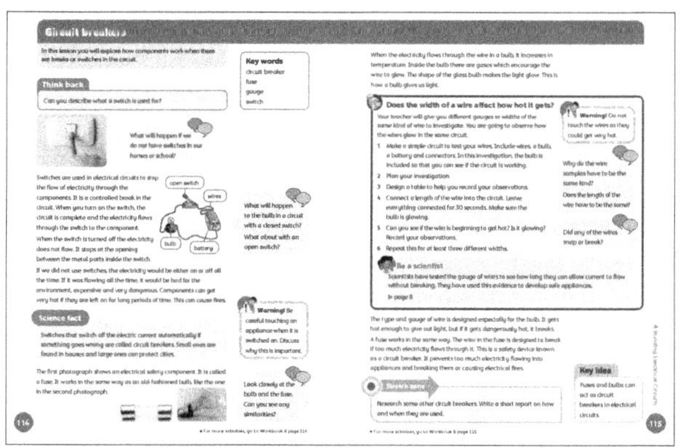

Supporting activities are in Workbook 6 pages 114–115.

Getting started

In this lesson students will explore how some components can break a circuit and stop the current. Students learn that these are called circuit breakers. We can use circuit breakers to control the flow of electricity around a circuit. A switch is an example of a component that stops the flow of electricity. This acts as a controlled break in the circuit. Fuses are also controlled circuit breakers. Students also explore how components can get hot when electricity flows through them.

Language support

Some of the key words in the lesson might be unknown to most students. They might have heard of 'fuse' and 'switch' but not fully understand what they mean. Some of the other words are not simple and they are not easy to spell, for example 'gauge' and 'circuit breaker'. A strategy you can use is to write the words on the board but leave some gaps so letters are missing. Students can copy the words and then discuss which letters are missing. They can self-check by referring to the key words box on page 114 of the Student Book. They can then share examples of when they have heard the words before so you can elicit prior understanding. Encourage students to make sentences up using some of the key words. Ask for volunteers to read out the sentences they have constructed.

Resources

Student Book: batteries; wires; connectors; bulbs; nichrome wire in pieces of different thickness.

Workbook: batteries; wires; connectors; bulbs; nichrome wire in pieces of different thickness.

Key words

circuit breaker   fuse   gauge   switch

Other words in the lesson

appliance   automatically   components

Scientific enquiry key words

Plan and/or carry out enquiries to answer questions

Make observations

Record data and results

Analyse data, notice patterns and group or classify things

Report and present findings

Draw conclusions and give explanations

Lesson at a glance

The key teaching points for students in this lesson are:

- circuit breakers control the flow of electricity
- switches and fuses are commonly used circuit breakers.

In the next lesson, students will draw circuit diagrams using circuit symbols.

Think back: Can you describe what a switch is used for?

Ask students to recall their prior learning by discussing the question about switches. Ask them to sit quietly and think back to earlier work on electrical circuits. Remind them that they made various circuits in Year 4. They can share ideas with a partner and then a volunteer can share their ideas with the class.

Possible response: Students should recall that when a switch is open there is a break in the circuit and electricity cannot flow. When it is closed the circuit is complete and electricity can flow through to the next component.

What will happen if we do not have switches in our homes or school?

Ask students to discuss the switch shown on page 114. You can also allow students to move carefully around the room and observe, but not touch, some switches and discuss what they turn on and off.

Possible response: The lights would be on all of the time. This would be expensive and could also be a fire hazard.

Ask students to look at the diagram of the circuit and read the information to the left of it. Ask them to discuss how the open switch could be called a controlled break in the circuit.

What will happen to the bulb in a circuit with a closed switch? What about with an open switch?

Ask students to work with a partner to discuss the questions. Ask them if they can link how the switch on the wall works with the ones they have used in their investigations.

> *Possible response: When the switch is closed the circuit is complete and the bulb would light up. When the switch is open there would be a break in the circuit so the electricity would not flow and the bulb would not light.*

Science fact Switches that switch off the electric current automatically if something goes wrong are called circuit breakers. Small ones are found in houses and large ones can protect cities.

Read out the Science fact or ask a volunteer to read it out. Ask students to discuss how this could keep us safe. They can suggest ideas about what might happen if a house did not have circuit breakers.

Warning! Be careful touching an appliance when it is switched on. Discuss why this is important.

Ask students to read the warning information. Discuss as a class why they think this is important. They might have noticed that monitors, smartphones and TVs can get very hot when they are in use. Touching them could result in burns.

Look closely at the bulb and the fuse. Can you see any similarities?

Ask students to read the information at the bottom of page 114 and study the photographs closely. Ask if they can see the wires.

> *Possible response: Students should see that there is a fine wire in both the bulb and fuse.*

Next, ask students to read the information at the top of page 115. Ask them what would happen if the wire in the bulb got too hot. You can point out that gases inside bulbs – such as neon and argon – do not react or burn easily and some bulbs have very little gas inside. This helps the wire to last longer and not burn up. Ask students to imagine what would happen to a light bulb filled with oxygen.

Investigation: Does the width of a wire affect how hot it gets?

The worksheet on page 115 of the Workbook supports this investigation. Arrange students into groups of three or four. Start by reading out the warning information and asking students to discuss why this is so important.

Warning! Do not touch the wires as they could get very hot.

Explain that the wires can get so hot that they can set fire to paper if it gets too close to them. Hand out different gauges or widths of the same kind of wire to investigate. Explain that students are going to test and observe how the different wires glow in the same circuit. Allow them to make a simple circuit to test the wires and remind

them that they need a bulb in the circuit so that they can see if the circuit is working. Students can follow the instructions to test the wires and record their findings. Discuss with them why they should repeat this for at least three different widths. During the investigation students can discuss the discussion questions.

> *Possible response: Students should find that the wider wires do not get as hot as the thinner wires. This is because thicker wire is able to let more electricity flow through it (it has less resistance).*

Why do the wire samples have to be the same kind? Does the length of the wire have to be the same?

Ask students to think about their investigation method. Ask them to think about the variables before they answer the questions.

> *Answer: The wires need to be the same kind otherwise too many variables would change. If they are investigating the gauge, only that should change so the length should be the same.*

Did any of the wires snap or break?

Students should discuss this in their group. If a wire did break, ask students how they could improve their investigation to avoid this happening.

> *Possible response: Some of the wires might have overheated so much that they broke. If students leave the wires connected for 30 seconds only, this is unlikely to happen.*

Be a scientist: Scientists have tested the gauge of wires to see how long they can allow current to flow without breaking. They have used this evidence to develop safe appliances. (Student Book, page 8)

Ask the investigation teams to discuss this information. This information is important when designing appliances. Let students continue to read the text at the bottom of page 115. Check understanding by asking students how a bulb and a fuse work in the same way.

Stretch zone: Research some other circuit breakers. Write a short report on how and when they are used.

Computing link: Students will need access to the internet to find out about other circuit breakers, or you can print off information sheets for them. Explain to students they need to share their findings as a written report.

> *Possible response: There are different types of circuit breakers including pole circuit breakers and AFCIs (arc-fault circuit interrupters), for example. These work by cutting the circuit to prevent systems overloading.*

Key idea

Fuses and bulbs can act as circuit breakers in electrical circuits.

Summarise the lesson by asking students to read the key idea. This will remind students of the uses of circuit breakers in electrical circuits. Ask students to describe to their partner what would happen if we didn't have these devices in electrical circuits. Ask them to describe to their partner how a fuse and a bulb have a similar design.

Workbook activities

How does a filament bulb work? (page 114)

Explain to students that they are going to find out about filament bulbs. Ask them to start by drawing a diagram of an old-fashioned filament bulb. They can use the photograph at the bottom of the page, but remind them that a scientific diagram is not an artistic drawing. You could also hand out some examples for students to study and draw. They should then explain how the bulb gives out light. They will find useful information on page 115 of the Student Book. Also remind students to use their investigations to help them.

Next, students should look at the photograph and explain what could have happened to the bulb and why it will not work even when it is connected to an electrical supply.

> *Answer: The filament wire in the bulb heats up as the electricity flows through it. This reacts with the gases in the bulb and it gives out light. The wire in the bulb at the bottom of the page has broken. This means that the electricity cannot flow through it so the bulb will not light.*

Does the width of a wire affect how hot it gets? (page 115)

This activity supports the investigation on page 115 of the Student Book. Ask students to explain what a circuit breaker is and draw a diagram of a fuse. They should then use the information on pages 114 and 115 of the Student Book to help them to explain how the fuse acts as a safety device in a circuit.

Students can use the table in this worksheet to record their results from their investigation and then write in their conclusion.

> *Answer: Students should describe a circuit breaker as a device that results in a controlled break in a circuit. When too much electricity flows through the wire in the fuse it gets so hot that it breaks. This prevents electricity flowing through it to the next part in the circuit. Students should have observed the wire getting hot during their investigation. This is why they only left them connected for 30 seconds. The width of the wire allows more electricity to flow through it so it doesn't get as hot.*

Review and reflect

Encourage students to find a quiet place to think about their learning in this lesson. Ask them if they found any part of the lesson challenging. Remind them that when they overcome a challenge, they are making their brain work harder. This is like going for a run to make your lungs work harder. Ask students how they felt when the learning was challenging and how they overcame the challenge. Ask them to record the strategies they used as this will help them in future situations. If all students found some part more difficult than others, you might want to review this as a class before moving onto the next lesson.

Conclude the lesson by asking students to complete questions 3 and 5 in the 'What have I learned about building electrical circuits?' activity on page 124 of the Student Book. (See also the teaching notes in this Teacher's Guide, pages 167–169.)

Extra activities

 1 **Maths link:** Students survey the space at home to find how many old-fashioned bulbs they have. They draw a diagram and explain to the people at home how the bulb gives us light.

2 Students research modern bulbs and suggest why these are better than filament bulbs.

Differentiation

Supporting: Hand out unbroken and broken bulbs for students to observe the filaments.

Consolidating: Demonstrate adding batteries to a circuit with a single bulb until the bulb 'blows' to show students how the circuit is broken.

Extending: Students write a paragraph to explain how a circuit breaker works and use the evidence from their investigation to support or refute this.

Differentiated outcomes	
All students	should be able to describe how a switch breaks a circuit
Most students	will be able to describe examples of circuit breakers
Some students	may be able to explain how the width of a wire can be used to make a circuit breaker

Using circuit diagrams

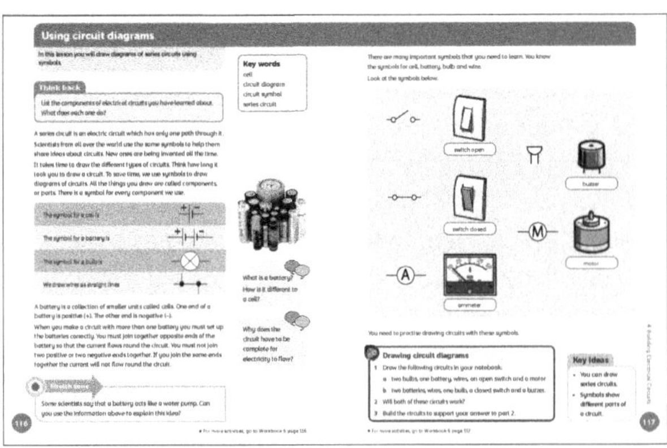

Supporting activities are in Workbook 6 pages 116–117.

Getting started

In this lesson students are introduced to symbols that are used to represent the basic components and how these are used in circuit diagrams. Students are given the opportunity to practise drawing diagrams. Students explore how using circuit diagrams is quicker and easier and also means that other scientists can use the same circuits.

Language support

Study the key words at the beginning of the lesson and have a class discussion about some of the unfamiliar words. Show students common symbols they may have seen in school or in the community. Point out that these are simple diagrams designed to explain without using words.

Key words

cell   circuit diagram   circuit symbol   series circuit

Other words in the lesson

battery   buzzer   component   motor   negative
positive   symbols

Scientific enquiry key words

Plan and/or carry out enquiries to answer questions

Make observations

Analyse data, notice patterns and group or classify things

Draw conclusions and give explanations

Lesson at a glance

The key teaching points for students in this lesson are:

- electrical components can be represented by symbols
- these symbols can be combined together to represent a circuit. These are called circuit diagrams.

In the next lesson, students will investigate series and parallel circuits.

Think back: List the components of electrical circuits you have learned about. What does each one do?

Ask students to make a list of all the things they have used when building circuits. This could be done as a class competition. See who can list the most or who can list the component that is the most uncommon. The idea is to construct a list of several components.

Hand out sheets of paper and ask students to draw a picture of one of the components such as a bulb or a battery. Point out that this is time consuming and not very clear. Show the images on page 116 of the common conventional symbols (cell, battery, bulb, wire). There is a symbol for every component that is used worldwide so everyone understands it.

Ask students to read the text and study the symbol drawings on page 116. Ask them to read the information about circuit symbols and how they are used. Discuss how these make drawings clearer as all scientists use the same ones.

What is a battery? How is it different to a cell?

Ask students to discuss in small groups what a battery is. The image of the symbol may help them work it out. Read out the text about batteries below the table of symbols. This explains that batteries are made of cells. One end is positive and the other negative. You could demonstrate what happens if the ends are joined positive to positive or negative to negative.

> **Answer:** *A battery is a chemical store of electricity. A cell is one part of the battery.*

Why does the circuit have to be complete for electricity to flow?

Ask students to discuss in their groups what will happen to the flow of electricity if the circuit is not complete. You could demonstrate using a broken circuit. Make sure students understand that electricity cannot cross the gap. The electricity needs wires and components to travel through.

> **Answer:** *If there is a break in the circuit, the flow of electricity stops. Electricity cannot flow over a break: it needs components to travel through.*

 Some scientists say that a battery acts like a water pump. Can you use the information above to explain this idea?

Allow students to work with a partner. Ask them to think about how a water pump would work.

> *Possible response: A water pump pushes water out of the main source to other places, usually along pipes. This can be compared to the flow of electricity from a battery along wires in a circuit.*

Lead a class discussion about the value of circuit diagrams. Point out that they are very useful but they must be drawn accurately. On the board, draw a circuit diagram with a gap in the circuit. Ask, 'What is wrong with this circuit diagram?'

Point out to students the important symbols on page 117 that they will need to use when drawing circuits. Discuss the symbols they are unfamiliar with (open and closed switch, ammeter, buzzer, motor). Then ask them to carry out the investigation activity.

Investigation: Drawing circuit diagrams

The purpose of this investigation is to allow students to apply their knowledge of component symbols in drawing complete circuit diagrams. Ask them to look at the symbols shown on pages 116–117 and then draw the two circuits described. Remind them that wires are drawn as straight lines so circuits are shown as rectangles. You can provide more support by letting students complete the activity on page 116 of the Workbook as this shows four circuit diagrams. Finally, ask students to predict if both of the circuits would work and then allow them to build the circuits to test their predictions.

> *Answer: Check that students have drawn the correct symbols for each component and that they are all joined together without any gaps or breaks in the circuit. Circuit **a** will not work because the switch is open.*

Key ideas
- *You can draw series circuits.*
- *Symbols show different parts of a circuit.*

Summarise the lesson by asking students to read the key ideas and then test their understanding by asking them to close their books and draw all of the symbols they can remember. They can then check their drawings against the ones on pages 116–117 and discuss any that they missed out.

Workbook activities

Using circuit diagrams (page 116)

Students can complete this activity individually or with a partner so they can share ideas. Ask them to study each circuit diagram. Point out how the components are shown spread out around the circuit and how they are positioned centrally along the wires rather than above or below them. Stress that a ruler has been used to draw straight wires with 90° angles at the corners. This makes circuit diagrams easier to follow. Students can then answer the questions about each circuit diagram and record their answers with the explanation.

> *Answer: Circuit A: 1 no; 2 close the switch; 3 neither dim nor bright. Circuit B: 1 yes; 2 nothing; 3 brightly because there are three batteries and one bulb. Circuit C: 1 yes; 2 nothing; 3 dimly because there are four bulbs and one battery. Circuit D: 1 no; 2 close the switch; 3 brightly because there are three batteries and one bulb.*

Drawing circuits (page 117)

This can also be an individual task or carried out in pairs if you wish to promote discussion or offer support to some students. Ask students to draw the circuit diagrams described. Remind them to include all of the components listed and then circle the correct answer to state whether or not each circuit will work. For circuit 1c, ask students to decide if there will be a reading on the ammeter.

> *Answer: 1 Make sure the correct symbols are used to represent the components and there are no breaks in the circuit. Circuit **a** will work; Circuit **b** will work; Circuit **c** will not work as the switch is open. 2 There will not be a reading on the ammeter as the circuit will not flow.*

Review and reflect

Ask students to test a partner by writing down some components and asking them to draw a circuit containing all of the components. You can make this a hands-on task by handing out components and students can actually construct and test the circuits suggested by their partner.

Conclude the lesson by asking students to complete questions 1 and 7 in the 'What have I learned about building electrical circuits?' activity on pages 124–125 of the Student Book. (See also the teaching notes in this Teacher's Guide, pages 167–169.)

Extra activities

1 Remembering the symbols can be difficult for some students, so they can play matching games with pre-prepared images of symbols and the words.

2 Ask students to represent specific series circuits with drawings and conventional symbols.

Differentiation

Supporting: Students practise drawing symbols when you call out the name of each component by referring to the Student Book and copying the symbol.

Consolidating: Display the symbols around the room so students become familiar with them.

Extending: Students can draw circuit diagrams from verbal instructions and then decide if the circuits will work.

Differentiated outcomes	
All students	should be able to recognise the electrical circuit symbols for bulb, battery, wire, switch, ammeter, buzzer and motor
Most students	will be able to draw a circuit diagram using circuit symbols, including bulb, battery, wire, switch, ammeter, motor and buzzer
Some students	may be able to correctly draw circuit diagrams using a number of symbols and predict if the circuit will work

Types of circuits

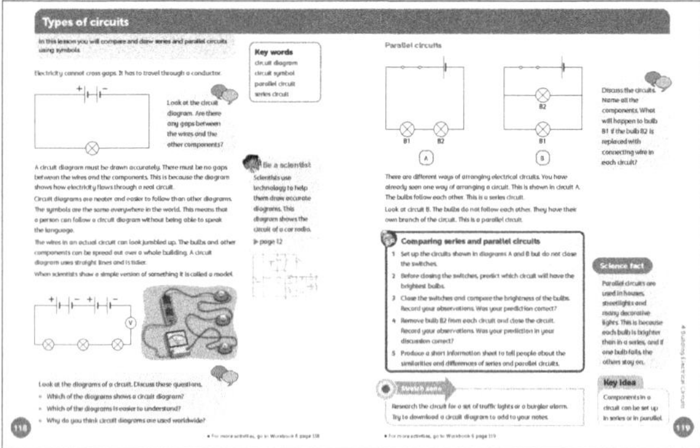

Supporting activities are in Workbook 6 pages 118–119.

Getting started

In this lesson students will be introduced to parallel circuits and how they are different from the series circuits that they have used. They will learn that parallel circuits are commonly used in buildings because if one component stops working, the others continue.

Language support

To help develop language skills you could use flashcards with the names of the different circuit symbols. Students have to name the symbol and describe how it is used. Then add diagrams of parallel and series circuits. Hold up the circuits and students have to say if it is a series or a parallel circuit.

Resources

Student Book: batteries; wires; connectors; bulbs; switches; materials to make information sheets.

Workbook: batteries; wires; connectors; bulbs; switches; materials to make information sheets.

> **Key words**
>
> circuit diagram   circuit symbol   parallel circuit   series circuit
>
> **Other words in the lesson**
>
> branch bulb component model switch technology

Lesson at a glance

The key teaching points for students in this lesson are:

- series circuits have the components in a line one after the other
- in a parallel circuit the components in parallel have their own branch of the circuit.

In the next lesson, students will investigate how changing voltage affects the brightness of bulbs and volume of buzzers.

Look at the circuit diagram. Are there any gaps between the wires and the other components?

Ask students to read the text at the top of page 118 and study the circuit diagram. Ask them to consider if the diagram is drawn correctly and discuss the question.

Answer: There are no gaps between the symbols in the diagram.

Ask students to continue to read the information on the rest of page 118 and to study the diagrams of a circuit – one using symbols and one using drawings.

Be a scientist: Scientists use technology to help them draw accurate diagrams. This diagram shows the circuit of a car radio. (Student Book, page 12)

Read out the Be a scientist information and ask students if they can identify any symbols in the radio diagram. This is very complex, but point out that it is evidence of how symbols are used and also that there are many more symbols used in electricity and electronics.

Look at the diagrams of a circuit. Discuss these questions.

- **Which of the diagrams shows a circuit diagram?**
- **Which of the diagrams is easier to understand?**
- **Why do you think circuit diagrams are used worldwide?**

Ask students to look at the two diagrams of a circuit. Ask them to discuss why we use circuit diagrams. Remind them that electricity is very dangerous, and elicit that

circuits must be set up correctly so that there are no accidents. Highlight the importance of internationally used symbols. Working from a photograph can be very confusing and could cause accidents. Explain that a circuit diagram is like an underground railway map – it is easy to follow. Railway lines may be at different levels and bend in many different directions but a map shows these in a simpler way. Show students a map if possible to demonstrate the similarity.

Possible response: The left diagram is a circuit diagram that uses symbols. Circuit diagrams are used worldwide because they are easy to understand and it doesn't matter what language you use as you can still use the symbols.

Parallel circuits

Ask students to look at the symbol circuit diagrams at the top of page 119 and read the text. Point out the difference in the ways the bulbs are connected in each circuit and ask students if they predict whether this will affect the way that the circuits work.

Discuss the circuits. Name all the components. What will happen to bulb B1 if the bulb B2 is replaced with connecting wire in each circuit?

Ask students to discuss each circuit with a partner. They can follow the line of the wires around the circuit with a finger and see what happens in the series circuit (A) and the parallel circuit (B).

Answer: There are two bulbs, a closed switch and one cell in each circuit. In both circuits the bulbs will light if B2 is replaced with a wire.

Then ask them to imagine a block in the circuits by placing a small object such as a pencil sharpener or eraser across the wire in circuit A between the bulbs. Ask them what would happen to the bulbs. Could the electricity flow? They then place the object between the bulbs in circuit B – near where the discussion box arrow reaches the circuit. Ask them again: What would happen to the bulbs. Could the electricity flow?

Investigation: Comparing series and parallel circuits

Explain to students that they are going to investigate the two types of circuits shown above – series and parallel. The worksheet on page 119 of the Workbook supports this investigation. Students can work in groups of three or four to plan and carry out the investigation. Once they have set up the circuits, they can predict which circuit will have the brightest bulbs and then close the switches and compare the brightness of the bulbs. Remind them to record their observations and check their prediction. Next, allow them to remove bulb B2 from each circuit and close the circuit. They should again record any observations and check their predictions. To share their conclusions they should produce a short information sheet to tell people about the similarities and differences of series and parallel circuits.

Answer: Students should find that both bulbs light in the circuits. If one of the bulbs in circuit A was broken, the other bulb would not light. If one breaks in circuit B, the other bulb would light as it has its own circuit.

Science fact Parallel circuits are used in houses, streetlights and many decorative lights. This is because each bulb is brighter than in a series, and if one bulb fails, the others stay on.

Read out the Science fact or ask a volunteer to read it out. Ask students to discuss what would happen if their school was wired with a series circuit.

Stretch zone: Research the circuit for a set of traffic lights or a burglar alarm. Try to download a circuit diagram to add to your notes.

Ask students to work with a partner or in a small group. Allow them to research their own circuit. This will give them some ownership of their learning. Check the circuit diagrams they have found and discuss how they would work.

Key idea

Components in a circuit can be set up in series or in parallel.

Encourage students to read the key idea and discuss it with a partner. This will remind students that there is more than one type of circuit. Ask students to write down one advantage of a parallel circuit over a series circuit and ask them why parallel circuits are used in buildings.

Workbook activities

Build circuits from diagrams and test them (page 118)

Arrange students into pairs so they can share ideas and discuss the tasks. Start by pointing out the Remember box. Ask students to follow the instructions to build the circuits shown and test each one. They can write down below each circuit what they changed and how bright the bulbs were. This is very useful practice in interpreting circuit diagrams and building series circuits.

Answer: Circuit A would work. To make the bulb brighter more batteries could be added. Circuit B would light the bulb brightly. Circuit C would not light the bulbs. The switch needs to be closed, then the bulbs would light brightly. Circuit D would light the bulb brightly and the ammeter would show a current reading. Circuit E would light the bulbs and the ammeter would show a current reading. To make the bulbs brighter more batteries could be added. Circuit F would not light the bulb. The switch needs to be closed. To make the bulb brighter more batteries could be added.

Investigating series and parallel circuits (page 119)

This activity supports the investigation on page 119 of the Student Book. Ask students to work in a group of three or four. They start by looking at the circuit diagrams and deciding which circuit shows a parallel circuit. Allow them to set up both of the circuits and investigate them using the prompt questions and instructions. Remind them to predict what will happen to the other bulb each time. They should then record their observations and determine if their were predictions correct.

Answer: The only difference is that one is in series and the other is in parallel; the components are the same. Circuit B is a parallel circuit. Students write their own predictions and investigation results. If bulb B2 is removed from Circuit A, then bulb B1 will not light because the circuit will be incomplete. If bulb B2 is removed from Circuit B, then bulb B1 will light because that loop of the circuit is still complete.

Students can then discuss the Stretch zone task and produce a short information sheet to tell people about series and parallel circuits.

Review and reflect

Encourage students to display their circuits and circuit diagrams and allow them to walk around to see what other students have produced. They can then have some quiet time to reflect on how they could modify their circuits to improve them. Point out that reflecting on their work and learning from others is a very important part of learning.

Conclude the lesson by asking students to complete question 6 in the 'What have I learned about building electrical circuits?' activity on page 125 of the Student Book and the fourth statement on page 124 of the Workbook. (See also the teaching notes in this Teacher's Guide, pages 167–169.)

Extra activities

1 Students take photographs of a range of circuits from their investigations. They ask the people at home to draw them. They then show how easy it is to use symbols.

2 Students research the symbols for a fuse and a diode. They describe what these components are used for.

Differentiation

Supporting: Allow students to refer back to the symbols on pages 116–117 when drawing series and parallel circuits.

Consolidating: Encourage students to practise drawing series and parallel circuits using instructions their partner gives to them.

Extending: Students find out where a voltmeter should be positioned as pre-learning for the next lesson.

Differentiated outcomes

All students	should be able to draw a parallel circuit using simple components
Most students	will be able to predict whether or not a bulb will light in series and parallel circuits if another bulb is removed
Some students	may be able to explain why a parallel circuit is used in modern buildings

Measuring voltage

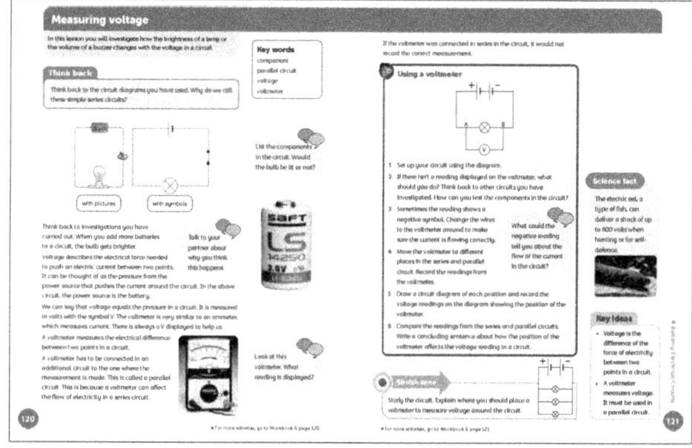

Supporting activities are in Workbook 6 pages 120–121.

Getting started

In this lesson students will study voltage and how it is the force needed to push an electric current around a circuit. They will investigate using a voltmeter. Students explore how the voltmeter must be placed in a parallel circuit to make accurate observations. Students find that voltage is the difference between the force of electricity between two points in a circuit.

Language support

To support language development, display a voltmeter and attach the symbol V to it. Explain what voltage is using different analogies, like a central heating system where the pump represents the voltage in an electrical circuit. Students could make word discs by taking a paper plate or round piece of card and drawing a picture of the voltmeter in one third, the symbol in another third and a brief description of what voltage is in the final third. These make attractive displays or students can test each other by covering up a third of their disc and asking their partner what is missing. They can do this for other components too if you wish.

Resources

Student Book: batteries; wires; connectors; bulbs; voltmeters.

Workbook: batteries; wires; connectors; bulbs; voltmeters; copper coins; aluminium foil; thin card; lemon juice or vinegar; salt; paper towels; sticky tape; LEDs with wires.

> ### Key words
>
> component   parallel circuit   voltage   voltmeter
>
> **Other words in the lesson**
>
> current   pressure   push   source   symbol

Scientific enquiry key words

Plan and/or carry out enquiries to answer questions

Make observations

Take measurements, using equipment accurately

Record data and results

Analyse data, notice patterns and group or classify things

Report and present findings

Draw conclusions and give explanations

Identify causal relationships

Lesson at a glance

The key teaching points for students in this lesson are:

- voltage is the difference of the force of electricity between two points in a circuit
- voltage is measured using a voltmeter
- increasing voltage increases the brightness of lamps and the volume of buzzers.

In the next lesson, students will use circuit diagrams to make predictions about circuits.

Think back: Think back to the circuit diagrams you have used. Why do we call these simple series circuits?

Ask students to work with a partner and read the Think back. Ask them to discuss the difference between a series and parallel circuit. Tell them if they put their finger on any point in a series circuit they can move through all of the components in one single loop. Ask them to try this with a parallel circuit. They will always come to a junction where they will have to decide which loop to follow.

Answer: All of the components are joined together in one single loop in a series circuit.

List the components in the circuit. Would the bulb be lit or not?

Ask students to study the two circuits at the top of page 120. Point out that one is shown by drawings and one is a symbol circuit diagram. Ask students to identify each of the components and decide if the bulb in the circuit would be lit.

Answer: The components are: one bulb, one cell or battery, one closed switch and wires. The bulb would be lit as the switch is closed and the circuit is complete.

Students should read through the first paragraph on page 120 below the circuit diagrams before talking about the question in the next discussion task. Ask them what happens to the brightness of the bulb when more batteries are added.

Talk to your partner about why you think this happens.

Point out that the text is explaining that adding more batteries makes bulbs brighter. They may recall this from work earlier in the unit. Ask them to think about the opposite scenario. What happens to a torch or other device if the battery runs down?

Answer: Batteries supply electrical power to the circuit so the more batteries there are, the more current will flow, so the bulbs will be brighter.

Ask students to read the rest of the text on page 120. This is challenging so you could ask volunteers to read a sentence each. To check understanding you can interrupt the text reading to ask the following questions: What do we call the electrical force needed to push an electrical current? What is the most common source of power in the circuits we use? What does V mean? What do we call the device we use to measure electrical force in a circuit? Should the device be connected into a circuit in series or in parallel? Why would adding more batteries make bulbs brighter in circuit?

Look at this voltmeter. What reading is displayed?

Ask students to look at the display on the voltmeter. Ask them to discuss the reading they see. The display shows a digital reading of the voltage in the circuit. Remind students to record the unit of measurement in addition to the reading displayed.

Answer: This voltmeter shows that the voltage in the circuit is 1.5 V.

Investigation: Using a voltmeter

Allow students to work in groups of three or four. Demonstrate how to set up the voltmeter and stress that it has to be set up in parallel – on its own branch. It is set up across components as it is measuring the difference between two points. More support on using a voltmeter can be found in the activity on page 120 of the Workbook.

Ask students to set up their circuit using the diagram to help them. If there isn't a reading displayed on the voltmeter, ask them what they should do. Suggest testing the components to check they have a complete circuit. If the circuit works but the voltmeter reading shows a negative symbol, then show them how to change the wires to the voltmeter around to make sure the current is flowing correctly.

Allow students to investigate what happens if they move the voltmeter to different places in a series and parallel circuit. They should record the readings from the voltmeter each time and draw a circuit diagram of each position. Ask them to compare the readings from the series and parallel circuits and write a concluding sentence about how the position of the voltmeter affects the voltage reading in a circuit.

What could the negative reading tell you about the flow of the current in the circuit?

Ask students to discuss this as they are carrying out their investigation as it can help with troubleshooting.

Science fact The electric eel, a type of fish, can deliver a shock of up to 600 volts when hunting or for self-defence.

Read out the Science fact or ask a volunteer to read it out. Ask students to study the picture and discuss how big a shock this would be. This is almost six times greater than most mains electricity.

Stretch zone: Study the circuit. Explain where you should place a voltmeter to measure voltage around the circuit.

Allow students to discuss the question with a partner or their investigation group. Ask them to study the components in the circuit. Ask them how many branches there are in the parallel circuit. Ask them to think about where the voltmeter should be positioned to be able to measure the voltage correctly. Remind them that the voltage is measured across a component so the voltmeter will need to be connected before and after a bulb, for example. It will be in parallel to the bulb. They can look at the diagram in the investigation box for a clue.

Key ideas

- *Voltage is the difference of the force of electricity between two points in a circuit.*
- *A voltmeter measures voltage. It must be used in a parallel circuit.*

Read out the key ideas or let students read them to a partner. Ask them to explain to their partner what voltage is and the units used to measure it. Ask them to talk about why we need to measure voltage and remind them about safety and electricity.

Workbook activities

Using a voltmeter (page 120)

This activity supports the investigation on page 121 of the Student Book. Ask students to set up the three circuits shown. Make sure they look carefully at how the voltmeter connects in each circuit. They should test each circuit, read the number of volts on the

voltmeter, and record this in the table. Ask students to compare the readings and decide if all of the readings are the same or if the position of the voltmeter changes the reading.

Making a battery (page 121)

Explain that students can make their own battery. Show them the diagram and then ask them to work with a small group to make their battery. They should follow the instructions and build the tower of repeated layers of coins, soaked card and aluminium and then make sure it is connected to a wire at the top and bottom. Ask students to find out what happens when they connect a voltmeter or an LED (light-emitting diode) to the wires. Explain that an LED is similar to a light bulb. When electricity passes through the LED it will glow. Point out to students that the different layers of material in their battery help to make an electrical current. The negative particles that form electricity (electrons) move from one metal to another.

Review and reflect

Encourage students to identify aspects they have not completed correctly and help them to identify improvements. This will help students to develop a positive approach to learning by understanding that learning is a process that will improve with practice and reflection. For example, they can walk around and look at drawings of parallel circuits to identify what they have done well and what they can learn from others as targets to improve their work.

Conclude the lesson by asking students to complete question 8 in the 'What have I learned about building electrical circuits?' activity on page 125 of the Student Book. (See also the teaching notes in this Teacher's Guide, pages 167–169.)

Extra activities

1 Students research how to make a battery using fruit or vegetables. They make a poster to show the method.

2 Students construct their fruit or vegetable batteries. The person who can record the highest voltage reading is the winner.

Differentiation

Supporting: Allow students to practise taking voltage readings using a voltmeter.

Consolidating: Encourage students to move the voltmeter around different circuits and record the reading.

Extending: Students can test voltage and current around series and parallel circuits with ammeters and voltmeters.

Differentiated outcomes	
All students	should be able to explain what voltage measures
Most students	will be able to measure the voltage in a circuit
Some students	may be able to compare voltage and current around a circuit

Using circuit diagrams to make predictions

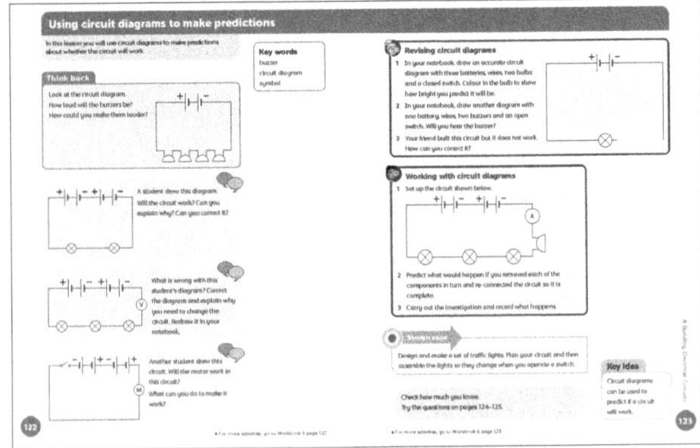

Supporting activities are in Workbook 6 pages 122–123.

Getting started

In this lesson students practise using circuit diagrams. They study the diagrams and predict what will happen to the components in the circuits. They also practise drawing circuit diagrams from memory. They will need to recall the symbols for components to be used in the diagrams from descriptions given. Students study circuit diagrams and offer suggestions to solve problems with them. They also study diagrams and predict what would happen if these were constructed. They test their predictions and record their observations.

Language support

To support language development you could create a Word wall covering some of the more difficult key words from this unit. You could also ask students to add words and definitions to their glossaries. A large circuit diagram using as many symbols as possible could be useful. Students could label this with the names of the components represented by the symbols.

Resources

Student Book: writing materials; batteries; wires; connectors; bulbs; ammeters; buzzers.

Workbook: string; paper plates or pieces of card; scissors; sticky tape.

> ### Key words
> buzzer   circuit diagram   symbol
>
> **Other words in the lesson**
> batteries   bulbs   buzzers   motor   switch   wires

Lesson at a glance

The key teaching points for students in this lesson are:

- analysing circuit diagrams can help us to predict how the circuit will work
- circuit diagrams can be tested by building the circuits and testing if the components work.

In the next lesson, students will review their understanding of building electrical circuits.

Think back: Look at the circuit diagram. How loud will the buzzers be? How could you make them louder?

Use the Think back diagram to encourage students to reflect on their learning and understanding. Ask them to discuss the diagram and then the questions asked about it.

Answer: The buzzers will be very quiet or more likely silent. More batteries would need to be added to the circuit to make them louder.

Ask students to work with a partner to study each of the diagrams and address each problem linked to them.

A student drew this diagram. Will the circuit work? Can you explain why? Can you correct it?

Remind students to be logical when fault finding or testing a circuit. They should start at the battery or batteries and then follow the circuit round to make sure all of the components are in the correct place and are linked by wires with no breaks.

Answer: This circuit would work. The circuit is connected, and there are two batteries and two bulbs.

What is wrong with this student's diagram? Correct the diagram and explain why you need to change the circuit. Redraw it in your notebook.

Again, recommend the logical approach and remind students that some components have to be set up in a special way. Check that they can remember what each of the component symbols represents.

Answer: Students should identify that the voltmeter should not be placed in a series circuit. They should redraw this circuit with the voltmeter in a parallel circuit.

Another student drew this circuit. Will the motor work in this circuit? What can you do to make it work?

This time ask students to list all of the components they can see to check that they recognise the batteries, switch and motor especially. If they follow the circuit around looking for gaps, they will soon spot where the circuit has a break. They should also notice that the batteries are the wrong way round.

Answer: The motor would not run because the switch is open and the batteries are connected the wrong way round. To make it work they would need to close the switch and turn the batteries around.

Investigation: Revising circuit diagrams

Students can work individually on this task or you can ask them to work with a partner. Ask them to draw an accurate circuit diagram containing three batteries, wires, two bulbs and a closed switch. They should colour in the bulbs to show how bright they predict they will be. Next, they draw another diagram with one battery, wires, two buzzers and an open switch and decide if they will hear the buzzer. Finally, ask them to imagine that a friend built the circuit shown but it does not work. Ask them to fault find and decide how it can be corrected.

Answer: The circuit diagram should be drawn accurately using the correct symbols for steps 1 and 2. For step 3 the circuit would not work because there is a gap in the circuit. They would need to properly connect the bulb to the wire.

Investigation: Working with circuit diagrams

Explain to students that in this investigation they are going to interpret a circuit diagram and make a circuit. Ask them to set up the circuit shown and predict what would happen if they removed each of the components in turn and re-connected the circuit so it is complete. Allow them to carry out the investigation and record what happens.

Answer: If the connections were correct, the components would continue to work. If they removed one battery, the voltage would be reduced and the bulbs and buzzers would not work.

◉ **Stretch zone:** Design and make a set of traffic lights. Plan your circuit and then assemble the lights so they change when you operate a switch.

Students might have found a circuit diagram for traffic lights in an earlier Stretch zone. They plan the components that they would need and how to construct the circuit. Ask them if any of the components should be in parallel. Ask them what would happen if the red bulb stopped working, for example. Would the green and yellow bulbs still light?

Answer: Students connect three different coloured bulbs in parallel, each with its own switch. When they switch on the circuit with the red bulb, for example, the red bulb should light.

Key idea

Circuit diagrams can be used to predict if a circuit will work.

Read out the key idea and remind students that circuit diagrams are often used to make predictions about whether a real circuit would work correctly. Ask students to name some examples of symbols that they have used in diagrams.

Workbook activities

🕹 Model circuits (page 122)

Explain to students that they are going to make models of circuits using string and paper plates or card. They can work with a partner or in a small group to model the circuits shown in the diagrams. Ask students to start by drawing the symbol for each component on a paper plate or piece of card (one symbol on each). Explain that the string is being used to represent the wires so they should cut lengths of string and stick them between the paper plates or card. The models can be displayed and students can then study them and decide which circuit each model is representing. For further practice in drawing circuit diagrams, students can then interpret each model and change it into a circuit diagram. Check that the symbols are correct and they are joined in a way that would make them work.

Circuit wordsearch (page 123)

🔤 This can be an individual activity. Explain that students must find some missing words that are important words from this unit. They are hidden somewhere in the wordsearch. Ask them to circle each word in the grid when they find it and cross out the word in the word box so they can keep a check on what is found and what is still missing. Explain that one example, which is a two-word phrase, has been done for them. They should explain to a partner what each word means. Finally, ask students to draw the symbol for each of these components: ammeter; buzzer; battery; closed switch; bulb; open switch.

Answer: Students should find and circle the following words: ammeter, battery, bulb, buzzer, conductor, insulator, metal, plastic, switch, symbol, wire. The symbols for each component should be correctly drawn.

Review and reflect

As this is the final lesson of the unit, unless you include a review lesson, you can ask students to look back through their notes and any models and posters they have made to select two pieces of work they are keen to learn more about. They can set this as an individual learning target.

Allow students a few moments of quiet reflection so they can enjoy the feeling of satisfaction from learning so many new things and working so well with others.

If you wish, you can ask students to revisit and discuss all of the questions on pages 124–125 of the Student Book and all of the statements on page 124 of the Workbook.

Extra activities

1. Students could prepare a two-minute presentation to tell the people at home the key things they have learned in this unit. This will also support revision.

2. Students make a card sort with symbols and names on separate cards. They play a matching game with the people at home.

Differentiation

Supporting: Set out each circuit on a table so students can link the diagrams with the actual components.

Consolidating: Give students practice in studying diagrams by asking them to regularly set challenges for each other by making lists of components their partner has to combine into a circuit diagram.

Extending: Set out complicated faulty electrical circuits so students practise fault finding and repairing them so that all of the components work.

Differentiated outcomes	
All students	should be able to identify components from their symbols
Most students	will be able to predict if circuits will work from studying circuit diagrams
Some students	may be able to identify and repair faults in complicated circuits

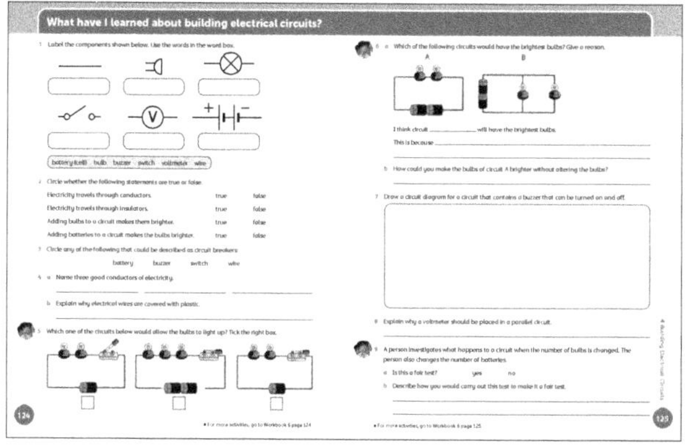

Supporting activities are in Workbook 6 pages 124–125.

Getting started

As you will know from previous units, the aim of this section is to encourage students to review their learning after all of the lessons in the unit. Remember that all lessons have questions and discussion tasks for students to answer and these can be used formatively during lessons. Use student responses to these to help you to assess student knowledge and understanding of the topic.

On pages 124–125 of the Student Book there are nine questions related to the content of the unit. As with other units, students can tackle these one at a time after each relevant lesson and specific advice on which questions are most appropriate is given under the 'review and reflect' heading. The questions within this section can also be answered as a single summative activity. This could be done by reading out the questions to the class and asking for volunteers to answer them, carrying out the activity as a group work task with students talking about each question or, if students can confidently read the questions, as an individual task. Whichever approach is adopted the questions are designed to give you and students feedback about progress and to help in identifying targets for development. The questions are arranged in increasing order of conceptual demand and not topic order.

On page 124 of the Workbook there is a section containing four 'review and reflect' statements for students to respond to. You can read through the statements and then ask students to tick one of the two self-review option boxes to show how confident they are about each topic. Some statements have been compiled to cover more than one lesson so you can use these after the last relevant lesson. As with the Student Book section, you can ask students to reflect on each statement separately after each relevant lesson or carry out the review as an end-of-unit activity.

Advice on which specific review and reflect statement is most appropriate is given under the 'review and reflect' heading in the relevant lesson. It would be very useful to do both and this is recommended, as you will be able to measure student levels of confidence throughout the unit to respond quickly and also gain a measure at the end of the unit to find out if this has changed for any topics. For example, time to reflect may make a student more or less confident or later learning may support understanding of a topic that had been causing problems for some students. As teachers, we know that not all learning takes place at the time the teaching takes place.

It is important students feel confident that their reflections will be listened to and dealt with sensitively so that they do not hesitate to report areas they are not confident with. This information is vital for the teacher to provide support strategies in the end-of-unit summative assessment.

1 Label the components shown below. Use the words in the word box.

   Point out the six symbols and remind students that the answers will be in the word box. No other words will be needed and the words are only used once.

   Answer: Top row from left: wire; buzzer; bulb. Bottom row from left: open switch; voltmeter; battery (cell).

2 Circle whether the following statements are true or false.

   Electricity travels through conductors.

   Electricity travels through insulators.

   Adding bulbs to a circuit makes them brighter.

   Adding batteries to a circuit makes the bulbs brighter.

   Warn students to read all of the statements carefully as they look very similar. Remind them of examination technique and make sure they understand to circle one of the two choices in each case and don't underline or cross out.

   Answer: true; false; false; true

3 Circle any of the following that could be described as circuit breakers.

   battery   buzzer   switch   wire

   Point out to students that if they do not know the answer, they can start by eliminating any they know to be incorrect to cut down on the possible correct answers. Again, they must circle any answers they think are correct.

4 **a** Name three good conductors of electricity.

  b Explain why electrical wires are covered with plastic.

Ensure that students know to write each example in the appropriate place on the line beneath each question. Remind them to write clearly so answers are readable.

5 Which one of the circuits below would allow the bulbs to light up? Tick the right box.

Remind students to read the question carefully and point out that an important word in the question is the word 'one' so they know only one of the three choices is correct. Hint that they can follow the circuits with their finger to check for any breaks.

6 **a** Which of the following circuits would have the brightest bulbs? Give a reason.

  b How could you make the bulbs of circuit A brighter without altering the bulbs?

Ask students to use their observation skills to identify the two different types of circuit. Remind them that they have built and tested circuits like these. Point out that there will be credit for giving the answer and also the explanation.

7 Draw a circuit diagram for a circuit that contains a buzzer that can be turned on and off.

Remind students to draw the symbols clearly and use a ruler to ensure the lines for the wires are straight so the circuit diagram is neat and easy to follow.

8 Explain why a voltmeter should be placed in a parallel circuit.

Point out that there is not a lot of space for the answer so students will need to get their answer across in just a few words. If they need a prompt, ask them to look back at page 120 of the Student Book.

9 A person investigates what happens to a circuit when the number of bulbs is changed. The person also changes the number of batteries.

  a Is this a fair test? yes no

  b Describe how you would carry out this test to make it a fair test.

Explain to students that they will need to use their knowledge of investigations and fair testing to answer these questions. Remind them that fair testing is about variables – those they change, those they measure and those they keep the same.

Summative assessment

The questions in the 'What have I learned about building electrical circuits?' activity on pages 124–125 of the Student Book can be used to consider the progress of each student individually. You can also use the information to create summative reports – such as end-of-term reports – for each student. If you wish to allocate a score or mark for the questions, then the total number of marks you could allocate is 22 (question 1 = 6; question 2 = 4; question 3 = 1; question 4a = 3; question 4b = 1; question 5 = 1; question 6a = 1; question 6b = 1; question 7 = 1; question 8 = 1; question 9a = 1; question 9b = 1).

It may also be useful to keep a record of whole-class overall confidence levels to identify areas that may need revision later on.

This feedback can then be used to form support strategies to help students improve. Keep the recording and analysis of student self-evaluations simple. A general impression of the class's self-evaluation, not individual student records, is all that is required, e.g. 'Fifty per cent of the class were not confident about …'.

Additionally, you can ask students to complete the Quiz Yourself questions for this unit to help check their understanding. These are on pages 146–147 of the Workbook.

Investigate like a scientist

The 'Investigate like a scientist' task on page 125 of the Workbook is designed to encourage students to apply their investigative and creative skills and review key aspects of the content of the unit.

Can a pencil be used to complete a circuit?

Resources: batteries; wires; connectors; bulbs; pencils.

Provide students with the equipment for this investigation in which students work together as a team to investigate if a pencil could complete a circuit in the place of a wire. Students have made circuits and tested materials to find out if they are insulators or conductors of electricity throughout this unit. Ask the teams to predict whether or not the pencil will work. The teams can then discuss the problem and agree on a design to investigate and test this. Allow them to set up a circuit to test the pencil and ask them to research the properties of the pencil. Students can observe and record their results and then report on their findings. Ask them to decide if their results support the research findings. To share their ideas, students can present their investigation and findings as a news report.

Making symbols

Resources: internet and science books, materials to make booklets.

Ask students to work with a partner or individually for this task. Explain that they should find 20 symbols for components that they recognise or have heard of before. You could suggest some of the following to add to the ones they have used in lessons: antenna; transformer; resistor; fuse; diode; LED; amplifier; push button switch and loudspeakers. Point out that students are then going to be allowed to invent new symbols for these components. Stress to students that the symbols they invent should be easy to draw and also make sense to other scientists. If they are too elaborate, they will take too much time to draw and people will not use them. Suggest they might decide to make their symbols easier or more meaningful for a younger audience. Ask students to share their ideas by making a booklet to show their new symbols alongside the old symbol. Remind them to also describe the symbol with a brief description of what the component does.

5 Adaptation and Inherited Characteristics

In this unit students will:

- recognise that living things have changed over time and that fossils provide information about living things that inhabited the Earth millions of years ago
- recognise that living things produce offspring of the same kind, but normally offspring vary and are not identical to their parents
- identify how animals and plants are adapted to suit their environment in different ways and that adaptation may lead to permanent change.

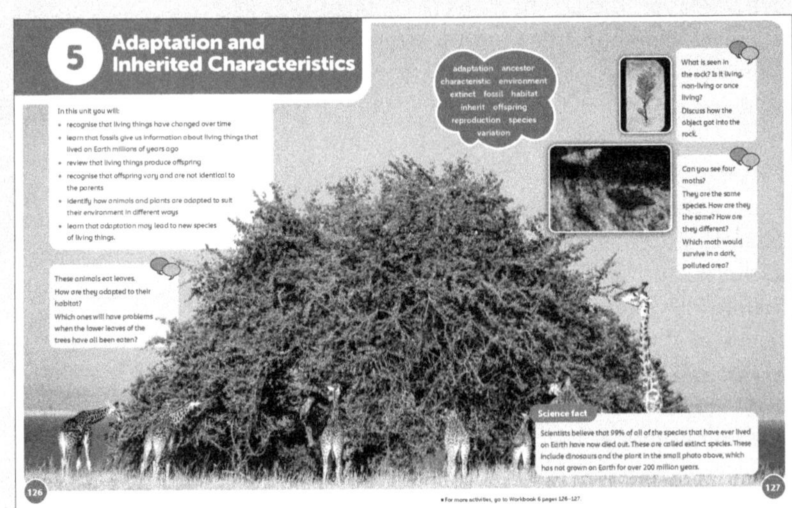

Supporting activities are in Workbook 6 pages 126–127.

Getting started

This unit explores the inherited characteristics of plants and animals. Students will review their knowledge of the types of adaptations seen in plants and animals. They will consider the formation and types of fossils and consider what fossils can tell us about living things and their environment in the past. They will also observe examples of some types of living things that have changed over time, others that have stayed the same, and some that have become extinct. Students will study how living things reproduce to produce offspring of the same kind but these offspring are not identical to the parents. They will learn that this variation within species is important in the survival of living things, especially if the environment changes.

Science in context

Use the lessons in this unit to encourage students to observe the adaptations seen in the living things around them. In addition to studying the environment around the school, carry out field trips and educational visits to farms, botanical gardens and zoological gardens and local forest and coastlines. Encourage students to meet people who use the breeding of animals and plants as part of their work and invite them into school to talk about why this is important. Students can also study the work of scientists who study fossils and, if there is a local natural history museum near you, arrange a visit for students to see fossils. You may even be able to organise a fossil hunt in rocks near the school, but take advice from local geologists on safety and what you are likely to find. Encourage students to take photographs, not samples, as these areas must be protected.

Scientific enquiry skills

Scientific enquiry skills for this unit focus on observation skills, surveys and the setting up of fair tests. Remind students that to plan a fair test they must be aware of the variables: independent (what they change), dependent (what they measure) and control (what they keep the same). Students will have opportunities to develop their skills of collecting, recording and presenting data. Encourage them to consider the accuracy and validity of the observations and measurements they take and ask them to evaluate their work. They will also be expected to interpret secondary data. Students will present their results in a variety of way and will have opportunities to use computer technologies to help in collecting and presenting data.

You can use the Investigation master sheet on pages 4–5 of this Teacher's Guide to support investigative work. This provides prompts and structure to support students in planning and carrying out fair tests and recording and drawing conclusions about their findings.

Resources

Student Book: selection of once-living things (e.g. shells; parts of a plant); dishes; modelling clay; water; plaster of Paris; paints; graph paper; samples of four different specimens of seeds and the plants they have come from; plant pots and labels; compost; access to the internet; books to research types of seeds / how an environmental factor can cause variation between the same species of animal / a plant that became extinct because of its inability to adapt to changing surroundings; items to prepare a computer or poster presentation; materials to make information leaflets and posters; writing materials; large sheets of paper; access to a local area to research plant adaptations; thin blue, green and red card; scissors.

Workbook: selection of once-living things (e.g. shells; parts of a plant); modelling clay; rolling pins; water; plaster of Paris; paints; labels; samples of four different specimens of seeds and the plants they have come from; plant pots and labels; compost; access to a warm dark, space and a dry, sunny space; items to prepare a computer or poster presentation; rulers; access to a local area to research plant adaptations; smartphones or digital cameras; thin blue, green and red card; scissors; plates or shallow dishes; large bowls; pebbles; pieces of wood with small holes; small seeds; rice; pieces of cork or polystyrene; tweezers; pliers; salad tongs; stopwatches or timers.

Key words for unit

Bold words are in the Word cloud and are included in the glossary.

adaptation ancestor breed cast
characteristic class cross breeding
environment extinct fossil habitat inherit
mould **offspring** reproduce **reproduction**
selective breeding **species** survival **variation**

Scientific enquiry key words

Plan and/or carry out enquiries to answer questions

Make predictions

Recognise and control variables

Make observations

Take measurements, using equipment accurately

Record data and results

Analyse data, notice patterns and group or classify things

Report and present findings

Draw conclusions and give explanations

Identify causal relationships

Language support

Place more responsibility on students and expect them to read and carry out work independently. They should be aware that they need to look up any words they find unfamiliar, so have scientific dictionaries to hand. As you will know if you have taught the other units in this book, it is very important to start the unit reading out the words in the Word cloud. Allow students to discuss each word and list those they are not familiar with. This monitoring of prior knowledge is vital and this activity will help students to recall prior work so it is fresh. Create a Word wall for this unit so students see the words often and can become familiar with them. Ask students to make their own word cards for display. A useful task would be to give each small group one or two words and ask them to make a word card that is bright and contains some relevant drawings. Try to keep repeating the words throughout the unit – do not just cover a word and then move on. Have a fun test every day that helps students to think about the words.

Remind students about the write-in glossary at the back of the Student Book and remind them to add definitions of key words throughout the unit. Plan in a regular end-of-lesson task or starter for following lessons to encourage recall.

The activities on pages 126–127 of the Workbook are designed to support language development as they encourage students to think about some of the key words. These can be used as a starter task, as an end-of-lesson task or for home learning. You could return to these later in the unit to help longer-term recall and understanding.

Remember that the eBook has examples of key words being pronounced. Use this at the start of lessons and return to it at the end.

Add to your science library. Collect unit-specific resources and download specific information about topics for each lesson or activity and make a booklet of these like a class magazine. Ask students to help in making small information booklets that include their own work. Seeing their work being used by a wider audience is very motivating.

Unit at a glance

The key teaching points for students in this unit are:

- to introduce the unit objectives
- to introduce the learning outcomes
- to engage students with the content of the unit
- to review and build on prior learning and understanding of the topics.

In the next lesson, students will learn about the fossil record and how fossils can provide information about things living millions of years ago.

The purpose of this introductory lesson is for students to start thinking about and reviewing prior knowledge of the characteristics of living things and how living things are adapted to their habitats. The introductory pages show a large photograph of giraffes feeding, a fossil plant and insects (moths) against a tree. The images are used as a starting point and prompt for discussions but they are also designed to create interest and curiosity.

Point out the unit objectives and read through each one – this will give students an idea of the learning they will be doing, but will also trigger recall of related work they have done in earlier years. For example, remind them that they studied life processes and the need for living things to produce offspring in Year 5 and adaptations linked to predators and prey in Year 4.

Read through the Word cloud words and then allow students to enjoy looking over the page before you start the sequence of discussion tasks.

Ask students to look closely at the photographs on the pages, and then ask them to talk about the discussion questions. A useful sequence is as follows.

These animals eat leaves. How are they adapted to their habitat? Which ones will have problems when the lower leaves of the trees have all been eaten?

Arrange students into pairs or groups of three or four for discussion work. Ask students to look carefully at the main photograph. They can discuss it and then agree on an answer to each question. Encourage them to talk about what the animals are and whether they are all the same. Elicit that some are adults and some are offspring. Volunteers can share their answers with the class.

Answer: The animals are giraffes. They are adapted to eat the leaves as the leaves are high up in the trees and giraffes have long necks and legs. Some students may suggest other adaptations from general knowledge, such as that giraffes have long tongues to grip the leaves and many flat teeth to grind the leaves. The smaller giraffes will have problems as they have shorter legs and necks and may run out of leaves to eat.

What is seen in the rock? Is it living, non-living or once living? Discuss how the object got into the rock.

Allow students to continue to work with their partner or small group to talk about the picture showing the fossil plant. Point out that they will be studying fossils and making their own versions in the next lesson. Ask them to consider if they have seen any living things like the one shown in the photograph. Remind them of the classification of living, non-living and once living and let some students share their ideas with the class so you can check answers and deal with any misconceptions. Listen to their ideas about how the object got into the rock. Don't correct any but tell students to remember their suggestions as they can check them soon and may change their minds.

Answer: A plant fossil is seen in the rock. It can be classified as once living. Students may suggest many theories, but the scientifically accepted one is that the plant died, was covered with sand and mud, and eventually turned to stone.

Can you see four moths? They are the same species. How are they the same? How are they different? Which moth would survive in a dark, polluted area?

Ask students to look carefully at the moths on page 127. Remind them to use their observation skills and to look carefully for four moths. Ask them to discuss the questions and then share their ideas with the class. Encourage them to think back to predator–prey relationships.

Answer: There are four moths – two are light grey and two are dark grey. They are identical except for the colouring (pigmentation). The darkest grey moth would survive in dark, polluted areas. Students may recall the word 'camouflage' here.

Science fact Scientists believe that 99% of all of the species that have ever lived on Earth have now died out. These are called extinct species. These include dinosaurs and the plant in the small photo above, which has not grown on Earth for over 200 million years.

Read out the Science fact or ask a volunteer to read it out. Ask students if they know the names of any extinct things other than dinosaurs. You can also ask them to think about what has happened to all of the living things that once lived on Earth. Point out that some of these will have left traces in the rocks for us to study.

Workbook activities

Key words (page 126)

This can be an individual or paired activity. Students are very familiar with wordsearch activities now so leave them to find the missing words. Point out that after finding the words they need to reflect on any of the words they have not used before and write them down. Ask them to keep coming back to this page to tick off words as they learn about and use them. You can hint that nine of the twelve words are in the wordsearch. The missing ones are characteristic, environment and habitat.

Answer: Students find and circle the following words: variation, adaptation, fossil, extinct, reproduction, ancestor, species, offspring, inherit.

Match the definition (page 127)

You can use this activity at the start of the lesson to gauge prior knowledge of these key words. Ask students to leave any blank that they do not know and return to them as they learn the words, or you can use it as an end-of-lesson summary. Explain that students should use a ruler and pencil to draw a neat line to link each word with its definition

Answer: adaptation – Any characteristic that helps a living thing survive in its habitat. ancestor – The living things that other living things developed from. extinct – A species that no longer exists. fossil – Evidence of living things preserved in rocks. inherit – The passing of characteristics from parents to their offspring. offspring – The young born to living things. reproduction – The making of new individuals by parents. species – Living things that are so similar they can make offspring together. variation – The differences between individuals in a species.

Extra activities

1 **Computing link:** Allow students to review their prior knowledge of adaptation by selecting a favourite animal and a favourite plant and using the internet to research how they are adapted to survive in their habitat. Students can download photographs and make a small display. Ask them to label the adaptations.

2 Ask students to create their own wordsearch or crossword puzzle using the key words for the unit. You can allow them to use one of the many free puzzle maker websites or hand out squared paper and let them make one the original way they were developed. Once completed you can copy them and make a puzzle book for students to try.

The fossil record

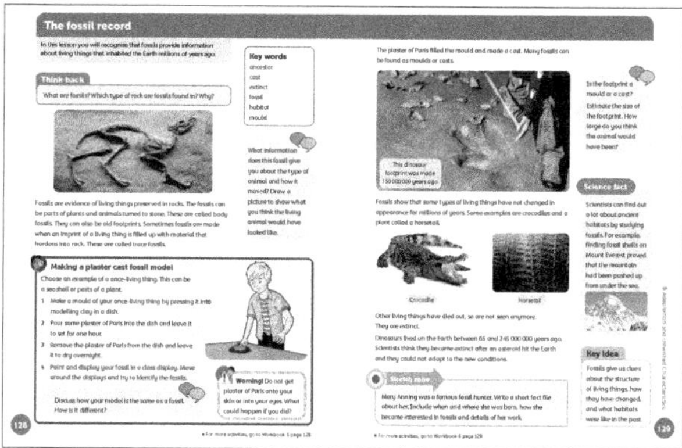

Supporting activities are in Workbook 6 pages 128–129.

Getting started

In this lesson students will learn that fossils are evidence of living things preserved in rock. They will review their learning of fossils from Year 3, study how fossils are formed, and learn that many are created as moulds and casts. They investigate by making their own fossil casts. They then consider examples of species of animals and plants that have not changed for millions of years and some that have died out or become extinct.

Language support

To help develop language skills and review key words you can write the word 'ancestor' on the board with some letters missing and replaced by dashes. Ask students to work out what the word is. Then they can make their own versions of partial words by selecting some of the key words for the lesson and writing them out with missing letters. Ask them to test a partner.

Resources

Student Book: selection of once-living things (e.g. shells; parts of a plant); dishes; modelling clay; water; plaster of Paris; paints.

Workbook: selection of once-living things (e.g. shells; parts of a plant); modelling clay; rolling pins; water; plaster of Paris; paints; labels.

> ### Key words
>
> ancestor   cast   extinct   fossil   habitat   mould
>
> **Other words in the lesson**
>
> asteroid   body fossil   dinosaurs   footprints
>
> imprint   preserved   trace fossils

Unit 5 Adaptation and Inherited Characteristics   173

Lesson at a glance

The key teaching points for students in this lesson are:

- fossils are evidence of living things preserved in rock
- fossils give us clues about what living things and habitats were like in the past.

In the next lesson, students will learn how fossils can help us to find out how some species of living things have changed over time.

Think back: What are fossils? Which type of rock are fossils found in? Why?

Ask students to think back to their work about rocks from Year 3. They can discuss the three main types of rocks (igneous, metamorphic and sedimentary) and then decide which type contains fossils. Remind them that if a rock is formed from molten rock, then living things will never have lived near it and also if any rocks are later heated up any fossils that might have been there are likely to be destroyed.

> *Answer: Fossils are evidence of living things preserved in the rock. They are found in sedimentary rocks – those formed from particles and shells settling in layers. Examples are mudstone, shale, sandstone and limestone.*

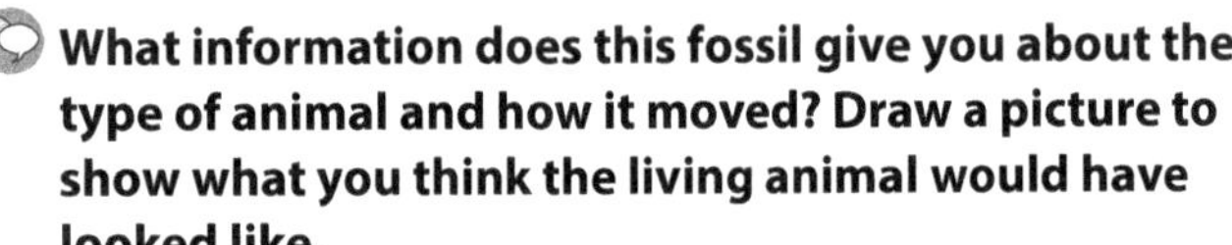

What information does this fossil give you about the type of animal and how it moved? Draw a picture to show what you think the living animal would have looked like.

Allow students to work with a partner to read through the text on page 128 and talk about the photograph showing the fossil bones. Encourage them to observe the shapes and sizes of the fossil bones and decide which animal group they belong to. Point out that the fossil will not be bone. Chemicals will have soaked into the bones and changed them to stone, or they will be the imprints of bones later filled in with rock. Students can draw on prior learning about the skeleton and the work they did on vertebrates in Year 3 and in Unit 1 earlier this year.

> *Answer: Students should know that the animal was a vertebrate and walked on four legs. It also had a tail. Students may recall their work on teeth from Unit 2 and may mention that they can see large molars for grinding food and incisors for cutting so the animal may have eaten a lot of plants.*

Investigation: Making a plaster cast fossil model

The purpose of this investigation is to allow students to explore one way that fossils are formed. Allow students to work in groups of three or four. If you wish to give more support, the activity on page 129 of the Workbook will be useful.

Warning! Do not get plaster of Paris onto your skin or into your eyes. What could happen if you did?

Start by reading out the warning information. Ask students to discuss it with their team and do not let them start until they can explain to you why they need to be careful with plaster of Paris and how they are going to prevent it getting onto their skin and into their eyes.

Students can work independently and follow the instructions. During the time they leave the plaster of Paris to set, you can ask them to complete the activity on page 128 of the Workbook or continue with the lesson in the Student Book. Once the cast is dry, ask students to remove it carefully and paint it. You can set up a 'fossil' exhibition.

Discuss how your model is the same as a fossil. How is it different?

Students can continue to work in their group. Ask them to compare their model to the photograph of the fossil at the top of the page. You could also display some other fossil examples.

> *Answer: Students should recognise that their model is the same as a fossil because the imprint or mould was the shape of the once-living thing and this was later filled in with something that sets hard. The difference is that students used plaster of Paris while fossils are formed from chemicals that trickle down through the rocks. Also, many fossils are not imprints but are bones or shells which are slowly replaced by rock.*

Is the footprint a mould or a cast? Estimate the size of the footprint. How large do you think the animal would have been?

Ask students to work with their partner to study the text and photograph at the top of page 129. This shows a fossil footprint. Point out that the photographer has made sure that a ruler or tape measure is included in the photograph. This is important for students to realise and bear in mind when they are taking photographs for science reports or posters.

> *Answer: As a footprint is made by an animal pressing down onto soft sand or mud, the initial footprint is a mould and then when it is filled in it becomes a cast. Explain that sometimes the cast is softer than the surrounding rock and so it can be washed away over time and leave the original mould shape. The size estimate is between 30 and 50 centimetres. The footprint is twice the size of the person's foot so the animal might be twice as large in terms of height – between 3 and 4 metres.*

Students can read the text beneath the photograph and study the photographs of the crocodile and horsetail. Stress that individual living things do not live for millions of years – it is the species that survives. Ask, 'How do scientists know that crocodiles have not changed for millions of years? When did dinosaurs become extinct? What is one theory about why this happened?'

Science fact Scientists can find out a lot about ancient habitats by studying fossils. For example, finding fossil shells on Mount Everest proved that the mountain had been pushed up from under the sea.

Read out the Science fact and ask students to think about how they would feel if they found a fossil shellfish high on a mountain. Challenge students by asking them what they could work out (deduce) about their area in the past if someone found a fossil whale, dolphin and coral in a hillside above where they lived.

Stretch zone: Mary Anning was a famous fossil hunter. Write a short fact file about her. Include when and where she was born, how she became interested in fossils and details of her work.

Computing link: Allow students to work individually or with a partner and allow them access to the internet to research Mary Anning. They can download photographs to help with their fact file. (These will be photographs of paintings due to the time that Mary Anning lived.)

Answer: Mary Anning was born in 1799 and died in 1847. She lived on the south coast of England and she contributed greatly to our understanding of fossils – especially fossil lizards and dinosaurs.

Key idea

Fossils give us clues about the structure of living things, how they have changed, and what habitats were like in the past.

Read through the key idea or ask a volunteer to read out the statement to the class. Ask students to write down the names of a species of animal and a species of plant that we know from fossils have not changed over millions of years. They can then turn to a partner and discuss how fossils form and then ask volunteers to share their ideas with the class.

Workbook activities

How are fossils formed? (page 128)

Explain that the pictures showing the formation of a fossil have been mixed up. As well as drawing on their work, this lesson reminds students that they studied the formation of fossils in Year 3. Point out that revisiting

work and then studying it in more detail is a vital part of learning. Ask them to study each diagram and then decide on the correct order. They record this as a timeline or flow chart at the bottom of the page. Point out the clues in the hint box.

Answer: The order is B, D, A, C.

 Making a plaster cast fossil model (page 129)

This activity supports the investigation on page 128 of the Student Book. It shows the method in a step-by-step sequence and has diagrams to make it easier to work out the process. Point out the warning information as in the Student Book version and then allow students to follow the numbered instructions as you supervise. Once the fossil exhibition is arranged, allow students to walk around to observe and try to identify the different once-living things.

Review and reflect

Encourage students to reflect on their own learning by pausing for a few moments and thinking about the way they undertook the investigations. Talk about how they managed this and how they worked as a team. Remind them that when they are doing tricky work they are training their brain and this will help their future learning.

Conclude the lesson by asking students to complete question 2 in the 'What have I learned about adaptation and inherited characteristics?' activity on page 140 of the Student Book and the first statement on page 140 of the Workbook. (See also the teaching notes in this Teacher's Guide, pages 193–195.)

Extra activities

1 Arrange a visit to see a local area that has fossils in the rocks, or take students to a local geological or natural history museum. Ask students to draw and label some examples of fossil animals and plants and record the dates the once-living things were alive. They can find this from information panels in museums or they can use books or the internet to find out about the fossils if they have visited local rocks.

2 **Computing link:** Ask students to work in pairs or small groups to research an example of an animal that has been extinct for millions of years. They should find out when it lived, what it looked like, what sort of habitat it lived in and why it became extinct. They can also make a note of any living animals that are similar to the extinct one. They can share their ideas by making an information leaflet or a page to put on the computer as a wiki or blog.

Differentiation

Supporting: Allow students to use the activity on page 129 of the Workbook to provide extra support for the investigation.

Consolidating: Display photographs of fossils around the room. You could split this into geological periods so students see which animals and plants were common at different times.

Extending: Ask students to research the geological periods, make a timeline of geological history from the Cambrian to the Tertiary, and place into the timeline one example of a fossil found during each period.

Differentiated outcomes	
All students	should be able to state that fossils are evidence of living things found in rocks
Most students	will be able to describe how fossils form and how they give us clues about how living things and conditions have changed over time
Some students	may be able to link specific fossils to specific time periods in the Earth's past

Changes over time

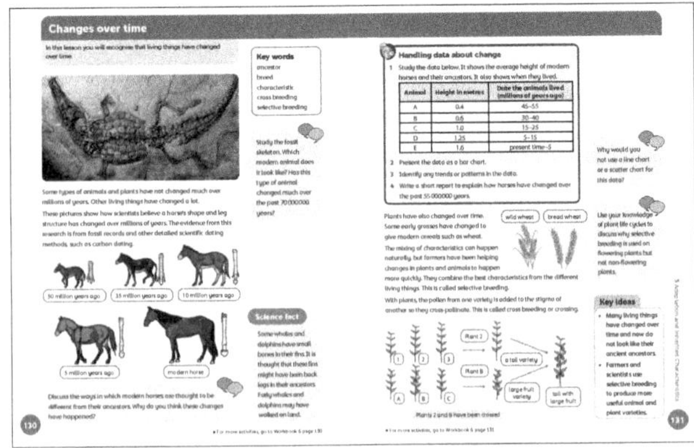

Supporting activities are in Workbook 6 pages 130–131.

Getting started

In this lesson students will learn that some types of animals and plants have not changed much over millions of years. They will study the changes in selected ancestors of the modern horse to observe fossil evidence that horses have grown progressively larger compared to the ancestor that lived 50 million years ago. Students will interpret data about this change and then consider some examples of changes in plants over time.

Language support

Students are now familiar with the words 'characteristic' and 'ancestor' but ask them to tell you what they mean so you can review this. Read out the other words in the key words box and explain that 'breed' means to reproduce – to produce offspring. Point out that cross breeding happens in plants and it means pollen from one variety of plant joins with the ovule of another to make a new variety. You can show them the diagram of this on page 131 as you explain this. With selective breeding, a person such as a farmer or scientist deliberately picks the varieties and then helps them to cross breed. Remind students that 'select' means to choose.

Resources

Student Book: graph paper.

> ### Key words
>
> ancestor   breed   characteristic   cross breeding selective breeding
>
> **Other words in the lesson**
>
> carbon dating   cereals   cross-pollinate
> dating methods   grasses   pollen   stigma   wheat

Lesson at a glance

The key teaching points for students in this lesson are:

- some types of animals and plants have not changed over millions of years
- many types of living things have changed over time and we can use fossil evidence to show this.

In the next lesson, students will learn that living things reproduce by producing offspring of the same kind.

Study the fossil skeleton. Which modern animal does it look like? Has this type of animal changed much over the past 70 000 000 years?

Students can work in pairs or small groups. Ask them to observe the photograph at the top of page 130. This shows a fossil. Ask them to think about any modern animal they have seen that looks similar. If they cannot think of one, ask them to turn back to page 129. They can share their ideas with the class.

Answer: Students should suggest the skeleton looks like a crocodile or alligator. They should state that it has not changed very much.

Ask a volunteer to read out the text on page 130. You can then ask students to study the pictures of the modern horse and its ancestors. Point out that these are only a select few from many but they all fit the same pattern.

Discuss the ways in which modern horses are thought to be different from their ancestors. Why do you think these changes have happened?

Students can continue to work with their partner or small group. Ask them to study each picture and compare each with the modern horse. They can make a list of any similarities and differences.

Answer: Students should note that the modern horse is larger than any of its ancestors and has longer legs with no evidence of toes, which are seen in most of the other ancestors. They may suggest the changes occurred as varieties that were larger and could run faster had more chance of surviving and so had offspring that were also larger.

Science fact Some whales and dolphins have small bones in their fins. It is thought that these fins might have been back legs in their ancestors. Early whales and dolphins may have walked on land.

Read out the Science fact and ask students if they have heard of the idea that whales and dolphins had ancestors that lived on land. Point out that the ancestors would have looked very different from how whales and dolphins look now.

Investigation: Handling data about change

Students can work with a partner so they can discuss their ideas. Ask them to study the table and present it as a bar chart. They can then analyse the data to identify any trends. Remind them that a bar chart has the categories (in this case, the date the animals lived) along the x (horizontal) axis and any number values (in this case, height) up the y (vertical) axis. Once they have completed their analysis ask them to produce a written report to explain the changes in horses over the past 55 million years.

If you wish to provide more support for students, they can complete the activity on page 130 of the Workbook.

Answer: Students should note that between 55 million years ago and the modern horse the ancestors have grown from an average of 0.4 metres to an average of 1.6 metres. The trend throughout is for an increase in height.

Why would you not use a line chart or a scatter chart for this data?

Maths link: Students can continue to work in their pairs to talk about the question. Ask them to think back to examples of line graphs and scatter graphs they have seen and used in science and other studies such as maths. You can ask them to look back at the bar, line and scatter graphs on page 11 to remind them. Ask volunteers to share their answers with the class or you can read out the answer and students can check to see if they were correct.

Answer: Students should know that the dates the animals lived are not linked – they are separate categories, like colours would be, so they are shown as bars. A line graph is used when both axes show numbers, and scatter graphs are useful when the points do not have a simple relationship. Point out that they can use a scatter graph when they survey shoe size later in the unit.

Point out the text on page 131 and ask volunteers to read out a sentence each. To check understanding you could ask students the following questions: What has wild wheat changed into? What is selective breeding? Why do farmers try to do this? In cross breeding, what is deliberately added to the stigma of one plant variety?

Use your knowledge of plant life cycles to discuss why selective breeding is used on flowering plants but not non-flowering plants.

Ask students to think back to their work on reproduction in flowering plants. You could show them some potted plants and flowers to remind them of the concepts they studied. You can hint by asking if non-flowering plants produce seeds. They may recall that, other than conifers, most do not.

Answer: Students should state that plants are called non-flowering plants for an obvious reason – they do not produce flowers. This means that most do not produce pollen so it isn't possible to deliberately place this on the stigma of another plant.

Key ideas

- *Many living things have changed over time and now do not look like their ancient ancestors.*
- *Farmers and scientists use selective breeding to produce more useful animal and plant varieties.*

Ask students to close their books and talk to a partner about what they have learned in the lesson. Then ask them to imagine that they are the authors of the book and they have to write what the key ideas should be. Let them write down their ideas and then they can open their books and check how close their version is to the one published.

Workbook activities

Handling data about change (page 130)

This activity supports the investigation on page 131 of the Student Book. Explain that the questions can be used as prompts to help students analyse the data in the table and a section of graph paper has been provided. This contains the axes to help in the design and production of the graph. Explain that, although the *x* (horizontal) axis has numbers, they are there to help them place the categories (date the animal lived) in the correct place and to get some idea of a timeline.

Answer: E is the modern horse. A is the smallest. The height of the horse ancestor living 16 million years ago was 1.0 metre.

Changes in plants (page 131)

Point out the drawings of the plants at the top of the page and explain that this is a black and white version of the picture on page 131 of the Student Book. Ask students to read the text and then complete the questions. They should use their knowledge of the results of cross breeding to predict the outcomes of various crosses.

Answer: 1 Plant 2 is tall and healthy and plant B has lots of fruit. Plant 3 is tallest but will take up more space. 2 For Plants 1 and A students should draw a short plant with very little fruit; for Plants 2 and A they should draw a tall plant with very little fruit.

Review and reflect

Ask students to look back at the Student Book and their activity pages for this lesson and tell them they are going to vote for the part of the lesson they found the easiest to understand and then vote for the part they found the most difficult. Set up two boxes at the front – one labelled 'easy' and the other labelled 'hard'. They write down both votes by writing the topics onto small pieces of paper and dropping them into the easy and hard boxes. Collect the paper from both boxes and see if the same topics occur in each. If all students find one aspect hard, you could go over it again.

Conclude the lesson by asking students to complete question 6 in the 'What have I learned about adaptation and inherited characteristics?' activity on page 141 of the Student Book. (See also the teaching notes in this Teacher's Guide, pages 193–195.)

Extra activities

1 Visit a garden centre or botanical gardens and arrange for students to have a talk from a scientist who works to cross breed plants to create new varieties. Students can learn about the techniques and even try some back in the classroom by cross-fertilising flowers and later collecting and planting the seeds.

2 **Computing link:** Ask students to research the ancestors of animals other than the horse. They could choose their own or you could allocate one of the following: lion; whale; eagle; elephant. They can download diagrams and drawings and provide information about when the ancestors lived and how they have changed over time.

Differentiation

Supporting: Support the investigation by providing graph paper with the axes already labelled for students. There is an example on page 130 of the Workbook.

Consolidating: Ask students to measure out the heights of the modern horse and its ancestors on a wall and fix drawings and labels to show how the heights have changed over time.

Extending: Ask students to find out about some of the other horse ancestors that are not included in the information in the Student Book. They can check to see if the pattern of increasing height over time is true for them all.

Differentiated outcomes	
All students	should be able to state that ancestors of the horse were smaller than modern horses and describe a general trend in the development of the horse over time
Most students	will be able to describe how cross breeding can result in different varieties of plants
Some students	may be able to describe a wider range of horse ancestors and determine if the general trend of increasing in height over time still holds true

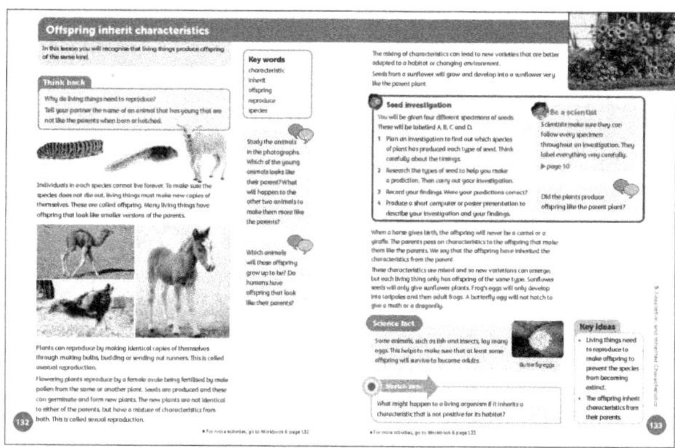

Offspring inherit characteristics

Supporting activities are in Workbook 6 pages 132–133.

Getting started

In this lesson students will review their knowledge that living things produce offspring of the same kind. They will discuss the need for living things to reproduce and consider offspring that are smaller versions of the adults and offspring that are very different. The latter allows them to review their prior work on the life cycles of the butterfly and frog. Students learn that asexual reproduction results in identical copies of living things and sexual reproduction offers the chance of new characteristics emerging. They will investigate how planting sunflower seeds results in new plants that are very similar to the parent plant.

Language support

To help develop language skills you could use flashcards with the key words written on. As you raise each card to show students, they have to find the word on pages 132–133 in the Student Book and then discuss with a partner any pictures that are associated with the word. They should then use each word in a sentence.

Resources

Student Book: samples of four different specimens of seeds and the plants they have come from; plant pots and labels; compost; water; access to the internet or books to research types of seeds; items to prepare a computer or poster presentation.

Workbook: samples of four different specimens of seeds and the plants they have come from; plant pots and labels; compost; water; access to a warm dark, space and a dry, sunny space; items to prepare a computer or poster presentation.

Scientific enquiry key words

Plan and/or carry out enquiries to answer questions

Make predictions

Recognise and control variables

Make observations

Take measurements, using equipment accurately

Record data and results

Analyse data, notice patterns and group or classify
things

Report and present findings

Draw conclusions and give explanations

Identify causal relationships

Lesson at a glance

The key teaching points for students in this lesson are:

- individuals in a species cannot last forever so
  they have to make new versions of themselves by
  reproducing
- these new versions are called offspring
- the offspring can be a mixture of characteristics
  from the parents and so the offspring are unique
  and may have new characteristics.

In the next lesson, students learn that offspring are not
identical to the parents and this leads to variation within
a species.

Think back: Why do living things need to reproduce? Tell
your partner the name of an animal that has young that
are not like the parents when born or hatched.

Point out the Think back task at the top of page 132.
Students can work with a partner to talk about why living
things need to reproduce. This will encourage them to
review prior work on life processes covered in earlier
years such as Year 3 and Year 5.

*Answer: Living things need to reproduce because
individuals do not live forever, so, if a species is
to survive, living things must make new copies of
themselves. Students could select many options,
including butterflies, moths and frogs.*

 **Study the animals in the photographs. Which of the
young animals looks like their parent? What will
happen to the other two animals to make them more
like the parents?**

Ask students to work with a partner to talk about
the questions. Tell them they should study the three
photographs at the top of page 132 and discuss the
questions. Ask them to start by trying to identify each
animal and then they can imagine the adult. You could
ask students to name the phases of the life cycles
shown by the left and centre photographs. Elicit larva
(caterpillar) and tadpole.

*Answer: The young sheep (lamb) looks like its parents.
The other animals will change to look like their
parents. The caterpillar will change inside a chrysalis
and the tadpole will grow legs and lose its tail.*

 **Which animals will these offspring grow up to be? Do
humans have offspring that look like their parents?**

Students can continue to work with their partner to talk
about the three photographs in the middle of page 132.
They should realise that the animals are examples of
offspring that look like small versions of the parents. They
can debate the bird as very young birds (chicks) have no
feathers but they soon develop them.

*Answer: Camel; chicken and horse. Yes, human
offspring generally look like their parents.*

Ask a volunteer to read out the text at the top of page
133 and point out the photograph of the sunflowers.
Explain that they are going to investigate the idea that
flower seeds grow into flower plants very like the adults.

 **Investigation: Seed investigation**

Arrange students into groups of three or four. Explain
that the purpose of the investigation is to grow seeds
from different flowering plants and then observe and
measure the plants that grow to identify which variety
of flower each set of seeds has come from. Show them
some potted plants of the same type as the seeds but
do not link the seeds to the plants. That is for students to
work out.

Point out that the seeds will need to be left for a number
of weeks as they germinate and the seedlings grow.
Ask students to read and follow the instructions. Hand
out four different sets of seeds from very different types
of flowers – different adult heights and flower shape
and colour. Use plant brochures, online information or
a local plant shop to help you to decide. Students can
review earlier work on growing plants and what they
need for healthy growth and then, when the flowering
plants have grown, they can compare them with the pot
plants you showed them at the start and decide which
seeds came from each plant. They share their ideas by
producing a short computer or poster presentation.

Be a scientist: Scientists make sure they can follow every specimen throughout an investigation. They label everything very carefully. (Student Book, page 10)

Ask students to read the information in the Be a scientist box as they are planning their investigation and explain to students that they should make sure they label everything carefully as they will be observing the seeds and plants for a long time.

Did the plants produce offspring like the parent plant?

Students can continue to work with their investigation group. Ask them to talk about the question to help them to arrive at their conclusions. Remind them to only use the results they see in their investigation to support any conclusions.

Ask students to work with a partner and take it in turns to read out a sentence each from the text on page 133. This explains that offspring are always specific to the adult and living things only make offspring like themselves. Ask, 'Can a camel have a horse as their offspring? Can a butterfly egg hatch to give a tadpole?'

Science fact Some animals, such as fish and insects, lay many eggs. This helps to make sure that at least some offspring will survive to become adults.

Read out the Science fact or ask a volunteer to read it out. Ask students to discuss why laying lots of eggs can help a species to make sure individuals survive. Remind them of food chains and ask them to think of some animals that might eat the eggs. Also ask them to think about how else eggs could be damaged. They should realise that laying many eggs means that even if some are eaten or damaged by storms or fires, then there is a chance some will survive and hatch.

Stretch zone: What might happen to a living organism if it inherits a characteristic that is not positive for its habitat?

Allow students to work in their small group. Point out that for this activity the emphasis is on a living thing that might not be well adapted to its habitat. Ask students to list some examples of adaptations and then think about what would happen if a living thing did not have these adaptations. For example, what might happen if the offspring of a polar bear did not have a thick fur coat?

Key ideas

- *Living things need to reproduce to make offspring to prevent the species from becoming extinct.*
- *The offspring inherit characteristics from their parents.*

Read out the key ideas or ask volunteers to read them out. Ask students to turn to a partner and tell them the names of two animals that have offspring unlike the parents and two animals that have offspring that are small versions of the parents. They can then list two characteristics that they have inherited from their parents.

Workbook activities

 **Seed investigation (page 132)**

This activity supports the investigation on page 133 of the Student Book. It provides more support by offering step-by-step instructions, a reminder to predict the outcomes and a blank results table for recording findings. Ask students to follow the instructions and complete the table.

Find the parent (page 133)

This is a useful activity to complete as an introductory task or to use if students are having problems understanding the concept of offspring either looking like the parent or not looking like the parent. Explain that they draw a line to link the parent to its offspring. Finally, ask students to think about how plants grow and develop and they can decide two characteristics that adult plants have that offspring do not.

Review and reflect

Ask students to stand by their computer or poster presentation and look at it carefully. They can then walk around and observe other presentations. When they return to their own, they can write down two things that they are really proud of and one suggestion for improving their presentation.

Conclude the lesson by asking students to complete question 3 in the 'What have I learned about adaptation and inherited characteristics?' activity on page 140 of the Student Book. (See also the teaching notes in this Teacher's Guide, pages 193–195.)

Extra activities

1 Ask students to download photographs of six animals and their offspring. They should select the animals so that three have offspring that are small versions of the adults and three that are unlike the adults. They can then make a large collage poster and label the characteristics that are the same and different for each animal.

2 Students could obtain family photographs and spread them out in the room with photographs from other students. They then look through the photographs and see if they can identify which photographs should be linked to each family. They have to use characteristics they have in common. If this is too sensitive with your students, you can download famous family photographs from around the world and do the same activity.

Differentiation

Supporting: Start the lesson with the activity on page 133 of the Workbook to help students grasp the concept of offspring and their parents.

Consolidating: Show students a photograph of a common animal from your local area and work with them to label five characteristics and then contrast this with another animal that has markedly different characteristics.

Extending: Ask students to research the work of Gregor Mendel and to find out about genes and how they help to explain how characteristics are passed on.

Differentiated outcomes

All students	should be able to state that offspring inherit characteristics from their parents
Most students	will be able to explain that asexual reproduction creates identical copies of a living thing but sexual reproduction allows new characteristics to emerge
Some students	may be able to relate the passing on of characteristics to the passing on of genes

Variation in living things

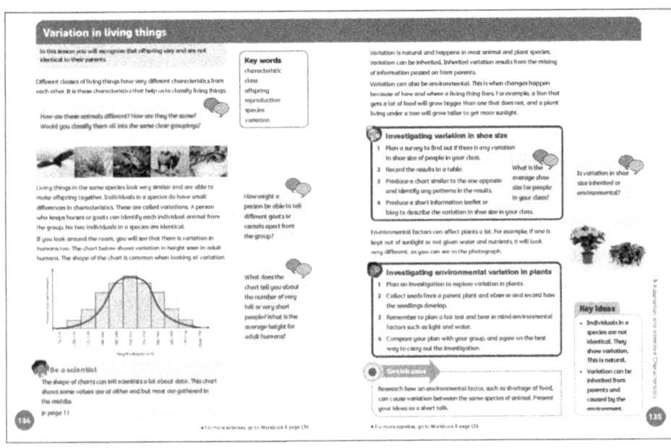

Supporting activities are in Workbook 6 pages 134–135.

Getting started

In this lesson students will learn that offspring vary and are not identical to their parents. They will observe similarities between living things of the same species, but also find out that there is a natural variation between members of the same species. They will analyse a normal distribution curve and investigate differences in hand size in students and investigate environmental variation in plants.

Language support

To support language development, you could refer back to the Word wall if you created one for this unit. Point out that two key words not found on the Word wall are 'variation' and 'class'. Explain that 'class' can have a number of definitions, but in this lesson it refers to a level of classification such as the major groups or classes of vertebrates – fish, amphibians, reptiles, birds and mammals. Ask students if they know any other definitions of the word 'class'. Point out that 'variation' means differences – how much something varies or is different.

Resources

Student Book: materials to make information leaflets; access to the internet or books to research how an environmental factor can cause variation between the same species of animal; seeds; plant pots and labels; compost; water.

Workbook: rulers; seeds; plant pots and labels; compost; water; access to a warm, dark space and a dry, sunny space.

> **Key words**
>
> characteristic   class   offspring   reproduction   species   variation
>
> **Other words in the lesson**
>
> environmental   hand span   identical   inherited   natural

Lesson at a glance

The key teaching points for students in this lesson are:

- living things of the same species look very similar but there are natural variations
- variation can be inherited and can be the result of environmental differences
- variations from environmental factors are not passed onto the next generation.

In the next lesson, students will identify how animals and plants are adapted to suit their environment in different ways.

Read out the text at the top of page 134 and remind students of their earlier work on classification from Unit 1 Classification and Habitats. You could write down the major levels of classification (taxonomy): kingdom, phylum, class, order, family, genus and species. Point out that as you work down through the levels from kingdom the living things are more and more alike.

How are these animals different? How are they the same? Would you classify them all into the same close groupings?

Ask students to discuss the questions with a partner and point out the photographs at the top of page 134. Suggest they will find the answers there if they use their observation skills. Ask students to write down their ideas and then volunteers can share their ideas with the class.

Possible response: Students should list differences such as body shape, presence or absence of wings, presence or absence of legs, number of legs, body coverings and having lungs or gills. They may recognise that each photograph shows a member of a different class of vertebrate – left to right they are: bird; fish; amphibian; mammal; reptile.

Ask students to read the text on page 134. This informs them that even living things in the same species show some variations. You can discuss identical twins but even these are not 100% identical – no two individuals in a species are identical.

How might a person be able to tell different goats or camels apart from the group?

Students can continue to work with their partner to talk about the question. Ask them to think about people they know. They are all in the same species but they can still tell people apart.

Possible response: Goats and camels will look very similar to other goats or camels but they will have slight differences in characteristics such as height, coat colour and facial features.

Explain the chart on page 134. It shows the distribution of heights in a community of people. This shape of graph is often called a natural distribution curve or bell curve (because it is bell shaped).

What does the chart tell you about the number of very tall or very short people? What is the average height for adult humans?

Students can work independently or in pairs or small groups. Allow them to study the chart and then ask them to look carefully at the *x* (horizontal) axis and ask them what it shows. Then ask the same about the *y* (vertical) axis. They can then talk about the questions and decide on the average height for adult humans.

Answer: Very few people in the population are very tall or very short. Most are nearer to the average. The average height, as shown in the chart, is between 150 and 154 centimetres.

Be a scientist: The shape of charts can tell scientists a lot about data. This chart shows some values are at either end but most are gathered in the middle. (Student Book, page 11)

Point out the Be a scientist information and ask students to think of any other examples of graphs that might be the same shape as the one on page 134. Tell them they can investigate to find out if hand size and shoe size follow the same pattern.

Allow students time to read the text at the top of page 135 and explain that inherited characteristics can be passed onto the next generation, but environmental ones cannot. They just affect the living thing at that time. Ask students to list one example of an environmental factor that can cause living things to vary from others in the same species. Elicit that lack of food, water, space or sunlight can affect plants and lack of food and water will affect animals.

Investigation: Investigating variation in shoe size

Allow students to work in groups of three or four. Explain that they are going to plan a survey to find out if there is any variation in shoe size of people in their class. Students should walk around and ask eight to ten people their shoe size, then record their results. Alternatively, you can make a class table and ask everyone to write in their shoe size. Remind students to design a table for their results or make a copy of the class table. They should then produce a chart similar to the one on page 134 and analyse it to identify any patterns. They can share ideas by creating a short information leaflet or blog to describe the variation in shoe size in the class or group of students they surveyed.

> *Possible response: Students should find that most people surveyed had shoe sizes that were at or near the average shoe size, so they may see a normal distribution curve.*

What is the average shoe size for people in your class?

Students can discuss this as they analyse their findings from the investigation. Ask them to use their chart to work out the average by finding a point where there are an equal number of people on either side.

> *Answer: Values will vary but students should be able to use the chart to work out the average shoe size.*

Is variation in shoe size inherited or environmental?

Allow the investigation groups to consider the question. You can provide clues by asking the following questions: What size shoes do people in your family wear? Do all of the adults have the same size shoes? If someone didn't have much food when they were a child, would they grow up with smaller feet?

> *Answer: Foot size is inherited but is affected slightly by environmental factors such as diet.*

Read out the text beneath the shoe size investigation on page 135. Point out the photographs of the two plants. They are the same species of plant and should look very similar. Ask students why they look so different. Elicit that one has wilted due to a lack of water. Wilting cannot be passed onto the next generation but it does affect how a plant looks.

Investigation: Investigating environmental variation in plants

Students can continue to work with their group or you could change groups around to encourage students to work with different people. Ask students to read through the instructions and then individually plan an investigation to explore variation in plants.

Allow students to collect some seeds from a parent plant and observe and record how the seedlings develop. You could buy seeds if it isn't possible to find plant seeds in the area. Remind students to plan a fair test and to bear in mind environmental factors such as light and water. Students should compare their plans with other groups and agree on the best way to carry out the investigation. Allow them to work independently as a research team but be ready to offer support when needed. The activity on page 135 of the Workbook can be used to provide additional information and structure if needed.

Students can share their findings by giving a short talk to the rest of the class.

> *Answer: Students should find that the seedlings grow into plants very similar to the parent plant but each seedling will not be identical. They will vary slightly in height and colour, for example.*

Stretch zone: Research how an environmental factor, such as shortage of food, can cause variation between the same species of animal. Present your ideas as a short talk.

Computing link: Students can work independently or in pairs or small groups. Allow them access to the internet and once they have gathered the information ask them to present their ideas.

> *Possible response: Students should find that environmental factors can cause differences between individuals of the same species. Living things that have plenty of resources will grow bigger and healthier than individuals that are short of resources – such as light, food or water.*

Key ideas

- *Individuals in a species are not identical. They show variation. This is natural.*
- *Variation can be inherited from parents and caused by the environment.*

Read out the key ideas or ask volunteers to read them out. Ask students to complete definitions for the words 'characteristic' and 'variation' in the glossary at the back of the Student Book, if they have not done so already. They can then tell a partner one example of an inherited characteristic that a young lion will have that a fish will not have.

Workbook activities

Investigating variation in hand size (page 134)

Allow students to work with a partner to share ideas. Explain that this activity is similar to the investigation on page 135 of the Student Book but they are going to investigate hand span instead of shoe size. Show them the diagram showing how to measure hand span so they take accurate measurements.

Ask students to plan their investigation and to save time they should survey only eight people in the class. They can record the results in the table and produce a bar chart or a line graph to show the results. Explain that either type of graph will work as both axes are numbers so a line chart would be suitable, but also students can produce a bar chart by grouping hand spans into categories or ranges – for example between 10 cm and 12 cm, between 12 cm and 14 cm, etc. and these can be shown as bars, with the number of people in each bar shown on the y (vertical) axis. Students can then work out the average hand span for the people they surveyed and determine how much variation in hand span they detected. You can combine the results for the groups to get a class table and ask them to draw another chart.

Possible response: Students should note that there is a normal distribution of hand sizes if the people are picked at random or a lot more people are surveyed.

Investigating variation in plants (page 135)

This activity supports the investigation on page 135 of the Student Book. You can use it if you do not want to leave the investigation open-ended or if some students require additional support.

Warning! Check with an adult before taking any seeds from plants. Why is this important?

Read out the warning information and ask students to discuss why the rule is so important.

Allow students to follow the instructions. They will observe and measure the seedlings on the days shown in the table. They should use the table to record their results and then interpret the data.

Possible response: Students should find that the seedlings were very similar to the parent plant but the seedlings will show some variation – such as slight differences in size or colour.

Review and reflect

Hold up word cards for the key words in the lesson: characteristic, class, offspring, reproduction, species, variation. Each time you hold up a word ask students to turn to a partner and use the word in a sentence. They should write down any of the words they are still unsure of and look these up.

Conclude the lesson by asking students to complete the second statement in the 'What I have learned about adaptation and inherited characteristics' activity on page 140 of the Workbook. (See also the teaching notes in this Teacher's Guide, pages 193–195.)

Extra activities

1 Allow students to review earlier work on plants and relate this to variation by allowing them to grow seedlings but changing the environmental conditions. For example, they could grow four sets of seedlings of the same type of plant – one in sunlight with water, one in sunlight without water, one in the dark with water and the final one in the dark without water. They can then consider how the environmental factors affected variation.

2 **Maths link:** Ask students to research examples of how the normal distribution (bell curve) applies to other scientific ideas. They can download examples and arrange a display. Possible examples are blood pressures, birth weights of babies, and shell height of limpets (a type of shellfish).

Differentiation

Supporting: Rather than having the variation in plants investigation open-ended, you can allow students to use the activity on page 135 of the Workbook.

Consolidating: Display photographs of groups of living things of the same species – such as herds of zebras and fields of wheat – so students can see similarities and variation.

Extending: Ask students to consider why the idea that giraffes have long necks because they reach up for high leaves is not correct.

Differentiated outcomes	
All students	should be able to state that individuals in a species are similar but not identical
Most students	will be able to describe how variation can be inherited but can also be the result of environmental factors
Some students	may be able to explain why inherited characteristics can cause variation in the next generation but environmental factors are not passed on

Adapting to the environment

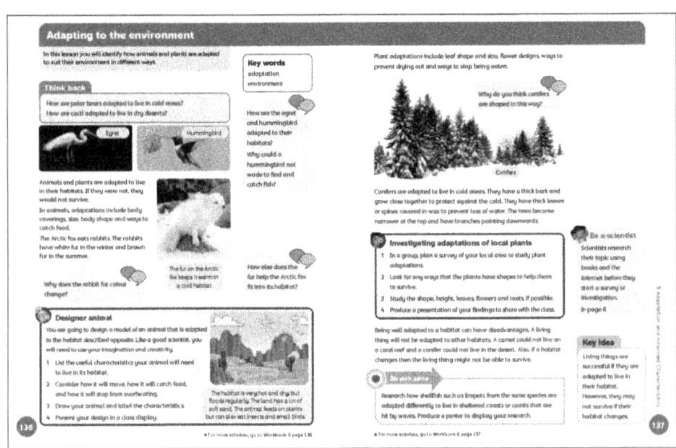

Supporting activities are in Workbook 6 pages 136–137.

Getting started

In this lesson students will review their understanding of how living things are adapted to live in their habitats. They will learn more about the adaptations that let animals and plants survive and will use their creativity to design an animal that is adapted to an imaginary habitat. Students will consider adaptations of conifers and then carry out an investigation to observe adaptations in local plants.

Language support

To support language development and to extend students' knowledge, ask them to cut pictures out of old wildlife magazines or download them from the internet to create a class display of some animal and plant species. Students should then add large labels identifying their different environments and habitats and label some adaptations to review these words.

Resources

Student Book: writing materials; large sheets of paper; access to a local area to research plant adaptations.

Workbook: access to a local area to research plant adaptations; smartphones or digital cameras; resources to make a presentation.

> **Key words**
>
> adaptation   environment
>
> **Other words in the lesson**
>
> characteristic   conifers   habitats

> **Scientific enquiry key words**
>
> Plan and/or carry out enquiries to answer questions
>
> Make observations
>
> Report and present findings
>
> Draw conclusions and give explanations
>
> Identify causal relationships

Lesson at a glance

The key teaching points for students in this lesson are:

- animals and plants are adapted to their habitats
- the better a living thing is adapted to its habitat the more successful it will be but if the habitat changes, it might not be well adapted to the new conditions.

In the next lesson, students will consider how adaptations can help plants and animals survive when faced with change.

Think back: How are polar bears adapted to live in cold areas? How are cacti adapted to live in dry deserts?

Ask students to work with a partner to talk about the Think back task. Remind students that they studied adaptations and characteristics in Unit 1 earlier in the year and actually discussed polar bears. After students have discussed possible answers, ask some of them to share their ideas with the class.

> ***Answer:** Students should recall that polar bears are adapted by having thick fur and a thick layer of fat to keep warm, the fur is white for camouflage and they have sharp teeth and claws for hunting. Cacti have spines instead of leaves to reduce water loss and they can store water. They have wide and deep roots to absorb what water there is from the soil.*

 **How are the egret and hummingbird adapted to their habitats? Why could a hummingbird not wade to find and catch fish?**

Students can continue to work with their partner. Ask them to study the photographs of the egret and hummingbird and then talk about the questions. Encourage them to use prior knowledge of animal adaptations as well as observations of the photographs to decide on their answers. Ask a volunteer pair to share their ideas with the class.

> ***Possible response:** Students should note the egret has long legs and a long beak for wading in water to catch fish. It has wings to fly from lake to lake. The hummingbird is small with a narrow beak so it can get close to flowers and drink nectar. Hummingbirds have small wings that beat quickly so they can hover. Hummingbirds cannot wade and catch fish as their legs are too small and their beaks are not adapted to catching and swallowing fish.*

Ask students to read the text on page 136 and study the photograph of the Arctic fox. They will need to use both in the following discussions.

How else does the fur help the Arctic fox fit into its habitat?

Still working in pairs, ask students to think about the Arctic fox and where it lives. Point out the snow clue in the photograph but also the name is a huge clue – they live in the Arctic region, which is often cold with a lot of snow. Read out the answer and ask students to check if they were correct. Allow them to alter their answer if they have written down an incorrect one.

Answer: Students should state that, as well as keeping the Arctic fox warm, the fur is white. This helps the fox to be camouflaged against the background to help it avoid predators and also catch its prey.

Why does the rabbit fur colour change?

After discussing the Arctic fox, the student pairs can move onto talk about the rabbits in the text. You can point out that the Arctic fox also eats animals that are very similar to rabbits – these are hares. Ask them to think about what will happen to the snow as the summer arrives in the Arctic region.

Answer: The rabbits and hares are white in winter to be camouflaged against the snow and brown in summer as the background is a brown colour.

Investigation: Designer animal

Students can work individually or you can form them into small design teams if you want to encourage them to collaborate. Explain that they are going to use their creativity and imaginations to design a model of an animal that is adapted to the habitat described in the investigation box. Ask them to start by listing the useful characteristics their animal will need to live in the habitat. Remind them that they will have to think about how it will move, catch food and keep cool enough to survive. Stress that the habitat is very hot and dry.

After students have drawn their animal, ask them to label the characteristics that are adaptations to help it to survive and then set up a class display.

Possible response: Adaptations should include ability to hide (so coat colour or being very small), ability to keep cool (so burrowing adaptations would be useful) and how to catch food (fast, good at hiding or creeping, sharp teeth and claws or poisonous fangs). Students may suggest wings to move from water source to water source and avoid floods.

After completing their designer animal, tell students they are now going to consider plant adaptations. Ask them to read the text and study the photograph on page 137. They can then talk about the discussion question.

Why do you think conifers are shaped in this way?

Ask students to look carefully at the photograph of the conifers on page 137. Ask them to talk to a partner about the shape of the trees. You could show them or ask them to think about the shape of roofs in wet countries and the shape of tents. Ask volunteers to share their answers with the class.

Possible response: The branches and spines of many conifers point down so snow does not collect on them and break them. The snow slides off.

Investigation: Investigating adaptations of local plants

Allow students to work in groups of three or four. You could bring pairs together to form the investigation groups. Explain that they are going to plan a survey of the local area to study plant adaptations. Warn them not to touch or eat any plants and allow them to discuss why, and then list any other safety rules they should consider during their survey. If you do not want the investigation to be so open-ended, you can provide more structure and support by asking students to use the worksheet on page 137 of the Workbook.

Ask students to read through the instructions and then take students out and allow them to study the various plants they can find. Remind them to record characteristics such as the shape and height of the plant, size and shape of leaves, type and colour of flowers and any roots visible. Ask students to produce a presentation of their findings to share with the class.

Possible response: Students will find many adaptations that allow plants to survive in their habitat – such as hairs and spines to protect against insects; large, scented flowers to attract pollinators; large leaves to capture more light; tendrils to hold onto objects and other plants; thick bark to prevent drying out; deep roots to search for water. A more comprehensive list is given as a checklist in the worksheet on page 137 of the Workbook.

Be a scientist: Scientists research their topic using books and the internet before they start a survey or investigation. (Student Book, page 8)

Remind students that books and the internet are examples of secondary sources. Ask them why it is important to do some research on secondary sources before starting an investigation or survey.

- **Stretch zone:** Research how shellfish such as limpets from the same species are adapted differently to live in sheltered coasts or coasts that are hit by waves. Produce a poster to display your research.

- **Computing link:** Allow students to have access to the internet and books to research the shell shape and thickness of limpets. Tell them that they may have a surprising result but that it will still fit into the main idea – animals and plants have to be well adapted to their habitat. Students can produce a short scientific paper about the limpets and include downloaded diagrams and photographs.

> **Possible response:** *Students should find that limpets on coasts that are hit by waves are lower than limpet shells from sheltered shores as the low shells let waves wash over them. However, although it might be expected that shell thickness on wave battered shores would also be greater than the thickness of limpet shells on sheltered shores to project from waves, the opposite is true. The reason is that there are more predator crabs on sheltered shores so the limpets here have thicker shells as protection. Students might find other adaptations – such as the width of the shells so they can hold onto rocks better and ridges in the shells to help waves wash over them.*

Ask students to read the text at the bottom of page 137 and ask them to think about how a highly-adapted living thing can have problems if the habitat changes.

Key idea

Living things are successful if they are adapted to live in their habitat. However, they may not survive if their habitat changes.

Read through the key idea or ask a volunteer to read it out. This will help students to review the main themes of the lesson. Ask them to suggest an example of an animal that would not survive if it were moved from one habitat to another. They should name both habitats.

Workbook activities

Adaptations for feeding (page 136)

This can be an individual or paired activity. Remind students that they studied the Galapagos finches in Unit 1 and considered the types of food each species of finch was adapted to eat. They also carried out an investigation to test different shaped beaks and the seeds they could pick up. Explain that they are going to extend this idea by looking at a wider range of birds. Point out that they need to draw lines to link the bird to the type of food they think it is most likely to eat. Hint that the shape of its beak will be the major piece of evidence.

- **Computing link:** Once students have completed the matching activity, ask them to think about the Stretch zone task. Allow them access to the internet to find out

the information they need about the pelican and then let them draw a pelican and write down how its beak is adapted to help it feed.

> **Answer:** *Matching of beaks is as follows: top bird = small animals; 2nd bird = insects buried in mud; 3rd bird = nuts; 4th bird = small seeds; 5th bird = small fish; 6th bird = nectar; 7th bird = insects in bark. Pelicans have a pouch-like beak used to scoop up fish.*

 Plant adaptation survey (page 137)

Explain that this activity supports the investigation on page 137 of the Student Book. Explain that students are going to survey different plants and make a presentation. They can work in small groups.

Explain that students should use the checklist to help them identify plant adaptations. They can tick any they find. Read out the warning that they may not find all of the listed plant adaptations in the area, but encourage them to try to find as many as possible.

Allow students to photograph or film the examples of plant adaptations they find and add these to their presentation.

 Review and reflect

Ask students to sit down for a few minutes and think about what they have learned from the lesson. They can then think of one idea from the lesson that they would like to investigate and find out more about. Ask them to make a plan about how they would do this. This planning time will allow students to think up some imaginative and original things to do and you could let them carry out these projects. They are highly motivating and reinforce learning.

Conclude the lesson by asking students to complete questions 1 and 4 in the 'What have I learned about adaptation and inherited characteristics?' activity on page 140 of the Student Book. (See also the teaching notes in this Teacher's Guide, pages 193–195.)

Extra activities

1. **Computing link:** Ask students to use the checklist on page 137 of the Workbook as the basis for a poster display. They can work within a small group to research examples of plants that show each adaptation and then design a poster. For example, they can download photographs of some plants with spines, stingers or spikes to prevent them from being eaten. They should do this for each adaptation listed or you could allocate each group four adaptations.

2. Students could review their earlier work in Year 4 on predator–prey relationships and use the Arctic fox and Arctic rabbits to show how the predator and the prey have to be adapted to their habitats. They could produce a short talk or computer slide show.

Differentiation

Supporting: Allowing students to use the activity on page 137 of the Workbook will support the plant adaptation investigation.

Consolidating: You can show film of animals and plants adapted to live in areas of the world that are unfamiliar to students. This will reinforce their understanding of adaptation but also provide a global perspective.

Extending: Ask students to research the main climate zones of the Earth and how these have an impact on the living things found there.

Differentiated outcomes

All students	should be able to describe some animal and plant adaptations
Most students	will be able to explain how adaptations allow living things to survive in their habitats
Some students	may be able to explain how adaptations mean that the distribution of living things in the world is not random but linked to many things, including the climate zones

Survival and change

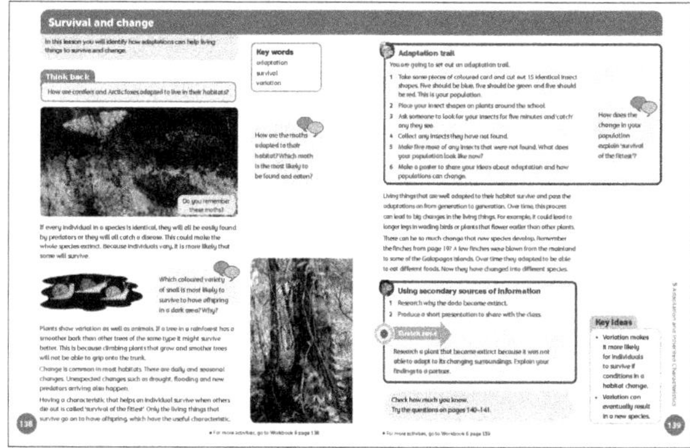

Supporting activities are in Workbook 6 pages 138–139.

Getting started

In this lesson students will learn in more detail how adaptations help living things to survive and also how this survival allows the living thing to reproduce and produce offspring. In this way useful adaptations are passed from generation to generation. Students will explore camouflage as an adaptation that is important for survival.

Language support

This lesson is the final one in the unit, unless you plan a review lesson (see next section). Encourage students to check their glossaries for completeness and to use key terms and words in their final presentation. Ask them to look back at the Word cloud on page 127 and make a note of any words they are still unsure of. They can use a dictionary or look through the Student Book to help them to recall any definitions they are unsure of.

Resources

Student Book: thin blue, green and red card; scissors; materials to make posters; resources to make presentations; access to the internet or books to research a plant that became extinct because of its inability to adapt to changing surroundings.

Workbook: thin blue, green and red card; scissors.

Key words

adaptation   survival   variation

Other words in the lesson

bark   capture/recapture   drought   flooding

generation   identical   polluted/unpolluted

population   predator   rainforest

Lesson at a glance

The key teaching points for students in this lesson are:

- if every individual in a species was identical, they might all become extinct if conditions change
- because individuals vary it is more likely that some will survive changing conditions.

In the next lesson, students will review their understanding of the topics and concepts covered in the unit.

Think back: How are conifers and Arctic foxes adapted to live in their habitats?

Ask students to work on their own initially and try to write down two adaptations for the conifers and two for the Arctic fox. They can then work with a partner to compare lists. Ask some students to read out their list to the class.

> *Answer: Conifers have waxy, spiny leaves to prevent water loss and the spines and branches point downwards to stop snow from building up. Arctic foxes have fur coats to keep them warm and are white in winter to camouflage themselves against the white background of snow.*

Point out the photographs of the moths at the top of page 138. Ask students if they have seen these moths before. Elicit that they were on the unit introductory pages. Tell them they are called peppered moths and this species has been studied a great deal. The moths have taught scientists a lot about camouflage, adaptation and survival. Explain that one variety of the moth has a dark melanin pigment and the other variety does not and so is paler in colour.

How are the moths adapted to their habitat? Which moth is the most likely to be found and eaten?

Ask students to continue to work with their partner to discuss the questions. Encourage them to look very carefully at the moths in the photograph and check that they can see four moths. Suggest they imagine they are a predator bird flying past the tree very quickly.

Which moths might they see? They can then share their answers with a nearby pair. Ask the fours to make a list to share with the class.

> *Answer: The pale peppered moth is well adapted to hide on pale surfaces and the dark variety is well adapted to hide on dark backgrounds. The moths that are pale grey against a dark background or dark against a light background are most likely to be seen and eaten.*

Ask students to read the text on page 138 until they reach the diagram of the snails. Then they should look at the diagram carefully and answer the discussion task. Explain that if an animal is found and eaten, it cannot have offspring and if an animal is not found and eaten, it can have offspring. This means that more and more of the variety of the animal best suited to a habitat are born. The predators help to remove those animals that are not well suited to a habitat. It is like a sorting process.

Which coloured variety of snail is most likely to survive to have offspring in a dark area? Why?

Ask students to stay with their group of four. They can talk about the different coloured snails and relate this back to their discussions about the moths. Remind them that snails can retract into their shells so they should look at shell colour mainly.

> *Answer: The central snail is best camouflaged and is more likely to survive to have offspring.*

Ask students to read the text on page 138 below the snail diagram and to look at the photograph of the tree. To check understanding, ask them to think up two questions about the text and pass these to a partner to answer. Ask volunteer pairs to share their questions and their answers.

Investigation: Adaptation trail

Allow students to work in small groups. Explain they are going to investigate camouflage by setting out an adaptation trail. Ask students to read through the instructions and set up their trail. If you wish to provide more support, the activity on page 138 of the Workbook provides more information and direction.

Hand each group some pieces of blue, green and red card and ask them to cut out 15 identical insect shapes. Allow them to place their insect shapes on plants around the school. Arrange it so members of other groups do not see where the insect models are being placed. Once all of the insects have been secretly placed, you can ask students to search for insects from other groups for five minutes and 'catch' any they see. After five minutes ask students to bring in any of their insects that were not found.

Maths link: Students can do some data analysis to compare the number of insects found within five minutes and the number of insects not found – and relate this to colour. They can produce bar charts to show their findings.

Ask students to make five more of any insects that were not found. This will show them how populations of better adapted animals will increase. Ask students to make a poster to share their ideas about adaptation and how populations can change.

> **Possible response:** *Red and blue insects placed against green plants will be found more easily than green ones. Red insects against red flowers, or blue insects against blue flowers will be less easy to find.*

How does the change in your population explain 'survival of the fittest'?

After completing the investigation, ask students to talk about the discussion question in their investigation groups. Encourage them to analyse their findings to help them draw a conclusion. They can share their ideas with the class.

> **Possible response:** *There will be more individuals of the variety of card insect that was the best camouflaged, therefore the best adapted fitted to the habitat.*

Investigation: Using secondary sources of information

Students can continue to work with their investigation group or you can set this as an individual challenge. Explain that students will need to use secondary sources to research why the dodo became extinct and then produce a short presentation to share with the class. Remind them that a secondary source is any source of information, from books, magazines, films and the internet, that they have looked up, not discovered themselves. You can ask students to present in different ways. One approach is to write alternative ways onto pieces of paper and then hand them out at random so the class experience a range of different presentation methods. You could include: information leaflet; poster; a stand-up talk; a play or poem; computer slide show; wiki; blog; or scientific article for a newspaper or journal.

> **Possible response:** *The dodo became extinct as it could not fly or move quickly and when people arrived in their habitat they were easy to catch for food. They were not adapted to survive the new predator.*

Stretch zone: Research a plant that became extinct because it was not able to adapt to its changing surroundings. Explain your findings to a partner.

Once the presentations about the dodo are finished, ask students to complete the Stretch zone challenge. After researching an extinct animal, they are now going to research a plant that became extinct. Remind them to include why it was not able to adapt to its changing surroundings. They can turn to a partner and explain their findings.

Key ideas

- *Variation makes it more likely for individuals to survive if conditions in a habitat change.*
- *Variation can eventually result in a new species.*

Read out the key ideas and ask students to turn to a partner and take it in turns to describe an example of an animal that is adapted to survive because it can hide from predators. They can then list two other dangers to the survival of animals and plants other than predators.

Workbook activities

Adaptation trail (page 138)

Explain that this activity supports the investigation on page 139 of the Student Book. Allow students to use the diagram to give them more clues about how to make their card insects and how to hide them. Point out that they should record their findings in the table provided and then answer the prompt questions to help with their analysis.

> **Possible response:** *Red and blue insects placed against green plants will be found more easily than green ones. Red insects against red flowers, or blue insects against blue flowers will be less easy to find. The best hidden insects are the most likely to survive to make offspring.*

Capture release data (page 139)

This is a deliberately difficult task so you may wish to use it as an extension challenge. Students can complete this as an individual activity but many will benefit from working with a partner to discuss ideas. Ask students to read the background text about Doctor Henry Kettlewell and his work. They should then analyse the data in the table, using the prompt questions to help. The main conceptual leap is for students to realise that recapture means survival. If a moth has been eaten, it will not be around to be caught again by the scientists.

Ask students to use the data to answer the questions. If they need more support, remind them that a polluted forest will have dark trees from soot and smoke and then let them turn back to the photograph of the moths on page 138 of the Student Book.

Review and reflect

As this is the last lesson in the unit, unless you plan a review lesson, ask students to look back through their notes and the Student Book and Workbook for the unit. They can then draw two large speech bubbles. In one they can finish the sentence, 'Two important things I have learned about adaptation and inherited characteristics are …' and in the other they can finish the sentence, 'Two things I am proud of in this unit are …'. Pin these to a wall so students can look at each other's answers and achievements.

Conclude the lesson by asking students to complete question 5 in the 'What have I learned about adaptation and inherited characteristics?' activity on page 141 of the Student Book and the third statement on page 140 of the Workbook. (See also the teaching notes in this Teacher's Guide, pages 193–195.)

Extra activities

1 Ask students to design an adaptation trail at home to teach their family about camouflage. They can use coloured material, strands of wool or coloured card to represent small animals. Allow them to draw a plan of their house and plan where they will place the small animals. Ask them to carry out the trail and report their findings to the class the next day.

2 Place objects around the room and tell students to spend three minutes looking for them. You can use card or paper to make objects with a rough outline so there are no straight edges and try using speckled and striped colouring to see if this makes it easier or harder for students to find them.

 3 **Computing link:** Download photographs of camouflaged animals from the internet for students to study.

Differentiation

Supporting: Place some small, coloured sweets (candy) onto a multi-coloured cloth and ask students to find them to help them understand camouflage. They can count which they find most easily. It is up to you if you let them eat what they find, but remind them of the need to eat only small amounts of sugary food!

Consolidating: Allow students to watch clips of wildlife films and documentaries to identify any adaptations of animals and plants.

Extending: Use the activity on page 139 of the Workbook as an extending challenge.

Differentiated outcomes	
All students	should be able to state that living things that are well adapted to their habitat have a greater chance of producing offspring than those that are not
Most students	will be able to describe how predators are an important way that the survival of the fittest happens in habitats
Some students	may be able to use experimental data from secondary sources to support the idea of the survival of the fittest

What have I learned about adaptation and inherited characteristics?

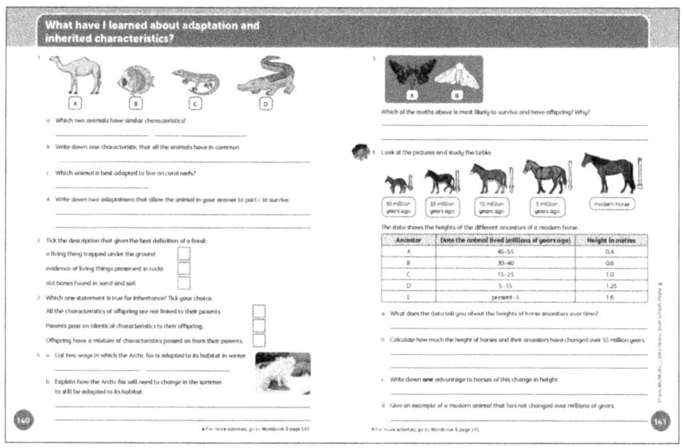

Supporting activities are in Workbook 6 pages 140–141.

Getting started

As with other units, the aim of this section is to encourage students to review their learning after all of the lessons in the unit. If you have taught the other units in this book, you will be familiar with the various ways you can use the 'What have I learned about … ?' sections and you may wish to move directly to the questions and answers below. If this is the first time you have completed a unit, then the following information will be useful to you.

All the lesson activities provide you with opportunities to assess students formatively and make decisions about how well students are coping with the concepts and tasks. Now you have reached the end of the unit, you will find that on pages 140–141 of the Student Book there are six questions related to the content of the unit. Students can tackle these one at a time after each relevant lesson and specific advice on which questions are most appropriate is given under the 'review and reflect' heading in the relevant lesson.

The questions within the 'What have I learned about adaptation and inherited characteristics?' section in the Student Book can also be answered as a single summative activity. This could be done by reading out the questions to the class and asking for volunteers to answer them, carrying out the activity as a group work task with students talking about each question, or as an individual task. Whichever approach is adopted, the questions are designed to give you and the students feedback about progress and to help in identifying targets for development.

The questions are arranged in increasing order of conceptual demand and not topic order.

On page 140 of the Workbook there is a section containing three 'review and reflect' statements for students to respond to. Ask students to tick one of the two self-review option boxes to show how confident they are about each topic. Some statements have been compiled to cover more than one lesson so you can use these after the last relevant lesson. As with the Student Book section, you can ask students to reflect on each statement separately after each relevant lesson or carry out the review as an end-of-unit activity. Advice on which specific review and reflect statement is most appropriate is given under the 'review and reflect' heading in the relevant lesson. It would be very useful to take both approaches.

What have I learned about adaptation and inherited characterisitcs? answers

1 a   Which two animals have similar characteristics?

   b   Write down one characteristic that all the animals have in common.

   c   Which animal is best adapted to live on coral reefs?

   d   Write down two adaptations that allow the animal in your answer to part **c** to survive.

Point out the four animal pictures and ask students to study each in detail. They should answer on the lines provided and ask them to check carefully if one or two answers are required for each section.

Answer: a C and D; b any one from mouth, eyes, head, tail; c B; d any two from gills, fins, smooth scales, streamlined shape.

2   Tick the description that gives the best definition of a fossil:

   a living thing trapped under the ground

   evidence of living things preserved in rocks

   old bones found in sand and soil

   Remind students to read the question carefully and note that the word 'best' means that only one answer from the possible three options is needed.

Answer: evidence of living things preserved in rocks

3   Which one statement is true for inheritance? Tick your choice.

   All the characteristics of offspring are not linked to their parents.

   Parents pass on identical characteristics to their offspring.

Offspring have a mixture of characteristics passed on from their parents.

Again, by reading the question carefully, students should pick out the crucial word 'one' so they are looking for only one correct answer from the choices given.

Answer: Offspring have a mixture of characteristics passed on from their parents.

4 **a** List two ways in which the Arctic fox is adapted to its habitat in winter.

   b Explain how the Arctic fox will need to change in the summer to still be adapted to its habitat.

Point out that students can use the photograph of the Arctic fox to help them to recall the adaptations of this animal. This time the crucial word in part **a** of the question is 'two' so remind students they will need to list two adaptations in this part.

Answer: a any two from: warm fur coat, fat layer, white fur for camouflage (accept hiding); b The snow will melt so the fox will need to be brown to blend into the darker background.

5 Which of the moths above is most likely to survive and have offspring? Why?

Ask students to imagine the background is a dark piece of tree bark and remind them that other animals eat the moths. Point out the word 'why' in the question and stress this always means they have to give a reason.

Answer: Moth A is most likely to survive. It is camouflaged (hidden) and so is less likely to be eaten and can go onto have offspring.

6 Look at the diagram and study the table. The data shows the heights of the different ancestors of a modern horse.

   a What does the data tell you about the heights of horse ancestors over time?

   b Calculate how much the height of horses and their ancestors has changed over 55 million years.

   c Write down **one** advantage to horses of this change in height.

   d Give an example of a modern animal that has not changed over millions of years.

Ask students to use both the diagrams of the horse and horse ancestors and the table of data to help them to answer the questions.

Answer: a height has increased over time; b 1.2 metres; c any one from: bigger to fight off predators, longer legs to run faster, taller to eat vegetation from a wider range of places; d crocodile or alligator (any other if suggested)

The questions in the 'What have I learned about adaptation and inherited characteristics?' activity on pages 140–141 of the Student Book can be used to consider the progress of each student individually. You can also use the information to create summative reports – such as end-of-term reports – for each student. If you wish to allocate a score or mark for the questions, then the total number of marks you could allocate is 17 (question 1 = 6; question 2 = 1; question 3 = 1; question 4a = 2; question 4b = 1; question 5 = 2; question 6a = 1; question 6b = 1; question 6c = 1; question 6d = 1).

You can also keep a record of individual and class responses to the 'review and reflect' statements in the Workbook. This lets you reflect on the overall confidence levels of students to identify areas that may need revision later on.

By reviewing responses to the questions and the self-review statements you can tailor specific interventions to help students improve. Keep the recording and analysis of the student self-evaluations simple. A general impression of the class's self-evaluation, not individual student records, is all that is required, e.g. 'Fifty per cent of the class were not confident about …'.

Additionally, you can ask students to complete the Quiz Yourself questions for this unit to help check their understanding. These are on pages 148–150 of the Workbook.

Investigate like a scientist

The 'Investigate like a scientist' task on page 141 of the Workbook is designed to encourage students to apply their investigative and creative skills and review key aspects of the content of the unit.

Investigating beak adaptations

Resources: plates or shallow dishes; water; large bowls; pebbles; pieces of wood with small holes; small seeds; rice; pieces of cork or polystyrene; tweezers; pliers; salad tongs; stopwatches or timers.

Warning! You must not touch the food or water with your hands. Do not pick up food from the table.

Start by reading out the warning and asking students to discuss why the rules are so important.

Allow students to work in groups of four so they can share out roles and ideas. Ask them to set up the four model habitats:

- marsh – a plate of water with floating objects such as cork or polystyrene
- pond – a large bowl of water with pebbles at the bottom
- forest – the piece of wood with holes in filled with small seeds
- grassland – a dry dish with dry rice in the bottom.

Then students can try to pick up each piece of food from the habitats using the different model beaks. They should count how many pieces of food they pick up with each model beak and record their results in the table provided. Finally, ask them to discuss what the results tell them about beak adaptations and obtaining food.

> **Possible response:** Students should find that it is not only the type of food that a beak has to be adapted for, but also where that food is found.

The questions are found on Workbook 6 pages 142–150.

1 Classification and Habitats

1 Across: 3 recycle; 5 conservation;
6 greenhouse effect; 8 waste; 9 acid rain;
11 landfill; 12 environment

Down: 1 deforestation; 2 pollution; 4 litter;
7 habitat; 10 care

2 **a** Grey wolf; Adonis blue butterfly

b mammals; birds; reptiles; amphibians

2 Organs and Systems

3 **a** 1 heart; 2 stomach and intestines;
3 liver; 4 lungs; 5 brain; 6 kidneys

b Students' own answers showing position of
organs on human body outline (see Student
Book page 44 to check correct positions).

4 brain – second box; heart – third box; lungs –
first box; stomach and intestines – fifth box;
kidneys – sixth box; liver – fourth box

3 The Way We See Things

5 Students' drawings showing a ray of light
leaving the torch, hitting the mirror and a
reflected ray. The angle of incidence and angle
of reflection should be the same (see Student
Book page 83 as a reference).

6 **a** Students' drawings should show a bigger
shadow of the same outline shape.

b Students' drawings should show a smaller
shadow of the same outline shape.

4 Building Electrical Circuits

7 Students should draw the following symbols:

battery

open switch

bulb

8 **a** Worst to best conductor: graphite, mercury,
gold, aluminium, copper

b copper

c aluminium

d Copper is not as strong as aluminium and is
more expensive.

9 **a** circuit; insulator; conductor; battery

b circuit – a complete path that an electric
current can flow around

insulator – material that does not allow
electricity to pass through it

conductor – material that conducts
electricity, allowing it to pass through it

battery – an object containing chemicals
that produce an electric current when
connected to a circuit

5 Adaptation and Inherited Characteristics

10 Across: 4 variation; 6 offspring;
9 inheritance

Down: 1 species; 2 ancestors; 3 reproduction;
5 adaptations; 7 fossils; 8 extinction

11 **a** The farmer has carried out selective
breeding. They have combined the best
characteristics (tall and big fruit) from
different plants. The offspring of the cross-
pollinated plants will have the characteristics
of the parent plants.

b Students' drawings should show a short
plant (the height of plant 1 or A) with small
fruit (smaller sized fruit than in plant A).

12 **a** A – 3rd statement; B – 4th statement;
C – top statement; D – 2nd statement;
E – 5th statement; F – bottom statement

b Bird G is not matched to a food type.

c This species would die out (become extinct)
in the area if it could not move away to
find food.

Glossary

adaptation — the characteristics living things have that help them to survive in their habitat

ammeter — a device that measures electric current

ancestor — the living things that other living things are descended from

battery — an object containing chemicals that produce an electric current when connected to a circuit

beam — a group of light rays

bulb — part of an electrical circuit that gives out light when connected

buzzer — a buzzer or a bell makes a sound when it is in a complete circuit

characteristic — a characteristic is something that a living thing or object has that helps us to identify it

circuit diagram — a diagram that uses symbols to show the parts of an electrical circuit

circulatory system — the system of heart and blood vessels that moves blood around the body

classification — the grouping together of similar living things

component — part of an electrical circuit. Bulbs, batteries and buzzers are components

conductor — materials that conduct electricity allowing it to pass through them

conservation — trying to stop animals and plants from dying out. Conservationists are people who try to stop living things becoming extinct

defence mechanism — any way the body protects itself from disease-causing microorganisms

deforestation — cutting down trees on a very large scale

digestive system — the body system that allows food to be broken down and taken into the body

drug — a substance that has an effect on the body

environment — the conditions in which a living thing exists. Soil, climate, and other living things all count as part of the environment

extinct — when a type of living thing no longer exists

fossil — the remains of an animal or plant buried in rocks

function — the job that something does

greenhouse effect — the way that the Earth's atmosphere traps heat from the Sun

habitat — the area where an animal or plant normally lives

infectious disease — a disease caused by microorganisms such as bacteria or viruses

inherit — when characteristics are passed from the parents to their offspring

insulator — an object that does not allow electricity to pass through it

key — a way of identifying animals and plants by using questions

kingdom — one of the very large groups of living things, such as animals, plants and fungi

lifestyle — the way people live their lives

light intensity — how much light is being transmitted

light source — the place or object that light comes from

medicine — chemicals we take to make us feel better or to cure a disease

microorganism — a living thing that is too small to see without using a microscope

mirror — a sheet of glass or metal which reflects an image

nervous system — the system of the brain, spinal cord and nerves that helps to control the body

offspring — a new living thing produced by parents

opaque a material through which light cannot pass

organ parts of the body that do a particular job

parallel circuit an electrical circuit where the components are connected on different branches of the wires

pollution anything that spoils the environment

ray a line of light

reflect to bounce back light; we see objects because they reflect light

reproduction the process by which living things make copies of themselves

series circuit when the components of a circuit are all joined one after the other with no side branches

shadow a dark shape made when light is blocked

silhouette a dark picture or image made on a pale or white background

species a small group of living things. Members of the same species are closely related and can mate to produce

switch a device for making and breaking an electrical circuit so other devices are turned on and off

translucent a material that lets light through but is not completely see-through

transparent a material that lets light through and is completely see-through

urinary system the system, containing the kidneys, bladder and tubes, that helps to remove water and some harmful substances from the body

vaccine a dead or weakened pathogen injected to help protect a person from an infectious disease

variation the differences between individuals in a species, such as height

voltage the difference in electrical energy between two parts of a circuit. It is the amount of push from the power source

voltmeter a device that measures voltage in an electrical circuit